International Conference on Humanities and Social Sciences (ICHSS 2018)

SHS Web of Conferences Volume 53 (2018)

Kuala Lumpur, Malaysia
13 – 14 August 2018

Editor:

S. L. Chong

ISBN: 978-1-5108-7306-3

Printed from e-media with permission by:

Curran Associates, Inc.
57 Morehouse Lane
Red Hook, NY 12571

Some format issues inherent in the e-media version may also appear in this print version.

Printed by Curran Associates, Inc. (2018)

For additional information, please contact EDP Sciences – Web of Conferences at the address below.

EDP Sciences – Web of Conferences
17, Avenue du Hoggar
Parc d'Activité de Courtabœuf
BP 112
F-91944 Les Ulis Cedex A
France

Phone: +33 (0) 1 69 18 75 75
Fax: +33 (0) 1 69 28 84 91

contact-edps@webofconferences.org

Additional copies of this publication are available from:

Curran Associates, Inc.
57 Morehouse Lane
Red Hook, NY 12571 USA
Phone: 845-758-0400
Fax: 845-758-2633
Email: curran@proceedings.com
Web: www.proceedings.com

TABLE OF CONTENTS

https://doi.org/10.1051/shsconf/20185300001

Preface

This collection of papers was first presented in the World Engineering, Science and Technology Congress 2018 (ESTCON 2018) where the International Conference on Humanities and Social Sciences 2018 (ICHSS 2018) was held in Kuala Lumpur Convention Centre, Malaysia, from 13th – 14th August 2018.

Representing work conducted by researchers located in Australia, Indonesia, Japan, South Korea, Malaysia and United Kingdom, the papers in this collection raise questions related to the following themes:

1) ASEAN Community
2) Communication and Media Studies
3) Language Use and Concerns
4) Religions, Science and Morality
5) Teaching and Learning

The research represented in this collection attempts to consider complex sociological issues from the theoretical, historical, empirical and philosophical perspectives. Because these issues affect and shape human lived experience, the papers in this collection contribute towards extending important conversations that are current in this millennium. The aim of this collection is for readers to carry on the conversations.

This collection is made possible with the full support and generosity of Universiti Teknologi PETRONAS, Malaysia, who was the main organizer of ESTCON 2018 and ICHSS 2018. Special acknowledgement must also be given to the many reviewers from Australia, Belgium, Malaysia and Singapore who provided useful and timely feedback to the authors of the papers in adherence to the blind review process.

It is hoped that the International Conference on Humanities and Social Sciences 2018 (ICHSS 2018) will be a springboard to further research work that continues to confront issues in Social Sciences and Humanities.

Editor,

Dr. *Su Li* CHONG

edp sciences

Web of Conferences

Statement of Peer review

In submitting conference proceedings to *Web of Conferences*, the editors of the proceedings certify to the Publisher that

1. They adhere to its **Policy on Publishing Integrity** in order to safeguard good scientific practice in publishing.
2. All articles have been subjected to peer review administered by the proceedings editors.
3. Reviews have been conducted by expert referees, who have been requested to provide unbiased and constructive comments aimed, whenever possible, at improving the work.
4. Proceedings editors have taken all reasonable steps to ensure the quality of the materials they publish and their decision to accept or reject a paper for publication has been based only on the merits of the work and the relevance to the journal.

Title, date and place of the conference

INTERNATIONAL CONFERENCE ON HUMANITIES AND SOCIAL SCIENCES (ICHSS 2018),
KUALA LUMPUR CONVENTION CENTRE, KUALA LUMPUR, MALAYSIA
13th - 14th AUGUST 2018

Proceedings editor(s):

Dr. CHONG, Su Li

Date and editor's signature

27/09/2018

17, avenue du Hoggar - PA de Courtabœuf – BP 112 - 91944 Les Ulis Cedex A (France)
Tél. : 33 (0)1 69 18 75 75 - Fax : 33(0)1 69 07 45 17 – www.edpsciences.org

Addressing inequality in Malaysia through corporate social intrapreneurship

Asrif Omar Che Yusoff[1,*]

[1]Candidate, Master of Studies in Social Innovation, Judge Business School, University of Cambridge, United Kingdom

Abstract. Inequality has been a long standing issue in Malaysia, although the situation has been statistically improving over the past 40 years. From a Gini coefficient of 0.51 in 1970, the government has done considerably well to bring the figure down to 0.39 in 2016. Efforts toward improving the situation are aplenty, but there is room for improvement in terms of the coordination and collaboration of initiatives that are carried out within the public, private, and social sectors. This paper explores the idea of corporate social intrapreneurship as a potential vehicle to mitigate inequality in the country for the long term. Through the analysis of existing literatures and data on the subject, the aim is to first of all, provide a historical and global context on how the roles of corporation have evolved over the years, discuss the transformative views on social intrapreneurship against traditional corporate social responsibility, and offer considerations to further corporate social intrapreneurship initiatives through public-private partnerships in Malaysia.

Before any discussion can be made on the idea of corporate social intrapreneurship, it is important to form a common understanding of why corporations exist, and how they have influenced society as it is today. This analysis could then provide context on how early corporations or business organisations were originally formed, why were they established, and how they have evolved over time. The past, present, and future perspectives on the functions of corporations in the social, economic, and political context are crucial indicators on their importance in shaping the world as we know today. As the engine of world economy and vehicle of capitalism, it is natural to form a hypothesis that supports the significant role that corporations play in influencing the accumulation of global wealth, and the inequalities that occur as a consequence. This presumption, however, needs to be further justified and contextualised.

The observations and researches that have been made on corporations have led to suggestions on how they should operate, both in the traditional and progressive sense. A holistic assessment of these opinions can facilitate the formation of ideas to be adopted towards the ideal and sustainable operation of corporations. For instance, proponents of social intrapreneurship assert that the concept provides a platform for individuals from within the corporate setting to make global impact or change for the better. The philosophy behind this idea can transform the roles that corporations play and redefine the values that they create. At the end of the day, any inequality that corporations have a hand in creating can never be fully undone. But minimising this factor and reversing its impact is worth exploring, even if it might take some time.

1 Capitalism: A historical context

There is wide body of work surrounding the origins of capitalism, which provides the foundation for the formation of corporations. The conventional view on capitalism history prior to its modern-day form, broadly chronicles it to the periods of agrarian (14th century), mercantilism (16th century), and industrial (18th century). Together with this evolution, is the development of differing functionalist views on the roles that capitalist organisations play in shaping society. From the division of farmed land, to early international trade, to the emergence of systematic production lines, the precursors of corporations as we know today have undergone massive shifts in serving individuals, governments, and society as key stakeholders and beneficiaries.

Despite these significant transformations over the past eight centuries, one common denominator across all iterations is the objective of generating wealth. This traditional view remains largely true in present day business organisations, and in its simplest form, capitalism can be universally understood today as a system that facilitates the process of turning of investments into profits (Appleby, 2010). Arising from this monetary motivation, is another 'derived' common denominator which is societal impact, both positive and negative. While agrarian capitalism restricted wealth generation to land owners only, and mercantilism led to global imperialism and war, the Industrial

* Corresponding author: aoc23@jbs.cam.ac.uk

Revolution created job for the masses -- although this development did not come without its own set of challenges such as variances in working conditions.

While the functionalist view focuses on the dynamic of roles within society in shaping the world, the conflict view on the matter is skewed towards the tension created between different social groups. This perspective provides the foundation to the theories of Karl Marx, whom together with Friedrich Engels, published 'the Communist Manifesto' in 1848 which discussed the dynamic between bourgeois (middle class) and proletariat (working class). These 19th century ideas can be viewed as the framework to explain the inequality that is being observed today. Capitalism, with corporations as its vehicle, is seen as the root cause behind the prominent aphorism, "the rich getting richer and the poor getting poorer". Nevertheless, while Marx and Engels envisaged a proletariat triumph in the future, Jeffries (2012) argues that the exact opposite is happening today. "The proletariat, far from burying capitalism, are keeping it on life support," he says.

2 Contemporary views on the role of corporations

The literature on the role of the modern corporation is wide and diverse. As capitalism is commonly seen as the driving factor behind global inequality, corporations are becoming more susceptible to unfavourable critique and judgments. Despite this preconceived observation on the corporate world, the awareness on corporate citizenship seems to be gaining momentum over the past decades. This development signals the effort among corporations to move away from the negative views on them, and be more involved with the communities where they operate. To facilitate discussion, Danley (1994) adopts the view of two common ideologies in discussing the roles of corporations: Classic Liberalism (economic freedom) and Managerialist (organisational responsibility).

To elaborate further, Classic Liberalism can be related to the concept of 'maximising shareholder value', which is commonly attributed to Jensen and Meckling (1976). The idea revolves around increasing returns to shareholder investments, and is one of the most widely adopted conventional concepts globally. This view limits the goals of a corporation towards creating monetary value only, for the benefit of those who have invested in the organisation. In recent years, however, this belief has received its fair share of criticisms. Stout (2013) attempted to debunk the shareholder value myth through historical, legal, and empirical perspectives. Kramer (2012) highlighted the irony of the concept as it "enriches senior executives at the shareholders' expense". At a more extreme end, former GE CEO Jack Welch once famously cited shareholder value as the "dumbest idea in the world" asserting that it is a result instead of a strategy.

On the other hand, the Managerialist view can be related to the concept of Shared Value (Porter and Kramer, 2011) which focuses on the "connections between societal and economic progress". There are the internal responsibilities of individuals within a corporate organisation that managerialism advocates, but the Shared Value perspective views the corporation as an ecosystem that interacts with society and the environment in its operations thus requiring it to ensure that the value chain is run efficiently, responsibly, and sustainably. As much as encouraging and progressive the idea appears to be, it does not escape criticism. While acknowledging the appeal of Shared Value, Crane, Palazzo, Spence and Matten (2014) claim the idea to be unoriginal and naive about the challenges of business compliance, among others.

Opposing views on the role that corporations should play have certainly added colour to discussions on the topic. And if the volume of emerging literature related to the subject is any indication, there is a long way to go before any consensus is reached. These strong views, however, suggest that the ideal corporation should strike a balance between the two schools of thought. A stable and sustainable corporation should be able to manage the expectations of its internal and external stakeholders, while constructively coexisting with the communities within the environment that it operates in. The question that corporate leaders need to address is how to embed these complex elements across the many facets of their organisations.

3 Global wealth, inequality, and corporations

In the 2017 edition of its Global Wealth Report, Credit Suisse indicates that total global wealth today has reached GBP 207 trillion, a 27% growth over the preceding decade at the onset of the global financial crisis. During the same period, 8.7 million new millionaires were created, and the global mean wealth per adult reached a new record high of GBP 41,839, an 8.5% increase from 2007. The first 12 months of 2017 recorded the fastest growth in total global wealth at 6.4%, faster than the rate of increase for world population. These indicators reflect how global wealth remained on a growth trajectory despite a tumultuous start at the back of the financial meltdown of 2008, with some figures indicating 'pre-crisis' trends.

It is important, however, to have a holistic view on this perceived progress of the global economy. In the same report produced by Credit Suisse, it was highlighted that the wealthiest 1% of world population owns over 50% of all household wealth; an increase from the 45.5% recorded at the turn of the millennium. These data provide a high-level indication of how lopsided global wealth is becoming. In its 2016 report on income inequality, the Organisation for

Economic Co-operation and Development (OECD) suggests that arising from the 2008 global financial crisis, economic rivalry has not reduced inequality. This assertion was made based on the uptrend of average worldwide Gini coefficient of disposable household income.

In his 2014 publication, 'The Drivers of Economic Inequality', Dr. Nicholas Galasso of Oxfam America outlines a range of factors that influence inequality around the world. Derived from a range of past and current literature, Dr. Galasso categorises the drivers as geography, technology, globalisation, wage structure, and politics, among others. Across these diverse groups, linkages to industries and corporations can be observed. The literature cites Kusnetz (1955) who argues that technological change leads to wage inequality between skilled and unskilled workers. The same can also be observed through globalisation as foreign direct investments favour workforce with existing skill sets which waive further expenditures on training and development. In terms of political influence, the paper argues how rules controlling the 'revolving door' between government and the private sector can be "weak or ill defined".

The understanding of emerging trends in global economy and linking them to key drivers to inequality provide a map that can facilitate the identification of roles that corporations play in wealth distribution. The influence appears to be sporadic across the fabric of society, with varying degrees in terms of direct and indirect impact. Nevertheless, it can be observed how economic inequality can be attributed to corporations through the creation of the wage gap, employment imbalance, and unfair advantage gained through political influence. These outcomes might differ in terms of significance between developed and developing countries, but common traits exist in defining ways of intervention, mitigation, and solution.

4 Social intrapreneurship and the CSR dilemma

The origins of social intrapreneurship can be tracked down to the municipal socialism and public enterprise movements in the UK and North America respectively between the 19th and 20th century. One of the most famous proponents of these movements was Joseph Chamberlain whom British Prime Minister Theresa May regarded as a "political hero" and Winston Churchill called "the man who made the weather". As the mayor of Birmingham in late 19th century, Chamberlain saved the city from "choking on its own filth" (Goodall, 2016) by taking over the gas and water supplies. Together with his housing and education reforms, he transformed Birmingham into an advanced city, which provides the foundation for it to be claimed as one of the second cities of the United Kingdom.

These transformations made by Chamberlain indicate early influence of structured and systemised management of basic civil requirements to society. Through effective delivery of basic amenities to the public, the overall quality of life enjoyed by a certain community can be viewed as being more equal. Such an improvement is a reflection of how the philosophy behind social intrapreneurship can lead to a seemingly more just distribution of wealth, both in tangible and intangible forms. It can be argued that these services are expected of an entity with authoritative powers, but it is worth appreciating that during its time, the changes brought forward by Chamberlain can be considered as progressive if compared to his contemporaries.

In the modern context, the definition of social intrapreneurship covers a broader range of activities. The term is attributed to Pinchot (1985) who refers it to the formation of new ventures from within an established organisation. Although potentially similar in nature, Stewart (2013) strongly argues that social intrapreneurship should not be confused with Corporate Social Responsibility (CSR), which she claims only serves to "protect the company from risk or improve its brand". The debate over the definition of CSR, in itself, is a subject of high contention among both academics and professionals. As a result of this subjectivity, the definitions and subsequently, action plans of CSR programmes across different business environments remain broad and expansive. It can also be argued that the term 'CSR' is generally more widely known by the public compared to 'social intrapreneurship'.

The debate over the validity and legitimacy of CSR can drag into an exhausting discourse. Opinions seem to be strong to support the importance of CSR as a vital component of an organisation's competitiveness (Porter and Kramer, 2006). On the contrary, there are equally strong views against the categorisation of CSR as a nobility but instead as a form of 'greenwashing' (Alves, 2009). Within this perspective, CSR is merely a widely misleading reputation management tool to balance out any unfavourable view held by the public on a corporation. A view that fits the narrative of corporate entities being inherently evil, or, as pop culture commonly perceives it, "a supporting player in the war against individualism" (Allan, 2016).

Social intrapreneurship provides an alternative perspective for organisations to do good to society and the environment. Furthermore, CSR excellence should firstly be attained through responsible business practices. From the provision of conducive working environment, to sustainable sourcing and procurement, elements of highly responsible corporate citizenship should already be embedded within a corporation's operations. These practices are fundamentals that are expected to be strengthened from within an organisation, before they even reach out to external stakeholders via extravagantly branded CSR programmes that might not last beyond three to five years. Some of these programmes even change almost overnight with the change of leadership. Focusing on social intrapreneurship, as a discipline, can facilitate sharper focus and line-of-sight in bringing change to society and the environment. But it should be done only after core and basic business practices are conducted with the highest compliance to environmental standards and formal or informal social regulations.

5 Social intrapreneurs as change agents

The body of literature on theories and concepts of social intrapreneurship is growing as it gains further interest from academics. Pinchot (2011) who has been attributed to coining the term, published 'The Intrapreneur's Ten Commandments' which include, among others, 1) Come to work willing to be fired, 2) Circumvent any orders aimed at stopping your dream, and 3) Do any job needed to make your project work regardless of your job description. These commandments suggest the determination and dedication required of social intrapreneurs in executing their respective projects. It also indicates the resistance that they might face in going against the grain of traditional corporate culture.

Further to these challenges, Patel and Gong (2017) assert that social intrapreneurs are bound to be asked questions in their corporate journey. Tough questions like whether their time is effectively used, or how their efforts are contributing to the company's bottom line. It can also be queried if these differences are down to generation gap, as corporations are commonly comprised of individuals from across different age groups, from Baby Boomers to Millenials. This conflict is further amplified by the wide array of definitions that companies have on the meaning of CSR, some of which, use the term interchangeably with social intrapreneurship. These scenarios signal how it can be a mountain to climb for social intrapreneurs in convincing management to permit their causes to be operationalised, and provide the necessary funding and resources.

One hurdle that is limiting social intrapreneurs from gaining support from within the organisation is the intangible nature of their causes. Social impact is not easily measured even if it can be quantifiable in terms of value created for the brand or company reputation. This challenge is further aggravated when they are faced with traditionalists in management who only believe in bottom line figures as the main indicator of a company's wealth. It is therefore crucial for social intrapreneurs to reach out and collaborate with the public and social sectors so that their efforts are better consolidated and coordinated. Before any programme is implemented, it is crucial for a screening of existing initiatives by the public or social sectors to be done, to avoid any duplication of approaches on the same issue.

Realising the gap that exists between corporate changemakers and their management, Lodge and Wilson (2006) propose that with the resources that they have, multinational corporations have got the opportunity to restore their legitimacy by playing a role in reducing poverty in developing countries where they operate. Legitimacy, in this context is not limited only to compliance to the rule of law, but also acknowledgement of public opinion over how ethical the company is operating. Corporations are expected to not only adhere, but also listen to the feedback that society have towards their ways of doing business. This perspective solidifies the view that corporations are best partnering with the public and social sectors if they aim to effectively executive their social goals. In the modern business arena, corporations of all scales are carefully and meticulously implementing their respective strategies through various partnership models.

An understanding of social intrapreneurship among the people in a corporation would then provide the platform for the formation of corporate social innovations or CSI. As proposed, the terms CSR and CSI are still considered interchangeable by certain quarters in the corporate world, which calls for a more streamlined distinction between the two. One proposal is to view CSI as a deviation from the traditional method of charitable acts by corporation. It can be viewed as an alternative that moves away from simply giving to the needy. Corporations should approach CSI as a mean of better integrating with communities, given that they have been historically viewed as a key factor behind the shaping and segregation of society.

6 Malaysia: A case study

To illustrate the interrelation between corporations and social wealth inequality, a localised view can be taken on Malaysia, a developing country in the heart of Southeast Asia.

Inequality has been a long standing issue in Malaysia, although the situation is statistically improving over the past 40 years. From a Gini coefficient of 0.51 in 1970, the government has done considerably well to bring the figure down to 0.39 in 2016. Despite this empirical proof of inequality reduction, Dr. Muhammed Abdul Khalid (2014), one of the foremost Malaysian scholars on the subject, the wealth gap between the urban and rural communities in the country has not changed since gaining independence from Britain in 1957. Further, he alerts how the gap, if left unchecked, can lead to social and political instability and stifle long-term economic growth. Further, Dr. Khalid discussed how complex the inequality situation in Malaysia is, stemming out of colonial policies, political drives, as well as cultural differences. These varying reasons indicate how inequality can be viewed as a 'wicked problem' in Malaysia.

Several others have voiced similar views on the subject. Kana (2015) argues that while overall Gini coefficient has decreased, the Gini of the Bumiputera (Malay indigenous) majority is lower compared to that of the Chinese and Indian population. This observation can be attributed to the many advantages that are given to the selected group in terms of education, housing, and working opportunities. He suggests that an aggravation of this seemingly marginal difference can lead to racial tension if not managed carefully by the government. Similarly, Faizli (2016) warns that if inequality is not addressed comprehensively, the country is at risk of "dragging its feet into the future". Coming out of the 1998 Asian Financial Crisis, the country has also undergone widening political divide, which can compound on the pre-existing socioeconomic gap.

While critical views over inequality management in Malaysia are aplenty, the opposing end of the argument is just as strong. Grewal (2016) attributes the government's achievement in reducing poverty and inequality to date is down to government-led micro credit programmes for the 'Bottom 40%' group of the country, together with cash assistance provision to the hardcore poor. This notion is further supported by Lee (2016) who opposed the narrative of worsening inequality in Malaysia arguing marginal income gap between the rich and the poor. He also asserts that the gloomy perspective is merely driven by perception, compounded by poor communication by the government on the effectiveness of their poverty eradication programmes. The advent of social media, while seen as an effective tool for government advocacy, can also be a medium that negates efforts put in by the authorities in clarifying their position on poverty eradication.

Within the Malaysian public sector, the reduction of wealth gap is under the purview of the Ministry of Rural and Regional Development. Among its core functions are to develop infrastructural and logistical projects for rural development, and to formulate upskilling programmes to equip rural communities, mostly underserved, with the skills required to enter the job market. There are some overlaps observed between efforts that are put forth by the Ministry of Rural and Regional Development, with those led by other ministries namely the Ministry of Education, the Ministry of Youth and Sports, and the Ministry of Human Resources -- especially in the area of human capital development. One example is the large number of vocational institutions run by many different federal and state agencies that can potentially confuse the public. The initiatives are nonetheless noble, but there is room for streamlining.

In between the public and private sectors in Malaysia, there exists a hybrid sector which comprise of government-linked companies (GLC). GLCs are mostly initiated and backed by the government, and formed to stimulate economic growth while providing social services to the public. They are mostly run with the governance and system of a private entity, and the vision of nation-building. For the purpose of discussion, focus shall be given on a company that has been at the forefront of social intrapreneurship towards empowering society, PETRONAS.

7 PETRONAS: Leading social intrapeneurship in Malaysia

Established in 1974, PETRONAS is the national oil company of Malaysia formed as custodian of the country's petroleum resources. Since its founding, PETRONAS has grown to become an integrated energy company which operates both upstream exploration and downstream processing and refining activities. As a business entity, PETRONAS contributes to the government in the form of annual dividend. In terms of social investment, the company has been focusing primarily in the education sector. Since 1975, the company has sponsored over 20,000 students from primary, secondary and higher education amounting to over RM 3 billion in total. Their focus on education and human capital development has been further exemplified through their network of learning institutions which includes a university, a vocational training institution, a marine academy, and a professional leadership centre.

PETRONAS' foray into social intrapreneurship began in 1992 when they initiated a programme that is now known as the Vocational Institution Sponsorship & Training Assistance (VISTA) programme. Through the VISTA initiative, PETRONAS collaborate with vocational colleges around Malaysia to enhance the delivery of their training programmes through sponsorship of equipment, curriculum development, and training of trainers. Leveraging on its wealth of experience in the fields of engineering, PETRONAS deploys its technical professionals to provide academic consultation and advice in delivering courses that are relevant to the needs of the industry. The VISTA programme is managed by the company's Education Division as its non-profit arm, and bridges the gap between vocational education and the industry which is crucial to ensure employability of the graduates.

Vocational education is viewed as a secondary alternative in the Malaysian education landscape. It is catered mostly for students who have not performed well academically during primary and secondary schooling. A high percentage of these students come from the underserved rural communities, some from families who live within the poverty line. Vocational education in Malaysia is seen as a fast route to acquiring skill and qualification for employment. To date, over 5,000 students have graduated from the VISTA programme, many of whom have joined the industrial workforce or began their own technical service firms, subsequently improving the lives of their families. VISTA is an example of how corporations can work together with the public sector in lending their expertise and resources in helping the underserved to get out of the cycle of poverty.

8 Conclusion

The diverse range of views over the roles that corporations should play in society signal a highly mixed and debatable stand on the subject. The rationale and reasoning for the formation of early corporations indicate how property division, imperial expansion, and national growth were the main drivers of history's pioneering companies. The evolution of these business organisations over time, however, led to the widely varied categories of corporations from public cooperatives to private entities. This expansive development has shaped the corporate landscape as we know today, together with the unique objectives that they have respectively as their reasons for being. And this diversity of functions marks greater challenges in pinpointing whether corporations are obliged, or otherwise, towards the betterment of society.

Zooming into the accumulation of global wealth, however, a clear link can be observed between the existence of corporations and their influence on how societies are shaped. From the benefits that are gained by modern corporations, through their utilisation (or exploitation) of resources, it can be argued that corporations are responsible in re-balancing the imbalance that they have created as a result of their pursuit of growth and profits. This attribution then raises the question of whether CSR is an add-on element to a company's operations, or an expectation that should be met through their business practices. After all, there are companies that consider improved working conditions, responsible sourcing, and environmental protection as CSR when these are basic requirements that they should already meet. Responsible business practices should already be incorporated within an organisation's DNA, which then allows social intrapreneurship to become the vehicle for corporations to make actual change globally.

The primary advantage that social intrapreneurs have in leading their causes is the access that they have to resources such as funding and manpower, being a part of their respective organisations. Commonly, the main challenge for them is convincing management to channel these resources towards the causes that they are driving. As most of the key targets of social intrapreneurship are social change, the key result indicators are less tangible and more qualitative. Therefore, the objectives of social intrapreneurship are often viewed as 'removed' from the company's bottom line or quantifiable value creation. In addressing this situation, it is imperative for social intrapreneurs to effectively link the roles of their organisations to societal motives such as human capital development, education, and poverty eradication.

It can be observed how through its VISTA programme, PETRONAS has demonstrated how a company's technical capabilities can be bridged with the education sector through social intrapreneurship. By deploying their highly qualified engineers to vocational colleges around Malaysia, PETRONAS has succeeded in equipping underserved students with the necessary skills to increase their employability, consequently making a difference in the future livelihood of their families. VISTA is an example of how a company utilises its strengths and resources in providing opportunities and opening doors for those in need. This model provides a framework that can be adopted by companies in other industries to identify and map the resources that they have with relevant community-building initiatives.

The PETRONAS example provides three considerations for social intrapreneurs in strategising and developing their respective programmes. First, a connection needs to be made between the core competencies of the corporation to the cause that it intends to support. This linkage can provide further justification in convincing management to approve the necessary funding or manpower required. Second, to ensure success, social intrapreneurs should reach out and collaborate with the public and social sectors. This is to ensure effectiveness in implementing a social impact programme, and to avoid overlapping of efforts. Third, the involvement of social intrapreneurs towards their causes should go beyond monitoring. The longevity of the VISTA programme can be attributed to the continued and frequent engagement between the benefactor and the beneficiaries.

History has shown how corporations and society interrelate in evolving into the forms that they are in today. The role of business organisations in global accumulation of wealth provides an indication of how they influence the demarcation of society. It is therefore vital for corporations to uphold their social contract in mitigating this situation, and social intrapreneurship is evidently an effective approach to meet this target.

9 References

A Corporate Solution to Global Poverty: How Multinationals Can Help the Poor

Abdul Khalid, M. (2014). The colour of inequality. 1st ed. MPH Group Publishing.

Appleby, J. (2011). The Relentless Revolution: A History of Capitalism. W. W. Norton & Company.

Alves, I. (2009). Green Spin Everywhere: How greenwashing reveals the limits of the CSR paradigm. Journal of Global Change & Governance, 2(1).

Allan, A. (2018). How the 'Evil Corporation' Became a Pop-Culture Trope. [online] The Atlantic. Available at: https://www.theatlantic.com/business/archive/2016/04/evil-corporation-trope/479295/ [Accessed 10 Dec. 2017].

Bernama Online. (2014). Bernama. [online] Available at: http://education.bernama.com/index.php?sid=news_content&id=1044531 [Accessed 5 Jan. 2018].

Crane, A., Palazzo, G., Spence, L. and Matten, D. (2014). Contesting the Value of "Creating Shared Value". California Management Review

Credit Suisse (2017). Global Wealth Report 2017. Credit Suisse.

Danley, J. (1994). The role of the modern corporation in a free society. University of Notre Dame Press.

Denning, S. (2016). Forbes. [online] Forbes.com. Available at: https://www.forbes.com/sites/stevedenning/2016/04/03/the-economist-defends-the-worlds-dumbest-idea/#8badf5826e3c [Accessed 8 Jan. 2018].

Deloitte. (2017). The Deloitte Millennial Survey 2017. [online] Available at: https://www2.deloitte.com/na/en/pages/about-deloitte/articles/millennialsurvey.html [Accessed 15 Dec. 2017].

Economic Planning Unit of Malaysia. (2016). Gini Coefficient by Ethnic Group, Strata and State, Malaysia, 1970-2016. [online] Available at: http://www.epu.gov.my/en/content/table-6-gini-coefficient-ethnic-group-strata-and-state-malaysia-1970-2016 [Accessed 10 Dec. 2017].

Faizli, A. (2016). Income disparity in Malaysia The Myth about the Rich and the Poor. [online] Astro Awani. Available at: http://english.astroawani.com/malaysia-news/income-disparity-malaysia-myth-about-rich-and-poor-121248 [Accessed 3 Jan. 2018].

Galasso, V. Nicholas, "The Drivers of Economic Inequality: A Backgrounder for Oxfam's Inequality Campaign," Oxfam America Research Backgrounder series (2014): www.oxfamamerica.org/publications/the-drivers-of-economic-inequalitythe-primer.

Goodall, L. (2018). Theresa May's unlikely political hero. [online] BBC News. Available at: http://www.bbc.com/news/uk-politics-37053114 [Accessed 21 Dec. 2017].
BBC News. (2018). Man 'who made the political weather'. [online] Available at: http://www.bbc.com/news/uk-england-28111051 [Accessed 21 Dec. 2017].

Inequality.org. (2018). Global Inequality. [online] Available at: https://inequality.org/facts/global-inequality/ [Accessed 15 Dec. 2017].

Jeffries, S. (2012). Why Marxism is on the rise again. [online] The Guardian. Available at: https://www.theguardian.com/world/2012/jul/04/the-return-of-marxism [Accessed 7 Jan. 2018].

Jensen, M. and Meckling, W. (1976). Theory of the firm: Managerial behavior, agency costs and ownership structure. Journal of Financial Economics

Kana, G. (2015). Inequality in Malaysia, is it too bad?. [online] Malaysiakini. Available at: https://www.malaysiakini.com/letters/313448 [Accessed 5 Jan. 2018].

Kramer, M. (2012). What's wrong with maximising shareholder value?. [online] The Guardian. Available at: https://www.theguardian.com/sustainable-business/blog/maximising-shareholder-value-irony [Accessed 13 Dec. 2017].

Kusnetz, S. (1955). Economy Growth and Income Inequality. The American Economic Review, 45(1).

Lodge, G. and Wilson, C. (2006). A Corporate Solution to Global Poverty: How Multinationals Can Help the Poor. Princeton University Press.

Ministry of Finance Malaysia. (2017). Government Investment Companies Division (GIC). [online] Available at: http://www.treasury.gov.my/index.php/en/contactus/faqs/gic.html [Accessed 3 Jan. 2018].

OECD (2016). Income inequality remains high in the face of weak recovery (pp. 1 – 6). [online] OECD. Available at: http://www.oecd.org/social/OECD2016-Income-Inequality-Update.pdf [Accessed 18 Dec. 2017].

Patel, A. and Gong, R. (2017). An Introduction to Social Intrapreneurship - Impact Design Hub. [online] Impact Design Hub. Available at: https://impactdesignhub.org/2017/05/17/an-introduction-to-social-intrapreneurship/ [Accessed 5 Dec. 2017].

PETRONAS. (2016). PETRONAS Reaffirm Efforts on Technical Education. [online] Available at: http://www.petronas.com.my/media-relations/media-releases/Pages/article/PETRONAS-MARA-MOU.aspx [Accessed 2 Jan. 2018].

Porter, M. and Kramer, M. (2011). Creating Shared Value. Harvard Business Review. [online] Available at: https://hbr.org/2011/01/the-big-idea-creating-shared-value [Accessed 10 Dec. 2017].

Porter, M. and Kramer, M. (2006). Strategy and Society: The Link Between Competitive Advantage and Corporate Social Responsibility. Harvard Business Review (pp. 63 – 77). [online] Available at: https://hbr.org/2006/12/strategy-and-society-the-link-between-competitive-advantage-and-corporate-social-responsibility [Accessed 14 Dec. 2017].

Pinchot, G. (2018). The Intrapreneur's Ten Commandments. [online] The Pinchot Perspective. Available at: http://www.pinchot.com/2011/11/the-intrapreneurs-ten-commandments.html [Accessed 20 Dec. 2017].

Pinchot, G. (1986). Intrapreneuring: Why You Don't Have to Leave the Corporation to Become an Entrepreneur. Harper & Row.

Stewart, E. (2018). How does a social intrapreneur add value to a business?. [online] the Guardian. Available at: https://www.theguardian.com/sustainable-business/social-intrapreneur-value-business [Accessed 18 Dec. 2017].

Stout, L. (2013). The Shareholder Value Myth. Cornell Law. [online] Available at: http://scholarship.law.cornell.edu/cgi/viewcontent.cgi?article=2311&context=facpub [Accessed 12 Dec. 2017].

SustainAbility (2008). The Social Intrapreneur: A Field Guide for Corporate Changemakers. [online] SustainAbility. Available at: https://www.allianz.com/v_1339502342000/media/current/en/press/news/studies/downloads/thesocialintrapreneur_2008.pdf [Accessed 18 Dec. 2017].

Singh Grewal, I. (2016). Economic inequality and the Malaysian experience. [online] The Star Online. Available at: https://www.thestar.com.my/opinion/online-exclusive/making-progress/2016/02/04/economic-inequality-and-the-malaysian-experience/ [Accessed 30 Dec. 2017].

Zahid, S. (2016). Is the income gap as big as people think?. [online] Malay Mail Online. Available at: http://www.themalaymailonline.com/malaysia/article/is-the-income-gap-really-bigger-only-perception-data-shows#Q5d6LxldDkEBf8eS.97 [Accessed 5 Jan. 2018].

Zainul, I. (2017). Petronas raises dividend payment to government. [online] The Star Online. Available at: https://www.thestar.com.my/business/business-news/2017/08/26/petronas-raises-dividend-payment-to-government/ [Accessed 3 Jan. 2018].

Data coding for indigenous language research: attaching local meanings in generating categories and themes

Roselind Wan[1*]

[1]Department of Management and Humanities, Universiti Teknologi PETRONAS, Seri Iskandar, 32610 Perak Malaysia

Abstract. Data analysis in qualitative research is the most complex phase of a study and covers a broad range of approaches with different variation in concepts, assumptions and analytic rules. It involves creating analytic categories reflecting the experiences of participants and highlighting the significance of cultural events happening in the research setting. Accordingly, thematic analysis of data is conducted to examine systematically the content of data to identify recurrent patterns in participants' response and their understanding of their social world. In this paper, I describe the development of categories and themes in a study designed to examine the perceptions on language and identity of Kayan elders in Sarawak, Malaysian Borneo. It demonstrates the analytic rigour in the stages of coding, and offers a rationale for initial coding in the indigenous language. This paper concludes with suggestion for considering indigenous language in coding as it enables the researcher to unpack layers of meanings that are embedded in the cultural context, thereby, enhancing richness of the analysis.

1 Introduction

In qualitative research, the data analysis process is considered the most essential aspect of the study. Analysis of data begins with reading and re-reading through the data corpus to make sense of the data (Hammersley & Atkinson, 1997), identifying analytic categories that reflect on the experiences of participants and henceforth, highlighting the significance of cultural events happening in the research setting. Various scholars consider qualitative coding of data as heuristic and an inductive process (Corbin & Strauss, 2015; Emerson, Fretz & Shaw, 2011; Richards, 2015; Saldana, 2009) and that, the coding process itself, is analysis (Miles & Huberman, 1994). In fact, the goal of qualitative coding is to retain the data until they are fully understood and "to learn from the data, to keep revisiting data extracts until you see and understand patterns and explanations" (Richards, 2015:104).

For an ethnographic study, it requires reading through interview transcripts and field notes, and taking the entire record of the field experience as a complete corpus for analysis. The heuristic nature of data analysis, hence, sees coding as an integration of sequences from observations; "a global reference which encompasses these observations and within which the different data throw light on each other" (Baszanger and Dodier, 2002:11). In which case, the researcher needs to have a certain perspective on observations made in the field so that the interpretations, and the field-notes, are in dialogue with each other (Ricoeur, 1992). This means, the researcher needs to understand the relationship between the how/what and the why of the phenomenon in the research. Nonetheless, qualitative research covers a broad range of approaches to data analysis with different scholars offer variation in concepts, assumptions and analytic rules based on their epistemological and ontological assumptions on research.

In this paper, I describe the development of categories and themes in a study designed to examine the perceptions on language and identity of Kayan elders in Sarawak, Malaysian Borneo. I present a rationale for using Kayan language in the initial coding, suggesting that coding done in participant's language may offer a viable alternative to deeper analysis of the data. Since the goal of ethnography is to "grasp the natives' point of view, his relation to life, to realise his vision of his world" (Malinowski, 1922: 19), coding the data in Kayan gives preference to participants' own words, and thus, retaining the essence of the native's meaning. For this paper, materials for data analysis are selected from the corpus of a larger ethnographic study that consists of data from interview transcripts, participant observation field notes, journals, memos, artefacts, songs, photos, folk tales and speeches. The aim is to demonstrate analytic rigour in qualitative data analysis and offers an alternative method on theme construction that can be used in similar studies. Readings from grounded theories (Corbin & Strauss, 2015; Emerson, Fretz & Shaw, 2011; Saldana, 2009) and ethnographic studies (Baszanger & Dodier, 2002; Brewer, 2000; Hammersley & Atkinson, 1997) informed my understanding and suggestion for theme development process in this analysis.

* Corresponding author: roselind_wan@utp.edu.my

2 Participants in the study

The participants are Kayan elders from a longhouse in the Baram region in Sarawak, Malaysian Borneo. They live in their traditional longhouses located in the rural regions of the Baram and in Belaga. The Kayan is a changing society. In the past, they are known as a communal society that lived entirely in the longhouse. However, in the past two decades or so, due to changes brought about by education and the pursuit of salaried jobs, many have left their longhouse to live in the towns. Although most still live in their longhouses, it is observed that large majority of Kayan in recent times have migrated to live in the towns. A 2010 census estimated the population of Kayan in Sarawak to be around 27,000, less than 0.04 per cent the total population of Malaysia. Kayan culture is generally deeply ingrained in the cultural landscape of Sarawak, and their traditional music and dance given notable attention in contemporary writings. It is unfortunate however, scholarly literature on the Kayan is still under developed. In particular, post-colonial writings on issues pertaining the Kayan language, and the social and cultural changes being experienced by the Kayan do not received sufficient attention from scholars. For this reason, my study hopes to fill the gap in research and contribute not just to our understanding of the Kayan but more importantly, to the much-needed scholarly literature on the Kayan.

The main sources in this study are participants from a Kayan longhouse located in the Baram district. I categorized them into two different groups of participants whom I refer to as elders (see Figure 1). For indigenous communities, elders are often older members of the community; individuals who are recognized as having spiritual and cultural wisdom and possess knowledge of traditional ceremonies, stories, and teachings from centuries past (Merculieff &Roderick, 2013). Based on Merculieff and Roderick (2013)'s contention, I consider those above fifty years of age as elders. The first group of elders, that is, Group 1 are those whom I classify as *daha aleng melo uma* [those who stay at home] and the second group of elders (Group 2) are *daha aleng melo ha'oh* [those who stay in the towns]. *Daha aleng melo uma* are elders that live their whole life in the longhouse. Elders in *daha aleng melo ha'oh* group were selected from those who work and live in the towns. Through my inquiries during fieldwork, this second group of elders has regularly returned to the longhouse over the past several years and they maintain close relationship with the Kayan in the longhouse. I chose these two groups as I anticipate that data from both may offer interesting insights on Kayan perception on their language and identity in Sarawak.

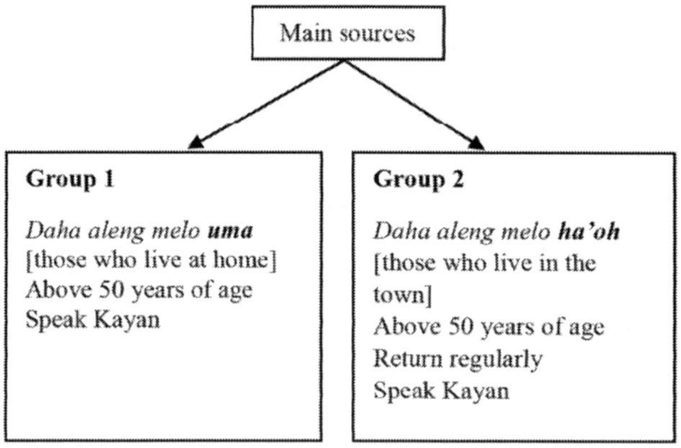

Fig. 1. Participants for the study

3 Understanding meanings in local context

It seems appropriate at this juncture to give a little explanation on the words *uma* and *ha'oh*. According to Fetterman (1998) and Spradley (1979), native words are rich in cultural meanings, thereby giving the English translations for the word *uma* [house] and *ha'oh* [downriver] in coding, may not fully represent their local meanings. For example, *uma* for Kayan is the longhouse but which also means home. In fact, *uma* unlike the English connotation, implies more than just house and home. According to Basso (1988), it is important to understand the environment, and perhaps the meanings and significance of the landscape, and that of the speech acts of the informants so that these could be interpreted in the manner that is fair and near to the meanings that locals attached to it. The elders have lived most of their lives in the longhouse. For them, *uma* suggests a connection to a sense of place, a place with whom the Kayan have an intangible spiritual bond (Basso, 1988), in that it conjures a feeling that is resonance to the mention of *asen* [root, origin], family, the longhouse and home.

Concomitantly, for those who live *ha'oh*, *uma* is often associated with *liveng* - a yearn for the past, for reflection into their memories of growing up in the longhouse. The Kayan in the second group may have a house in the cities, but home to them is always at the *uma* [the longhouse]. The word *ha'oh* in itself means 'downriver' (Southwell, 1990) which refers to any place downriver from the longhouse. In the distant past when travels were limited due to inaccessibility and having no means of communication between the longhouse and the towns, *ha'oh* means a place far away. The isolation of the longhouse from other communities made any place *ha'oh* as somewhat mysterious and out of reach. In contemporary reference, *ha'oh* is the town or cities where young Kayan *ilo hadui* [look for work] or where they *ilo urip* [look for life]. The significance of these terms is ingrained in the meanings that only members of the community know, although sometimes, and sadly so, these meanings are taken for granted (Basso, 1988). Understanding the meanings of words in their local context brings insightful layers to coding and enhance the qualitative analysis of a study.

4 Data Collection

Specifically, for this paper, data is gleaned from interview transcripts of only Group 1 (Table 1). In the extended analysis of the study, data from Group 1 is supplemented with data from Group 2, and from speeches and field notes taken during a study spanning over four years of intermittent field work. Two elders (*) passed away a few months after the interview. I have decided to retain their interview data as they hold valuable insights that complement the other data sets.

Table 1. Profile of elders in Group 1 – Daha aleng melo uma

Profile	*Usun*	*Jau*	*M. Ing**	*Mok Ya**	*Ulau*
Age	64	75	76	80	64
Gender	F	M	M	F	F
Education	Primary School	Primary School	Primary School	-	Primary School
Occupation	Farmer	Farmer	Farmer	Farmer	Grocery Shop owner
Language	Kayan	Kayan	Kayan	Kayan	Kayan
Location	Long Panai	Long Panai	Long Panai	Long Panai	Long Panai
Data collection type	Ethnographic interview	Ethnographic interview	Ethnographic interview	Ethnographic interview	Ethnographic interview

Data collection occurred in two phases (see Figure 2). In Phase One, I conducted ethnographic interviews lasting between 1 – 2 hours each for all the elders. As the elders are known personally to me, the nature of the interview is informal and although it is different from conversation, for ethnographic interviewing, the two would typically merge into one (Fetterman, 1998). Elsewhere in my thesis, I refer to this amalgamate as *tengaran*, a Kayan term for talk and conversation which can be both formal and informal. In Phase Two, data collection was through participants' observations and speeches. This includes field note observations, memos and recorded speeches that supplement data collected through interviews. The data corpus thus offers me insightful data on the social world of the community.

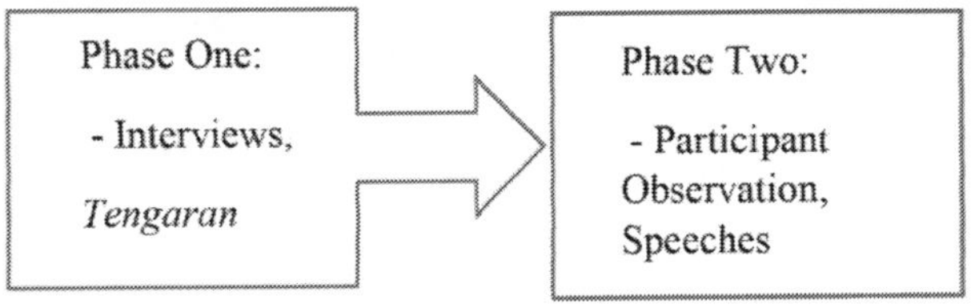

Fig. 2. Phases of data collection

5 Coding in Kayan language

Smith, Chen and Liu (2008) say it is helpful to develop a coding framework in the same language for qualitative interviews that are conducted in the local language. This is because original words, phrases and concepts are securely embedded in the local context and the risk of misinterpretation and loss of participants' intended meaning can be minimised. Proper management of the data ensures that the credibility, transferability, dependability and confirmability of data are not compromised. Identifying an initial coding frame requires skill in pinpointing recurring themes and concepts and developing meaningful labels for the data. To do this, I need to become familiar with the data through listening attentively to the audio recordings. From here, I was able to gather some general sense of emerging categories that indicate possible relevance to the research questions. The interview data was transcribed and then translated into English. Prior to analysis, I read each transcripts a couple of times and wrote analytical memo in Trello, an organizing app to help me have a visual overview of main ideas and emerging categories. I then started coding the English transcripts. Text segments (or meaningful units) were compared and contrasted and assigned inductive codes.

However, on hindsight, I discovered it would be much better to code in Kayan because some nuances of elders' meanings are not fully captured in the English version of the transcripts. For instance, as I highlighted in the earlier section, there are words and expressions from elders that given an English translation would lose some of their dynamics. Complexity also arise in translation, in particular when no equivalent word exists in the target language including the influence of the grammatical style on the analysis (Twinn, 1998). It is thus necessary for me to avoid problems of interpretation and to ensure accurate meaning of data. Also, when the coding is made in Kayan, the meaning will not be lost, the essence of the participants' voices and their cultural nuances are still being retained.

In my study, there are many such examples of words and phrases in the indigenous language that are embedded with local meanings, for instance, the phrases *dahok mahen* and *dahok murah* in the example below. Given a loose English translation, *dahok mahen* and *dahok murah* means 'expensive and cheap language'. This simple translation does not quite capture the layers of meaning that the Kayan word entails. The nuances of *dahok mahen* and *dahok murah* necessitates understanding the concept, and in what context the word is being used, as in Figure 3:

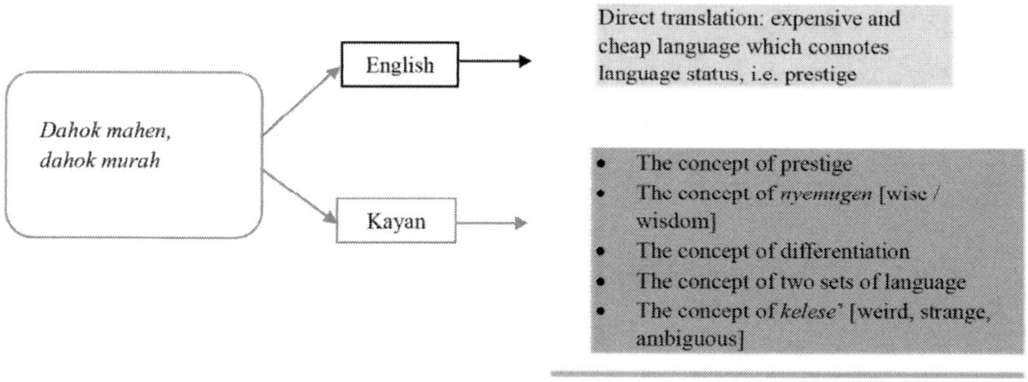

Fig. 3. The concept of *dahok mahen, dahok murah*

Using the Kayan transcripts for open coding allows for layers of meanings which an English translation would not have been able to capture. Here, given a coding in Kayan, the concept of *dahok mahen* and *dahok murah* encapsulates several layers of meaning. More significantly, it assumes the Kayan language to be of two types; old Kayan and 'contemporary' Kayan. That is, the old Kayan language is seen as a *dahok mahen* [prestige language], and often associated with the *dahok menuna* [language of the ancestors]. The *dahok mahen* is therefore considered pure and untainted, free from corruption. Further, their oral tradition *tekna'*, is sung in the old Kayan, where those who can sing the *tekna'* are often viewed as people who are knowledgeable and possess the wisdom of the ancestors. However, the contemporary Kayan language is *dahok murah* [cheap language] and *kelese'* [weird, strange and ambiguous]. The elders refer to it as *dahun nyam kere nih* [the language of the young people]. Consequently, there is the perception that those who do not know how to sing and understand the *tekna'* as quite illiterate in the *ju* Kayan [the Kayan way]. From this example, we can see how coding in Kayan offers voluminous potentials for meanings to be uncovered where *dahok mahen* and *dahok murah* illuminate the notion that 'old' Kayan is a language of prestige and tradition, whilst the contemporary Kayan as one which is contaminated and polluted due to language mixing.

6 Developing categories and themes

The text segment in Kayan transcript below (Table 2) is taken from an interview with one of the elders to illustrate initial coding process in the indigenous language. Making the analysis in the words of the elders allows for exploration of meanings as experienced by the elders themselves, therefore, a richer understanding on their cultural meanings can be achieved (Spradley, 1979). As can be seen, even from this short segment of data, it is possible to come up with a variety of codes from which relevant categories and themes will emerge. Descriptive and in vivo codes are applied to the text segment (Saldana, 2009). There are similar recurrent themes emerging from the data which I grouped into three categories: *meng melak, levah, asen* [do not abandon; the language will disappear or die; the language is our root or identity]. These categories would later be refined with other categories from other text segments to form higher level categories:

Table 2. Initial coding in Kayan

Talk data 5 **Source Mok Ya @ Aug 2014**	**Codes (descriptive and in vivo coding)**
bei, marung [1]dahun kayan nih, sayu lelan kah kurin, kenep ta re, nureh yah... [2]Nusi tam deng melak na. ita em jam ikam nyam nyam nih, ita anih tei pah na'a, daha anak kam anan usi nah ita jam. Pah si ya, ikam na na na ure' pah[3] iha seng te'. Nutih. [4]Nei pa ji nganak pa dahun uma beh, nei pah ji nganak pa dahun uma beh, dahun inggeris dahun melayu daha tih pah,[5]tam lak2 nah dahun tam jam na nipa. (avin na nun tam melak?) bei, avin [6]ha kajo dahok atih lo kereh, nuno, [7]daha duyah dahun tam yah, bi kurin ta anih sayu lelan dahun tam nih itam na pa'en. Kurin ta nih. Usi dahun tam nih seng tei [8]levah kurin ta nih. [9]Menangen tam na kah kurin ta nih, bi ha dap kerei nih pa, na dahun ha beh lelan daha men anak daha. [10]Anan pah ji tusah nih pah. Menangen kelo uk te dahok kayan anan [11]ja ha sayu kereh. [12]Ja ha em tei pah, [13]tei melak. [14]Bara asen tam atih ka sayu na. [15]tei ha melak [16]usi nah te dama asen tam la'an ujung [17]asen daha tua nah. Nure' pelemana kui reh. usi ta mun melak na urei ha em tei levah. [18]avin tada tam kayan atih pah, tei ha levah, ujung itung [19]dahun uma beh tua nah [20]usi gunah nutih, [21]bi te lan pu'un tam kayan anih, [22]meng na na levah, [23]na pa'en, nutih aleng pelemana kui reh [Regarding our language, it is really good, I think…it is not good for us to *melak* it. I do not know about you, young people, but when we are gone, I won't know what will happen to the language. May be it will be gone, finished. Unless you speak it. It is like that. It seems that, parents speak other languages (English or Malay) to their children, we have *melak* our language. We do not speak it. *Why do you think that we do not speak it?* Well, is it because the other languages are *kajo?* why is it that they do not want to speak our language, but to me, our language is *sayu lelan*, that we should speak it always. I would hope that our language will not *levah*. We should always speak it, speak it *menangen* to our children, grandchildren. That is the problem. We should *menangen* speak it. But it's different now. People just speak other languages to their children. This is the problem. You must *menangen* speak the Kayan language so that it is *sayu*. So that it will not *pah, levah*. Otherwise we will just speak other people's languages. It is *sayu* to show our *asen*. This is what I think about it. We should not *melak* it so that it will not *levah*. Because it is *sayu* to show our *asen*, a *tada* of our identity. If it is *levah*, then our *asen* is no longer as many, it would only be other people *asen*. This is what I think. Do not *melak* it so that it will not levah. Because it is our *tada* as Kayan. What is the point, if we end up just speaking other *uma ha beh* languages because we have our *pu'un* as Kayan? Do not let it *levah*, speak it always. This is what I think].	Dahun Kayan sayu lelan[1] Usi deng melak[2] Ikam nyam na na ure' seng te[3] Tei pah na'a[4] Tam lak avin pekunak dahun ha beh [5] Kajo sik daho ha beh[6] Daha duyah[7] Levah[8] Menangen tam na[9] Anan pah ji tusah nih pah[10] Ja ha sayu[11] Em tei pah[12] Tei melak[13] Bara asen[14] En tam lek dahun tam avin na dahun ha beh[15] Usi nah dama asen tam[16] Asen daha tua[17] Tada tam[18] Tei ha melak, usi te dama asen tam la'a[19] Usi gunah nutih[20] Te lan pu'un tam Kayan anih[21] Meng na na levah[22] Menangen na dahun tam[23] *<u>Kayan-English</u>* Melak: abandon, leave Kajo: beautiful, good Sayu lelan: very good Levah: disappears, gone Menangen: purposeful, persevere Pah: finish, gone Sayu: good Asen: origin, heritage Pu'un: origin, root Tada: symbol Uma ha beh: other people's

Dealing with the data and coming up with themes is not a straight forward process. In the earlier stages of my data analysis, interesting patterns emerged of possible loose categories (e.g. Christmas, Education, Intergenerational transfer, Land, Modernisation, Oral tradition, Social structure, Longhouse, Complexities, etc.). These loose categories form the basis for development of later categories and are helpful as they offer early analytic insights of what elders perceived as meaningful in relation to Kayan language and identity. The iterative nature of analysis (Fetterman, 1998) as such, makes data analysis in qualitative research both challenging and interesting. It demands that the researcher makes choices "between logical and enticing paths, between valid and invalid but fascinating data, and between genuine patterns of behaviour and series of apparently similar but distinct reactions" (pp 92).

Hence, developing categories from the codes previously identified was a time-consuming process. The categories of the data need to be clarified and developed in relation to one another, including specifying the links between the various concepts and indicators derived from the coding. In the text segment example presented above, 23 codes were identified. These codes need to be clarified with the other codes from the other data sets, which altogether for Group 1, there are about 300 codes. These were later categorized into 58 higher order themes. From the Kayan codes, categories were refined and regrouped into five categories and further refined into themes (See Figure 4):

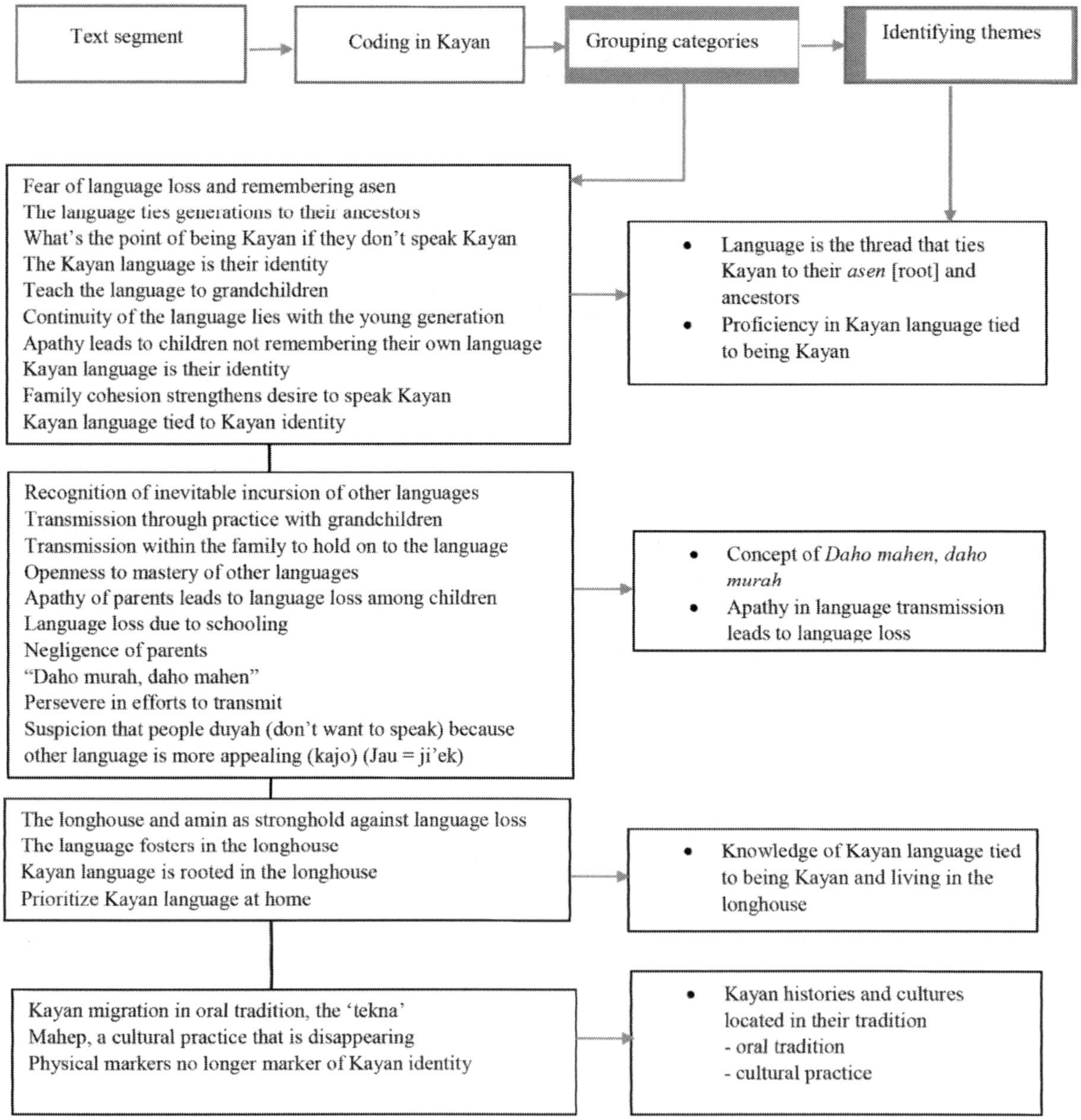

Fig. 4. Development of codes into categories and themes

7 Conclusion

Qualitative coding in the indigenous language open avenues of inquiry through identifying codes using the elders' own words without translating the text into English. Retaining the elders' words and coding in the original language enable the researcher to unpack layers of meanings that are embedded in the cultural context, thereby, enhancing richness of the analysis. It is possible that translating and coding in a translated form could have given a different interpretation to the text. Qualitative analysis is also enhanced when the researcher considers the real-world context of participants in the data analysis process. This paper has offered an alternative way of initiating the initial coding of a qualitative study in the Kayan language. It highlights that coding done in the original language of the elders could offer richness to the interpretation of the data while at the same time, making sure that the analysis still agree to the context from which data were collected.

8 References

Baszanger, I., & Dodier, N. (2002). Ethnography. Relating the Part to the Whole. In *Qualitative Research: Theory, Method and Practice*, pp 8-23. D. Silverman (editor). London: SAGE Publications Ltd.

Basso, K. H. (1988). "Speaking with Names": Language and Landscape among the Western Apache. *Cultural Anthropology, 3*(2), pp 99-130. Blackwell Publishing

Brewer, J. D. (2000). Ethnography. In *Understanding Social Research Series*. A. Bryman (editor). Buckingham, UK: Open University Press

Corbin, J. M., & Strauss, A. L. (2015). *Basics of qualitative research: Techniques and procedures for developing grounded theory* (4th edition). Thousand Oaks, CA: Sage Publications Ltd

Fetterman, D. M. (1998). Ethnography. Step by Step (second edition). *Applied Social Research Methods Series,*17. Sage Publications

Hammersley, M. & Atkinson, P. (1997). *Ethnography. Principles in Practice* (2nd edition). New York, NY: Routledge

Illarion L. Merculieff & Libby Roderick (2013). *Stop talking. Indigenous ways of teaching and learning and difficult dialogues in higher education.* University of Alaska Anchorage.

Malinowski, B. (2002). Argonauts of the Western Pacific. An account of Native Enterprise and Adventure in the Archipelagoes of Melanesian New Quinea. London: Routledge

Miles, M. B. & Huberman, A.M. (1994). Qualitative Data Analysis (2nd edition). Thousand Oaks, CA: Sage Publications.

Ricoeur, P. (1991) *From Text to Action: Essays in Hemeneutics, Vol. 2*. Evanston: Northwestern University Press.

Saldana, J. (2009). The Coding Manual for Qualitative Researchers. Thousand Oaks, CA: Sage Publications Ltd.

Smith, H. J., Chen, J., & Liu, X. (2008). Language and rigour in qualitative research: Problems and principles in analysing data collected in Mandarin. *BMC Medical Research Methodology, 8*, 44. Downloaded March 27, from http://doi.org/10.1186/1471-2288-8-44

Spradley, J. P. (1979). *The Ethnographic Interview*. New York: Holt, Rinehart and Winston, Inc.

Southwell, C. H. (1990). *Kayan-English Dictionary*. Kuching, Sarawak: Sarawak Literary Society

Twinn S. (1998). An analysis of the effectiveness of focus groups as a method of qualitative data collection with Chinese populations in nursing research. *Journal of Advanced Nursing, 8*, 654–661. doi: 10.1046/j.1365-2648.1998.00708.x

Providing Meaningful Interactions at KOSEN between Japanese and International Students

Richard Grumbine*[1], Natsuki Aka[1] and Riho Hirano[1]

[1] National Institute of Technology - Ariake College, Japan

Abstract. Internationalizing the Japanese students at KOSEN has proven difficult. While KOSEN welcomes international students with the hope of internationalizing the almost entirely Japanese student body, the challenge has been to get the Japanese students to interact with the international students and take advantage of the opportunity. Far too often international students live in a bubble and have meaningful contact with only a few students. This leaves the international students feeling isolated and the Japanese students not benefiting from the opportunity that KOSEN is trying to provide. Ariake KOSEN attempted to address this problem in two ways. A discussion class which included international students was created where the students were expected to discuss pre-assigned topics in small groups over the course of a semester. These groups often contained an international student. This forced exposure created a chance for Japanese students to communicate with international students. This communication was then seen continuing well beyond the discussion exercises. A survey was given at the completion of the one semester class to evaluate effectiveness. The results show that students found the opportunity to be helpful and in line with their own learning goals. Another approach was to create a more casual place for students to gather and interact with international students. An English Lab was created to give the students a place to interact outside of the traditional classroom atmosphere. This second effort is just in the beginning stages but the basic ideas are laid out in this paper. Anecdotal evidence seems to suggest that the Lab is working, with usage on the part of Japanese students being high, but the international students seem less inclined to use the room.

1 Introduction

The KOSEN system (Japanese National Institutes of Technology) is attempting to better internationalize and create more globally minded engineers. In fact, this is the third of three educational purposes espoused by KOSEN in their latest promotional literature, "To develop practical engineers equipped with a high professional ethics, a wide view, and international mind [sic]" (NIT-A School Catalog, 2017). To this end, KOSEN invite international students

[1] Corresponding author: richard@ariake-nct.ac.jp

to study at their campuses and make available many internships and exchange programs to both bring international students to KOSEN and give KOSEN students opportunities to travel abroad. Many of these students receive scholarships or other financial aid from the Japanese government to help in these internationalizing efforts, for example JASSO funds many such programs (JASSO, 2017). However, even with these programs and efforts in place, Ariake KOSEN usually only has about 5 full time international students and a small number of international interns who have much shorter stays. One of the reasons these international students are admitted is because of the need to internationalize the mostly Japanese student body and to help students build a more global mindset. But it was observed that the international and Japanese students did not interact very often. A survey was given to assess the reasons why Japanese students did not take advantage of the opportunity to interact with the international students (Hirano, 2018). No survey has as yet been given to the international students but is under consideration; however, the number of international students is so small as to likely yield statistically insignificant results. Concurrently to the giving of this survey two different English teachers tried two very different approaches to encourage greater interaction between the Japanese students and international students. One author created a discussion class featuring several international students and another author created an English Lab to encourage a more informal interaction between international students and Japanese students. Each will be discussed in turn.

1.1 Discussion Class

1.1.1 Introduction

In order to address the issue of international interaction, a selective discussion class was created including three international students. The class was made up of twenty-four Japanese fourth year KOSEN students (similar to freshman university engineering students) plus three international students. The Japanese students were divided into five groups according to their own preferences. Two groups contained a single international student while the other two groups shared an international student between them. The class met once a week with discussion topics pre-assigned. Before class the students were required to write a four-paragraph essay on the upcoming topic to help the students form opinions and to help them feel more comfortable during discussions. About 20 minutes of discussion time was allocated per meeting. An English teacher was also available to help facilitate the discussions and there were two TAs who helped the students write their essays, one a native Japanese but fluent in English and one an international student also fluent in English (both TAs had a similar class the year before with the same teacher giving them familiarity with the classroom mechanics).

1.1.2 Evaluation Survey

A five-question survey was given to the students at the end of the semester. The questions and results can be found below. Students seemed to enjoy taking the discussion class, since none answered they did not enjoy the class (Figure 1). As the discussion class was a selective class, most of the students took it to improve their English ability (Figure 2). The interesting point that was revealed by the survey was that while about 65% of the students found the class difficult or very difficult, and no one responded that the class was easy (Figure 3), they still felt that they would want to take a similar class again in the future (Figure 4). They also mentioned that the skills they would most like to improve in English are listening and speaking skills (Figure 5).

Q1. Did you enjoy the class?
Q2. Why did you take this class?
Q3. How difficult was the class?
Q4. Do you want to take a similar speaking and writing class again?
Q5. What skills do you want to improve?

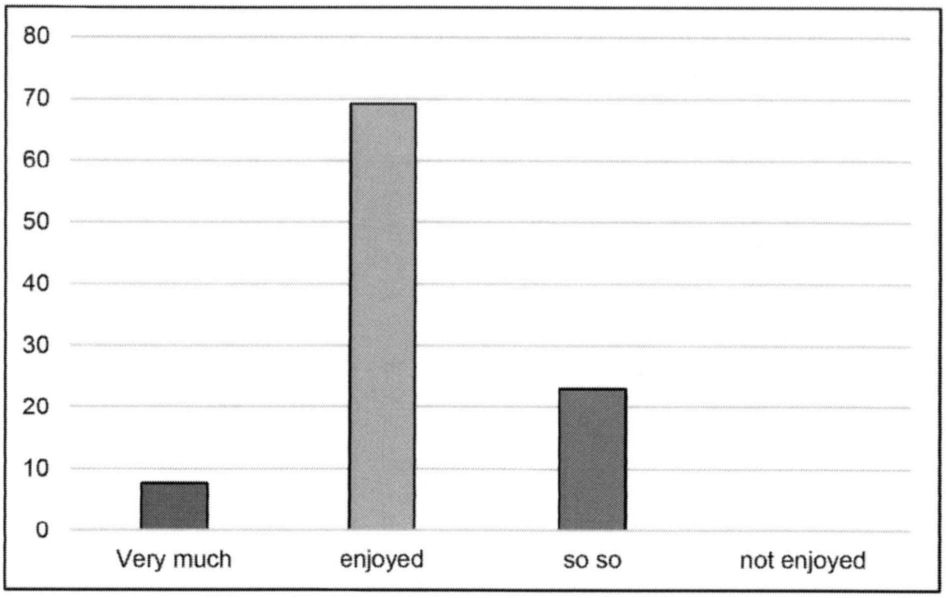

Fig. 1. The result of Q1: Did you enjoy the class?

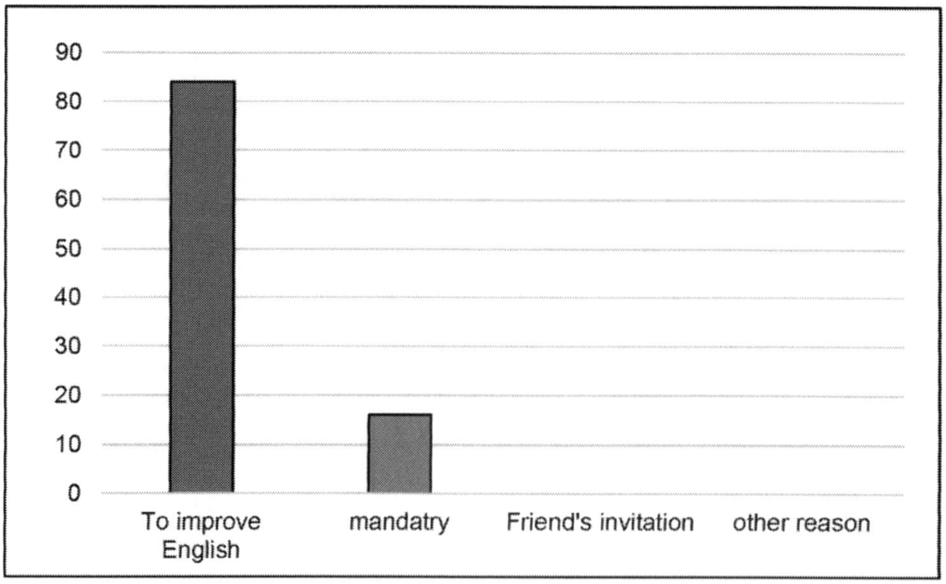

Fig. 2. The result of Q2: Why did you take this class?

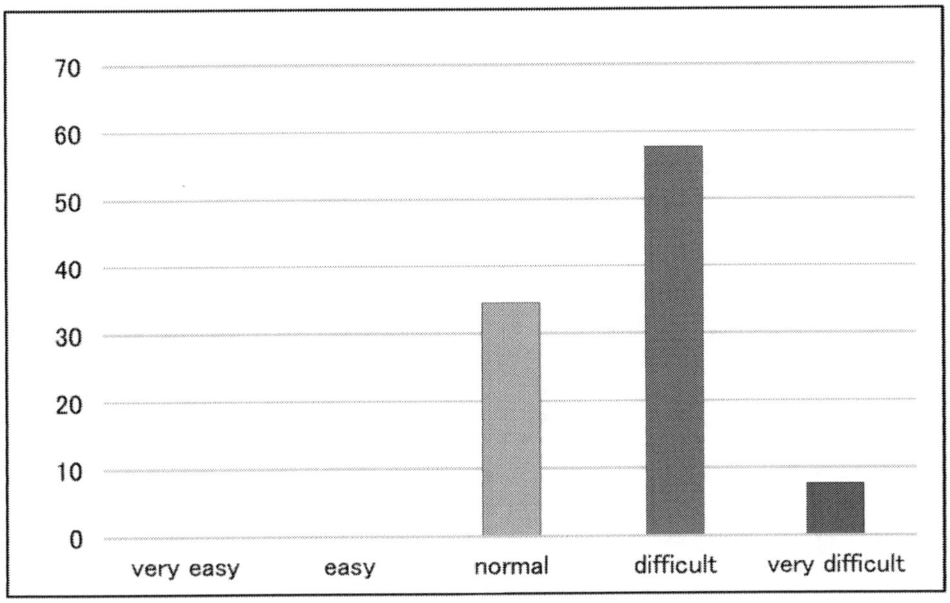

Fig. 3. The result of Q3: How was the class?

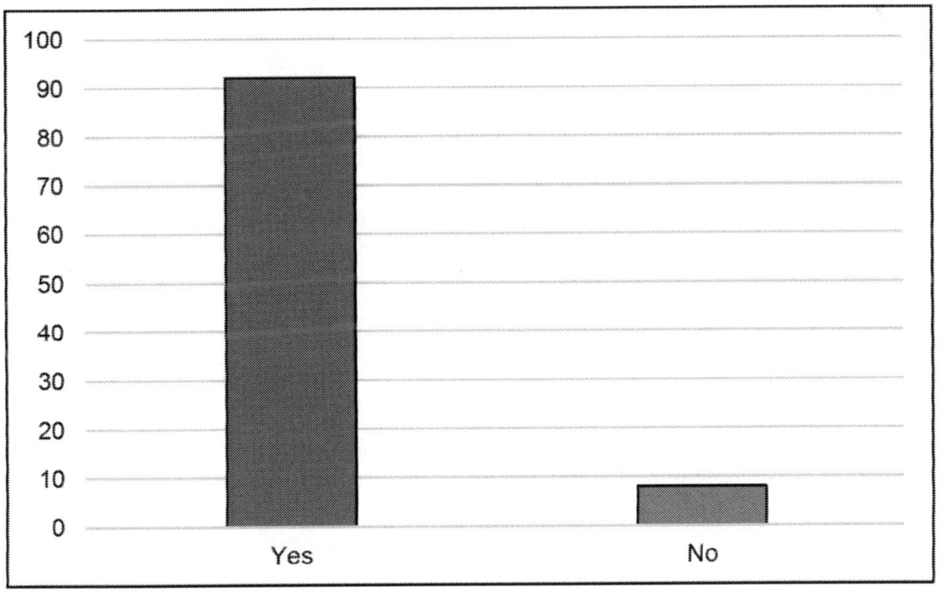

Fig. 4. The result of Q4: Do you want to take a similar speaking and writing class again?

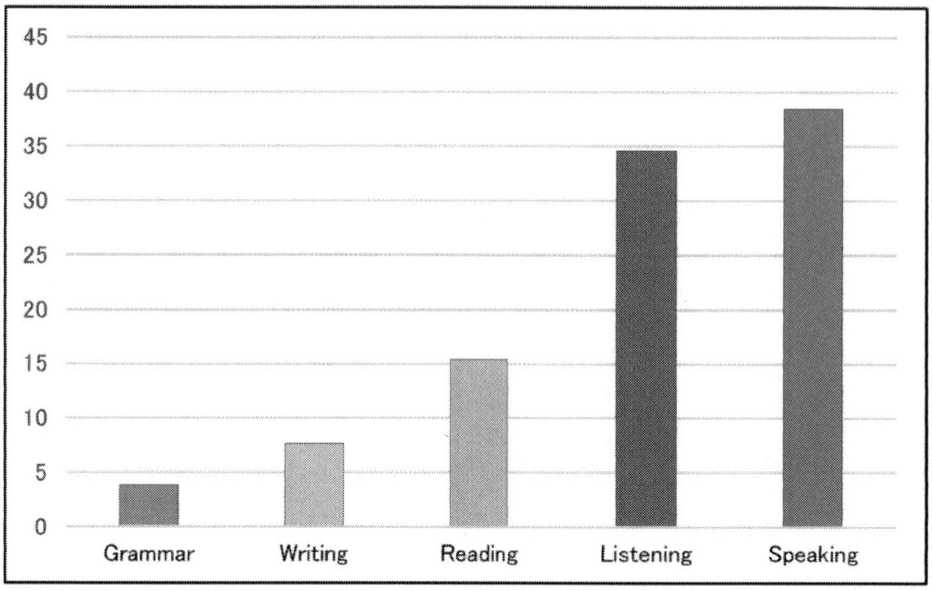

Fig 5. The result of Q5: What skills do you want to improve?

A follow up question was then asked, what had they learned in class. Below are the responses from students.

1. Even though I do not use difficult English words, and cannot speak fluently, I can convey what I want to say in English.
2. If I cannot understand what other speakers say, it is difficult for me to speak. I need to improve listening skills.
3. I could convey the message by using simple grammar.
4. It was a practice not only for speaking development, but also it enhanced my listening skills.
5. The important thing is to have my ideas first. If we do not know what to say in Japanese, then, we cannot express the thoughts in English.
6. I learned how to write an essay in English.
7. I found that if we take time, we can improve speaking skills. The class was beneficial for me and I found if I could speak more English, it would have been more interesting.

1.1.3 Discussion of results

According to the questionnaire, most of the students enjoyed the discussion class. One reason may be that the class was selective, therefore, mostly already motivated learners joined this class. Though the students felt it was challenging to prepare for writing and expressing their ideas in English, they said they would want to try this kind of class again. This is a good indication that they have learned something they feel is important through the discussion class. The most important point is they could have a chance to communicate in English. Japanese students tend to be very shy, always caring about mistakes (Saito and Eisenstein 2004). They feel making mistakes in front of other people is not good. English teachers need to address such self-imposed stereo-types concerning Japanese students. Conversation practice in small groups may help students to feel less pressure about making mistakes while talking to others in English. Such training will also help them realize that with simple English they can convey their message. The data also shows students hope to improve their speaking and listening ability, indicating they felt they want to improve these English skills even more after experiencing the discussion class.

The authors observed that groups with international students spoke in English more often and with more fluency than groups that did not contain an international student. Also, having an international student in the group allowed the students to explore issues of international differences. Discussions in these groups were also often seen continuing outside of class as well. These last two points show that this attempt to internationalize the Japanese students was somewhat successful. Students learned about foreign cultures and spent time talking with foreign students. But further research is needed to investigate this more thoroughly.

1.2 English Lab

1.2.1 Introduction

The English Lab was created with three purposes in mind. First and foremost, to help students with their English skills, especially conversational English and presentation skills. Secondly the LAB is meant to help faculty and staff with their English needs, primarily editing of academic papers and emails, and helping with academic presentations. And thirdly and pertinent to the issue at hand, internationalizing the students and giving them a chance to interact with people from other countries/cultures in a non-classroom setting. This last point was felt to be important by the authors as student anxiety in the classroom sometimes hinders learning (Horwitz, E., Horwitz, M., Cope J. 1986).

Although most KOSEN accept international students and perhaps employ a few international teachers (Ariake Kosen has roughly 1000 students and 80 faculty members, with five international students and three international faculty members from abroad), the Japanese students reported having difficulty interacting with the foreign students. A survey identified four areas of concern on the part of the Japanese students as reasons for not interacting with foreign students: shyness, lack of confidence, few chances for interaction, and difficulty finding a reason or topic for conversation (Hirano, 2018). The English Lab it was hoped could help address these concerns. The English Lab was envisioned to be a place where the international students could relax and interact with the Japanese students in a more casual atmosphere. (The school itself provides little in the way of community or common spaces on campus). The lab is small (seats about twelve people comfortably - three on a couch, and eight around a conference table plus one more chair) and is business casual in atmosphere (the donor room was a conference room and it still has to serve that purpose when helping students, faculty, and staff with presentations). Basic amenities to facilitate casual interaction and to provide a more relaxed atmosphere are provided by the Lab. These include: comfortable chairs, conference style table, quality audio and video systems, coffee and tea pots, a microwave, selection of books and magazines, phone charging station, a video game system, and a computer with internet access. It is hoped that by creating this space the students would find it an encouraging place to intact with one another. The lab was just completed at the end of last semester and this is the beginning of its first full semester of usage so little can be said about its effectiveness at present. Three weeks into the new school year, has seen the lab have regular attendance from many first-year students. About fifteen visitors per day (the lab is open four days a week, two hours after school Wednesday and Friday, and one hour at lunch time Tuesday and Thursday - so hours are quite limited). Most of the visitors are new students, though a few upper level students also come regularly. Only one intentional student currently comes with regularity. In the future, a paid system of attendance for international students may be tried to both encourage them to come to the room and to expand the hours of the Lab. A survey is also planned to ask students how they want to use the room and what amenities they would like to see provided, so that we can make it even more user friendly.

1.2.2 Challenges

There have been several challenges in creating, maintaining, and improving the lab. Acquiring a suitable space and enough funding for creating the lab were the first hurdles. In this case timing was good, as a semi suitable room became available, setting the stage for obtaining funding. A presentation was given to the school president, in order to obtain funds directly from the school (there was not enough time to pursue outside funding). In the end were given half of the monies we felt were required. This led to some cuts in amenities and a search for some outside funding sources in the future. Also, several English teachers donated some of their research funds to the project and or items that were needed. In all, about 500,000 yen was spent, with a few also items being donated. Once the lab was about to be up and running there were several other challenges that had to be dealt with. How would we get students to come to the lab? And how to get them to interact once they were there? Giving the lab an eye-catching exterior helped it capture attention (bright yellow vinyl wrap with the new room's name in big bold letters). Offering free coffee and tea also helped get students to explore the space and stay long enough to talk with one another and with the host (an American teacher at the school). On the down side, the international students this year have been off to a rough start due to some immigration issues beyond our control. So, no formal introduction for the three new incoming students this year has as yet been made to our students. And only one of the new international students has been to the lab so far. A returning international student however has been coming weekly and plans to come more often. Perhaps another problem is that this year most of the new international students do not speak much English and prefer Japanese, so the English aspect of the lab may be working against one of its purposes in some ways. This is a first-time issue, so not sure if it will be a continuing trend with future incoming students. A final challenge is staffing and paying for consumables because there is no current monetary allotment for these two things. The lab's only staff is also a full-time teacher at the school with most of the usual responsibilities of a full teaching load so there is not so much time for working in the lab, hence its rather short hours (the lab is currently only open six hours a week). Also, all monies for consumables comes out of that teacher's pocket at present. To address these last two issues, it is hoped that we can get the school to recognize the Lab as an official school committee or service and let its supervision take the place of other committee and service work on the teacher's part and find ways of getting some minimal funding for consumables.

1.2.3 Evaluation

The lab has so far only been partially successful in achieving its goals. While it has been heavily used by Japanese students and faculty members, there have been precious few visits by foreign students. The most recent TOEIC scores show that Japanese students who use the lab regularly scored very well. With five of the top ten scores being achieved by English Lab regulars and all five having shown great improvement in their scores over a short period of time (at least 100 points in a single semester). Likewise, faculty and staff usage of the lab's editing services has been quite frequent with at least one visitor per week. But precious few visits are made by the international students at the school. In fact, they only seem to occur about once every few weeks.

A survey given to regular Japanese attendees of the English Lab showed that 60% of the students would appreciate talking with foreign students in the lab and no students indicated that they would not like to interact with foreign students, so increasing foreign student attendance is a priority.

2 Closing remarks

These two attempts to internationalize the student body (integrated discussion class and English Lab) have been moderately successful but there is still plenty of room for improvement. It would be very helpful if the number of international students in the discussion class could be increased so that there was one international student per group. And in a similar vein, the English Lab needs better attendance by international students and more flexible hours so that it can be more available to Japanese students. The general lack of international students on campus makes solving these issues very difficult. The number at our KOSEN has been fixed at about five international students total in the student body. This number has remained unchanged for at least fifteen years. This is despite an upward trend (better than 10% increase year on year for the last three years) of international students enrolled in Japanese universities in general (JASSO, 2017). However, these numbers do not take into account the many temporary students visiting our KOSEN from abroad, and those numbers have been increasing over the last several years (12 in 2014, 16 in 2015, 18 in 2016 and 17, and at least 20 are expected this year). Although it is hard to integrate these temporary visitors into a discussion class it is easy to incorporate them into activities in the English Lab.

3 Suggestions

So how to improve the overall situation at KOSEN? Both of the afore mentioned attempts to improve globalization could benefit from greater numbers of international students being enrolled full-time at KOSEN. And in the case of the English Lab some small amount of funding to pay existing international students a stipend to attend the lab making the lab more available for use (from six hours now to perhaps twelve hours a week or so with additional hours being staffed by international students). Increasing the numbers of international students and the opportunities for interaction would only improve the chances of these programs having greater success.

4 Further research and other concerns

The issue of increasing international student enrollment will quite likely lead to another issue, how to adequately serve those international students. Anecdotal evidence seems to suggest that at least for Ariake, the environment is not too attractive for most international students. Also, the limited number of international students at the school makes meaningful research difficult as it is hard to expect any statistically significant results from such small numbers, so anecdotal evidence is about the best we can expect.

 https://doi.org/10.1051/shsconf/20185301003

5 References

Hirano, Riho (2018) Taking advantage of international students, *Japan Counsel of College English Teachers*, National Presentation Contest

Horwitz, E., Horwitz, M., Cope J. (1986) Foreign language classroom anxiety. *Modern Language Journal*, *70(2)*, 125-132.

JASSO 2017 (2017) International students in Japan 2017 retrieved from https://www.jasso.go.jp/en/about/statistics/intl_student/data2017.html

NITA School Catalog 2017 (2017) National Institute of Technology, Ariake College

Saito H., Eisenstein M. (2004) Seeing English language teaching and learning through the eyes of Japanese EFL and ESL students. *Foreign Language Annals*, *37(1)*, 111–124

Schweers, C. W. (1999) Using L1 in the L2 classroom. *English Teaching Forum*, *37(2)*, 6–15.

ASEAN Socio-Cultural Community: ASEAN Approach and Effort in Rural Development and Poverty Eradication

Fujica Anak Anggo[1,*] *and Louis Laja*[1]
[1]Institute of Design and Innovation, University Malaysia Sarawak, Malaysia

Abstract This paper examines the ASEAN Community approach and effort towards rural development and poverty eradication. The research by ASEAN Socio-Cultural Community which highlighted rural development and poverty eradication was driven by many factors. What has ASEAN planned to reduce poverty and increase rural development? Is this plan realistic? What approach was used? The World Bank defined rural development as a strategy designed to improve the economic and social life of a specific group of people; the rural poor. It involves extending the benefits of development to the poor among those who seek a livelihood in rural areas. "Human capital development should not be neglected as well as it can be the main engine of growth. We must leverage the existing human capital resources towards a holistic approach to rural development" (Bernama, 2017). In addition, according to Malaysia's former Deputy Prime Minister, Datuk Seri Dr. Zahid Hamidi, poverty eradication not only continues to be a priority in the ASEAN region as stated in ASEAN Socio-cultural Community Blueprints 2025 but also globally as outlined in the United Nations 2030 Agenda for Sustainable Development. This study will explain broadly what ASEAN Socio Cultural Community is building towards rural development and poverty eradication by its notion of people centricity.[1]

Keywords: rural development, planning, and poverty eradication

1 Introduction

ASEAN Community consists of three pillars: ASEAN Political Security Community, ASEAN Economic Community and ASEAN Socio-Cultural Community. ASEAN Political Security Community (APSC) aims to bring ASEAN's political and security relations to a higher plane. The second is ASEAN Economic Community (AEC), whose goal is of regional economic integration. The AEC intends to narrow the development gap among the member states and transform the region into a stable and prosperous community. The ASEAN Socio-Cultural Community (ASCC) intends to contribute to realizing an ASEAN Community that is people centric and socially responsible with a view of achieving enduring solidarity and unity among the nations and people of ASEAN by forging a common identity in building a caring and sharing society which is inclusive and harmonious where the well-being, livelihood and welfare of the people is enhanced (Amador, 2011). A short brief on ASEAN Socio Cultural Community - ASCC in consonance with the goal set by ASEAN Vision 2020, envisages a Southeast Asia bonded together in partnership as a community of caring societies. In line with the programs of action plan set by the 1976 Declaration of ASEAN Concord, the community shall foster cooperation in social development aimed at raising the standard of living of disadvantaged groups and the rural population, and shall seek the active involvement of all sectors of society, in particular the women, youths and local societies (Rodolfo, 2006).

Standing on fundamental premises and taking a long term perspective, one would say that the core of the ASEAN Community is the Socio-Cultural Community conceived as a vehicle for developing a sense of Southeast Asian identity, building a regional awareness and fostering mutual understanding among the people of ASEAN (Severino, 2006). The 10 countries of ASEAN have decided the right planning that benefits and engages the people. They cooperate to help each other to face arising issues. Moreover, when ASEAN Community was launched in 2015 it was a historic milestone and was seen as a culmination of ASEAN resilience and dynamism throughout a journey lasting nearly half a century, and gave a signal to the world of how far and how well the ASEAN Member States have achieved in coming together as one community (ASEAN Community, 2015). According to ASEAN Socio-Cultural Community Blueprint 2025, The ACSS 2025 opens a world of opportunities to collectively deliver and fully realize human development, resilience and sustainable development to face new emerging challenges together. Furthermore, ASEAN has developed an ASEAN Socio-Cultural Community Blueprint 2025 in order to lift the quality of life of its peoples through cooperative activities that are people-oriented, people centric, environmentally friendly, and geared towards the promotion of sustainable development to face the new and emerging challenges in ASEAN (ASEAN Socio-Cultural Community Blueprint 2025). By discussing the notion of ASEAN that is people -oriented and people centric, this paper provides important insights into understanding rural development and poverty eradication more deeply.

*Corresponding author: fujian_anggo@yahoo.com

2 Literature Review

Theoretically, sustainable development should be pursued in a coherent manner at global, regional, sub-national level, with nested objectives based on subsidiarity principal coordinated vertically (top-down and bottom-up) (Simon, 2015). Highlighting one of the ASEAN visions stated in ASCC Blueprint 2025, sustainability is defined as development that satisfies the needs of the present without compromising the capacity of future generations, guaranteeing the balance between economic growth, care for the environment and social wellbeing (Acciona, n.d).

As for the first approach, ASEAN countries are, according to United Nations.org, on September 25th 2015, countries that adopted a set of goals to end poverty, protect the planet and ensure prosperity for all as part of a new sustainable development agenda. There are 17 goals to transform the world - no poverty, zero hunger, good health and wellbeing, quality education, gender equality, clean water and sanitation, affordable and clean energy, decent work and economic growth, reduced inequality, sustainable cities and communities, responsible consumption and production, climate action, life below water, life on land, peace justice, strong institution and lastly, partnership for the goals. All 17 goals have been planned and are well aligned with the current and future needs of an ASEAN community. It is a practical agenda that works towards the goal of 2025. It is the main core of the planning to develop rural development as well as eradicate poverty.

Moreover, according to the Association of Southeast Asian Nation, inclusiveness of sustainable development through the promotion of rural development is a key strategy for ASEAN to eradicate poverty. The aim is to strive for an ASEAN community with equitable access to sustainable environment that can support its social development and its capacity to work towards sustainable environment. All ASEAN communities are free to access this accommodation and government planning equally in a way to increase the quality level of its social development. The social development among its people will impact the environment which will slowly head towards sustainable development goals in 2025.

There are several key result areas in sustainable environment that are lined up by ASEAN, Conservation and Sustainable Management of Biodiversity and Natural Disaster, Environmentally Sustainable Cities, Sustainable Climate and Sustainable Consumption and Production. All the strategic measures in each of the key result are aligned with the effort that leads to the 17 goals. These are huge effort that ASEAN community had planned, by putting their people first above all. Since ASEAN motto is one vision, one identity and one community, everyone is inclusive in ASEAN development planning. This will promote equitable access to opportunities for people. In summary, without looking at the level of income, skin colour, education level and etc., everyone will be treated equally and can become part of the ASEAN planning. Interestingly, its characteristic focuses on addressing the concerns of all people. It will reduce barriers, have equitable access for all, promotion and protection of human rights. The people is the main core of succession. One community, by putting the people as the main core of the succession plan will not take long to achieve goals. With this, the rural development will be implemented well according to plan since everyone is cooperative.

Resilience is part of the element in ASEAN Socio-Cultural Community Blueprint 2025. It is necessary to build resilient communities in the ASEAN region post-2015. Based on ASCC Blueprint 2025, resilience is an essential aspect of human security, and sustainable environment which is addressed by integrating policies, capacity and institution-building, stakeholder partnerships in disaster risk reduction, humanitarian assistance, and community empowerment among others. Resilience has to be inclusive, non-discriminatory and incorporate market, and technology based policies, including contribution from private sectors as well as the scientific and academic communities. Resilient community is able to withstand quickly from difficult conditions that happen in particular country. So this stability leads to massive growth in the economic, political and social areas in ASEAN. "The ASEAN experiences have shown that when countries in a region consult and cooperate, their chances of stability and success are greater. It is no accident that the ASEAN countries have such good records of economic growth and political stability" mentioned in a speech by the Malaysia Prime Minister, Tun Dr. Mahathir bin Mohamad at the 24thASEAN Ministerial Meeting in Kuala Lumpur, Malaysia on July 19, 1991. In Malaysia, the head leading the people become the liaison for the government to increase the quality of life for people in the rural areas. For example, in Kuching, Sarawak the people representative who has been voted and trusted by the people in a particular area will install electricity, increase the number of schools in the rural area, help excellent students to further their studies and help give seeds to those involved in agricultural activities. These are the efforts that are given by the government in rural areas for poverty eradication.

Agriculture is a great effort outlined by the government to increase the standard of living. For example, in Indonesia the government stresses on the efforts to grow agricultural industries in order to help its people and reduce poverty especially in rural areas. Renewed efforts towards agricultural and rural growth will be crucial to reduce the number of poor and food-insecure Indonesians. Rapid agriculture growth was a major contributor to the drastic reduction in poverty in the 1980s. In 2002 agriculture (wages and farming incomes) accounted directly for 43% of rural household income in two thirds of rural employment (ADB, 2006). Moving to Cambodia, according to Mr.Srun Sokhum, Deputy General

Directorate of Agriculture, Forestry and Fisheries, his country still considers agricultural sector as its main sector that contributes to economic growth, rural poverty reduction and food security. It is important as Cambodia has the highest number of farmers in Southeast Asia. In Lao-PDR, they addressed the issue of food-security not only in Lao but also in Southeast Asia where they highlighted the role of smallholder farmers in agriculture. According to the World Bank, Malaysia has had an inclusive economic growth rate of about 7 percent annually for the past 25 years. Since this growth has been inclusive, Malaysia has been successful at nearly eradicating poverty in the country (Paramasivm, 2018).

3 Methodology

It is significant to conduct an investigation to clarify the research information and data obtained. The most suitable methods used for this research is qualitative in nature. The two types of qualitative methods used for this study are observations and documentary analysis of publications that are published by the ASEAN organizations. The reason for choosing observation method is because it is a straightforward way to know what ASEAN people do under certain circumstances. The reason for choosing a documentary analysis method is to obtain data from existing documents without having to question people through interviews and questionnaires.

3.1 Observation Method

The observation is conducted to observe the behaviour of people towards the changes that were made by the government. Observation is a systematic data collection approach. Researchers use all their senses to examine people in natural settings or natural occurring situations (Foundation, 2008). Observations are conducted at places that has less access to facilities like clean water, electricity, less development and etc. Investigation was done at the place that has a bit of development done by the government to see the changes that shape the lives of people in that particular place. As the name describes, "observational" methods are all about observing the participants. The observation will help to understand the research better by looking into pre and post development in the area. For example, the government has given fruit seeds to people in rural areas in Sarawak to support the agricultural initiative. After two years of implementation, some people in the rural area have become entrepreneurs and they can afford to buy basic things. Observed behaviour will be recorded and analysed to match the research goals.

3.2 Documentary-Analysis

The documentary-analysis is conducted to analyse the source that is being published by the ASEAN countries to investigate deeply the planning and effort made that need to be implemented. The document can be found at the State Library, internet and some government offices. Document analysis is a form of qualitative research in which the document is interpreted by the researcher to give voice and meaning around assessment topic. It describes the nature and form of the document, outline the advantages, and limitation of document analysis, and offers specific examples of usage of documents in the research process (Bowen, 2009). The publications can be found via the internet through the particular website. For example, the ASEAN official portal is where all the publications can be found in the website and workshop proceedings. It is a genuine source which is up to date and suits the analysis process as well. Analysing documents incorporates coding content into themes like how a focus group or interview transcript are analysed (Bowen, et al. 2009). The types of documents used for this research was newspaper articles, journal articles, government policy, etc. The relevant source is taken and analysed to support the data, evidence, and information.

4 Results and Findings

To answer the main purpose of the research, the results were obtained. The research and the methodology method were aligned with the study. There is significant planning that ASEAN has outlined to make ASEAN a better region, peaceful, harmonious, with zero poverty and increase development in the rural areas. It is implemented to suit with the current situation and event that occur in that particular countries. How many years would it take to settle the problem? The answer is, it depends. ASEAN team members did cooperate actively and respond to the issues that were raised. The evidence is Laos, Cambodia, Myanmar, Indonesia and more are participating actively in agricultural sectors by organizing Agricultural Workshop Proceeding which is held at each country. The agricultural sector has highly impacted the quality of people's life. Many European countries are confident enough to cooperate not only to expand their businesses but also to invest in particular countries. After conducting the research study, it was revealed that all the ASEAN countries are supporting and implementing agricultural sectors as a main agenda in poverty eradication. The approach used by the government is realistic and practical.

One of the approaches is the role played by the head of the people in the particular areas. For example, at Kampung Buntal, Sarawak the role played by the head is providing the wet market for the fishermen to sell their marine catch, so the people living in the area can survive. Moreover, according to Sarawak Minister of Modernization of Agriculture and Rural Economy, Datuk Amar Douglas Uggah Embas at the National Smallholders and Commodities Entrepreneurs Conference (NSCEC) 2015, rural areas can be transformed through agriculture development to assist the community there to get out of poverty and the low-income trap. 'These transformations will enable rural folks to earn more and help the Penans escape from the clutches of poverty. This situation will lead to the growth of the economy and slowly increase the purchasing power of people in rural areas and reduce the poverty rate. 'We in ASEAN have enjoyed not only peace and stability but have also witnessed an era of strong economic growth throughout the region' said Malaysian Prime Minister, Tun Dr. Mahathir Bin Mohammad during Reflection on ASEAN, 2004. Most importantly, the approach taken must be realistic, so that all the government will implement the same planning and ways to reduce poverty issues and increase more development in rural areas. It is indeed the One Vision, One Identity, One Community motto.

5 Discussion

The findings of this study indicate that the ASEAN Community is working together in eradicating the poverty in the country. The agricultural effort is believed to be one of the ways that was found efficient in reducing the poverty level. Moreover, the planning outlined in the ASCC 2025 did mention its benefits and how it engages the people. It touches all the people in a holistic way. Not only for the people but also for the culture, beliefs, dreams, language and etc. It is clearly stated in the ASCC Planning that the main focus is sustainability, inclusivity and resilience. Each of the focus has a realistic approach and effort that can be put into action.

Rural development is also implemented well by the government. The people representative or head plays an important role in rural development because all the complaints and demands made are reported to the head. The head becomes the liaison between the people and government. So the development highly depends on the role played by the government. Southeast Asia cannot be enduring security community or an effective economic community. Indeed, it cannot be an ASEAN Community in its truest and deepest sense-without being a socio-cultural community (Rodolfo, 2006).

Lastly, the planning and effort in rural development and poverty eradication are well planned and cooperation is given actively by the ASEAN members for each other. They are responsive towards the issues that raised by the community members and are given immediate solutions. Again, the approach outlined in the ASCC 2025 that focuses on sustainability, inclusiveness and resilience is holistically practical. Aspiring to a greater notion of social integration will help transform the domestic political agenda and build on the achievement of the ASCC which should be a key force (Maramis, 2017).

The strength of the study firstly shows that sustainable development, resilience and inclusivity have positively impacted the planning of ASEAN Socio Cultural Community to increase the level of rural development and agenda to eradicate poverty. Each of the approach has its own agenda and focus to be implemented. Secondly, ASEAN countries are highly committed in agricultural sector to help the poor and the needs of its people. Agricultural sector is nurtured by given fruit seeds, fertilizer, and even programmes to help the people in rural areas. These research papers have strong evidence to support each of the main points. However, the weakness of this paper was it didn't mention the exact year the event occurred in the areas. Thus, the research only focussed on 10 countries in ASEAN instead of a single idea from outside of ASEAN. For example, the involvement of an economic agenda from Europe at ASEAN community.

6 Conclusion

It clearly shows that people are impacted by the ASEAN Community planning. Three key factors in ASCC highlighted sustainability, resilience and inclusivity being used by ASEAN community as an approach to make the 'people-centricity' vision a reality. Implementation is the only action that can make any approach and effort become reality. Smart strategy is worthless if not implemented (Bruce, 2016). Increased development has impacted the level of poverty specifically in the rural areas. The government in ASEAN countries are putting agriculture as the main core of the planning in the poverty eradication matter.

Based on this research, it is highly recommended that governments should be strict in rural development implementation especially when the budget has been allocated to raise the development. There a lot of rural areas that still do not have accommodation and are poor. A second recommendation is to plan more entrepreneurship programs to help the rural people to sell their agricultural product and marine catches. At the same time this will increase the income and purchasing power. The third recommendation is to develop a more realistic universal approach that suits all ages by respecting the differences in all ASEAN countries. Thus, the ASEAN countries have high potential to be successful because of its uniqueness, in terms of being multicultural, and strong determination to help its people.

7 References

Acciona. (n.d.). *Sustainable Development*. Retrieved from Acciona, Business as Unusualhttps://www.acciona.com/sustainable-development/#

ADB, A. I. (2006). Indonesia Strategic Vision for Agriculture and Rural Development. 3.

Bernama. (2017). *Malaysia urges for Asean poverty reduction to single-digit rate by 2025.* Kuala Lumpur: the Sun Daily.

Bowen, G. A. (2009). Document analysis as a qualitative research method. *Qualitative Research Journal*, 27.

Caballero-Anthony, M. (2005). Understanding Development Gaps in ASEAN. *IDSS Commentaries*, 1-3.

Cirt. (n.d.). *R Observational Method.* Retrieved from Center for Innovation in Research and Teaching: https://cirt.gcu.edu

Foundation, R. W. (2008). Observation. *Qualitative Research Guidelines project.*

H. E. Mr.Rodolfo C. Severino, S.-G. (2001, February 6-7). ASEAN: Building The Peace in Southeast Asia. *Association of Southeast Asian Nation.*

Maramis, L. (2017). *Buidling Asean Socio-Cultural Community & Nation Building* . Retrieved from United Nations Economic and Social Commission for Asia and the Pacific: https://pidswebs.pids.gov.ph/CDN/EVENTS/maramis_asean_socio-cultural_community_(road_traversed_and_journey_ahead)_revised_21-8-2017.pdf

Mohamad, M. (2004). *Reflections on Asean.* Putrajaya, Malaysia: Pelanduk Publications (M) Sdn Bhd.

Nation, U. (2015). *Sustainable Development Goals, 17 Goals to Transform Our World.* Retrieved from United Nation, Sustainable Development Goals: https://sustainabledevelopment.un.org/?menu=1300

Secretarial, T. A. (2017). Asean Framework Action Plan on Rural Development And Poverty Eradication 2016-2020. *asean.org.*

Secretariat, A. (2015). Asean Community. *Fact Sheet- Asean Community*, 1-4.

Secretariat, T. A. (2016). Asean Socio-Cultural Community Blueprint 2025. *The Association of Southest Asia Nations (ASEAN)*, 10-13.

Severino, R. C. (2006). *Southest Asia in Search of an Asean Community.* Singapore: ISEAS Publishing.

Simon H. Olsen, S. T. (2015). *ASEAN Community and the Sustainable Development Goals: Positioning Sustainability at the Heart of Regional Integration.* Researchgate.

V.Paramasivm. (2018). *Op.Ed: Why Malaysia is leading the way in Asean for poverty eradication.* Asian Corresspondent.

Workshop:Combodia, N. I. (2017). Agricultural Transformation and Market Integration in the Asia Region: Responding to Food Insecurity and Inclusiveness Concerns. *Workshop Proceedings*. Phnom-Pehn, Combodia: ATMI Combodia.

Digital preservation of language, cultural knowledge and traditions of the indigenous Semai

Sumathi Renganathan[1,*] and *Inge* Kral[2]

[1]Universiti Teknologi PETRONAS, Malaysia
[2]Australian National University, Australia

Abstract. In this paper we describe two community-based participatory research projects in an Orang Asli community that set out to document their local knowledge and culture. We describe how with the support of technology we are able to document indigenous oral traditions and practices that are on the verge of disappearing. The Semai are the largest Orang Asli community in Peninsular Malaysia and mainly live in the states of Perak and Pahang. Like in many other indigenous contexts, the Semai peoples' transition from an oral culture to a literate culture is relatively recent. In this paper we discuss how our long-term relationship has facilitated two projects using digital media technology that focus on the documentation of local knowledge and culture of the community members in a Semai-speaking village in Perak. Elders in this community, having local knowledge accumulated over generations through direct experiences and participation, were the main source of information for these documentation projects, while younger people assisted with film and audio recording, editing, as well as language transcription and translation. The elders in this Semai community recognise the value and importance of transmitting their local culture and knowledge to the next generation. The projects described in this paper led to the production of two short films in 2014, and a book project, which began in early 2017, is currently underway. Thus, this paper highlights how technology is an enabler in the preservation of endangered indigenous knowledge and culture.

1 Introduction

According to Ethnologue (Simons et al., 2017), the most extensive catalogue of the world's languages, there about 7000 distinct languages in the world. However, by the end of this century it is predicted that about half of this world's languages will be extinct as it is estimated that one language dies every two weeks (Kraus, 1992; Aulakh, 2013). Languages becoming extinct is not recent phenomenon, in fact every region in the world is vulnerable to this.

The UNESCO Atlas of the World's Languages in Danger (2011) cites that there are approximately 3000 endangered languages across the globe and those most at risk of extinction are Indigenous languages (UNESCO 2003). The loss of endangered minority indigenous languages represents a profound loss to humanity at a global level.

> Local and indigenous knowledge refers to the understandings, skills and philosophies developed by societies with long histories of interaction with their natural surroundings. For rural and indigenous peoples, local knowledge informs decision-making about fundamental aspects of day-to-day life. This knowledge is integral to a cultural complex that also encompasses language, systems of classification, resource use practices, social interactions, ritual and spirituality (UNESCO, 2018).

Language is the medium linking people to knowledge and therefore, a community's connection with its history, its culture and tradition and its base of specific knowledge are lost when a particular language has ceased

* Corresponding author: sumathi@utp.edu.my

to exist anymore. "The wisdom of humanity is coded in language," and thus when a language dies, the knowledge disappears with it (Campbell, L. cited in Wallace, 2010). Indigenous cultural knowledge refers to traditional practices and knowledge of the ways of life of that particular indigenous community which are passed down generation to generation that forms part of their unique identity and heritage.

For many indigenous minority communities everywhere, the threat of their language becoming extinct is very real. As indigenous communities come from a non-literate tradition, their knowledge and culture are passed from one generation to another orally. In addition, oral narratives not only pass on these forms of knowledge, but also the social values and the collective memory of that community. Thus when a language of a community is lost, traditional knowledge of the community is also lost. There are many reasons why a language becomes endangered or even extinct. In Malaysia, for minority communities such as the indigenous Orang Asli, learning and speaking the majority languages such as Malay and English are seen as economically advantages. These are the languages of education and it is perceived that mastering them will give speakers access to better jobs and successful futures. On the other hand, indigenous Orang Asli languages are perceived to be less prestigious and usually seen as useful only for personal uses at home or between indigenous community members. Thus, negative perception and low status are some of the major underlying causes for indigenous languages becoming endangered. According to Karim (2001), in Malaysia, many indigenous Orang Asli languages and cultural traditions face extinction, largely because many of these languages are not documented (Benjamin, 1976, 2012). Documentation is often regarded as extremely important for the continued survival of indigenous minorities' cultural heritage.

2 The Semai

We now turn to the projects focusing on one indigenous Orang Asli community, the Semai. These projects set out to document the language, cultural knowledge and traditions of the Semai speakers from a small village in Perak. The Semai indigenous people are mainly found in the states of Pahang and Perak in Peninsular Malaysia. A community that once depended on hunting and swidden farming, at present are mainly laborers in factories and oil palm and rubber plantation; or work for daily wages from available low–paid odd jobs.

Semai is a Mon-Khmer language belonging to the Aslian family of languages. Aslian languages are spoken in areas of Peninsular Malaysia and the southernmost part of Thailand. According to the Atlas of the World's endangered languages, Semai is categorised as severely endangered (Moseley, 2010; Theguardian, 2017) and is only spoken by about 15000 people. A language that is severely endangered implies that the language is spoken by the older generation including grandparents; while the parent generation although may understand the language, they do not speak to children or among themselves.

For a non-literate indigenous community like the Semai, languages live in stories, songs, and traditional cultural practices and like other intangible cultural heritage, oral narratives are threatened when modernization and environmental changes take place. Sometimes literate resources and practices such as books, magazines, radio, television and internet have had a damaging effect on oral traditions. Therefore, the digital documentation projects that we present in this paper are crucial in providing a record of Semai history, culture and traditions should they be lost if the Semai language itself is no longer spoken by the Semai community.

3 The Documentation Projects: Towards a digital preservation of Semai Heritage

One of the major difficulties working on documentation projects is having access to indigenous communities. In addition, for projects where the documentation of indigenous community cultural practices is central, a level of trust and accountability must exist between researchers and community members. This is one of the main reasons why community projects such as this are not only limited but are also difficult to carry out. However, the authors of this paper have worked closely with this particular Semai village since 2008 (Renganathan & Chong, 2009, 2010; Renganathan, 2016; Renganathan & Kral, 2017). Furthermore, in both the documentation projects we describe and discuss in this paper, we worked very closely with the indigenous village community members especially the *Tok Batin* (Village Headman). Both projects adopted an ethnographic approach and mainly utilised in-depth interviews and participant observations. In addition, both projects discussed in this paper focused on the concept of 'language as cultural practice' (Nicholas 2014; Schecter & Bayley 2002).

3.1 Project 1- Documenting the Semai cultural traditions

This project took place in the year 2014. The outcome of this project was 2 short one-hour films where the use of digital technologies and the film-making medium not only supported the documentation project but also initiated an intense interest in Semai language and culture especially among the youth in this village. This project took about 2 weeks of rigorous fieldwork in the Semai village working together with a specific team of Semai youth and other community members. The young people involved were volunteers who answered the *Tok Batin's* call to be part of this project. They consisted of four young women aged between 16-26 and one 16-year old adolescent male.

The content for the films was decided by the *Tok-batin* and the youth team members. They brainstormed and decided among themselves what, how and where the content for this project should be filmed. The young people were given two Kodak Playsport HD video cameras (similar to the size of a hand phone) to film whatever they themselves identified as important whereas the other aspects of the filming was directed by the Tok Batin and filmed by the second author using a professional video camera. This consisted of a professional quality Sony HXR-NX70P DV camera with wireless lapel microphone.

Once all filming was completed, the editing part of the project began. This was handled by two specific members from the Semai Youth team, Erna, a mother of two who was about 22 years old at the time this project was undertaken and Roomrid, a 16-year old school dropout who had completed 3 years of secondary school and was waiting at that time to start work in a nearby restaurant at washing dishes. The Tok Batin also assisted, especially in providing the necessary cultural authority and language input whenever needed. The editing took place in the village Community Centre where we used two computers, a Mac Book Pro and a PC laptop. We used the PC for the compilation of text documents, while the film was edited using Final Cut Pro (FCP) software on the Macbook. The young male, Roomrid, acquired the skills to independently edit the film after an initial demonstration on how to use the FCP software and minimal support. As the Semai are from a non-literate tradition, Roomrid was a natural visual learner who quickly remembered and mastered the sequence of actions demonstrated to him by the editing mentor by focusing on the visual codes. Later Roomrid taught Erna how to edit, by using similar aural cues he had learned from earlier. As both these young people were the ones who were familiar with the content, they became the experts in sequencing and editing the films and the narrative accompanying the visuals. This project demonstrated how technology enabled youth from a non-literate tradition to become confident in exploring and gaining competence in using complex technical editing functions. These actions not only empowered them, but also gave them control of telling their own stories as they were in charge of the oral, written text and the visual narrative of the films.

The digital documentation project enabled the young adults to participate in a meaningful role in their cultural traditional practices in a modern way and placed them in a crucial position as 'knowledge holders and conduit of cultural information for the next generation' (Kral & Renganathan, 2018, p.379). The two short films: *May Serak Temboh Bekeet: Tradisi May Sengooy Semai* (The traditions of the Semai) and *Ki Cerita Tentang Sinuy May Sengooy Semai* (Stories of the Semai) have now been archived at the Pacific and Regional Archive for Digital Sources in Endangered Culture (PARADISEC). Hence this data will now be accessible for future generations. Work is also underway to provide suitable bilingual subtitles in Semai and Malay, and is expected to be completed by end of 2018. A detailed account of this project is presented as a chapter in an edited volume published in 2018 (Kral & Renganathan, 2018).

3.2 Project 2 – Production of an edited volume on language, culture and history of the Semai

This second project again involved collaborative work with the Semai villagers and began in early 2017. This project involved the digital recording of a collection of stills and audio visual materials on the oral traditions, history and cultural practices of the Semai elders. The idea for this project was first conceptualized by the Tok Batin and a few elders living in the village. The project involved about 30 people from the village, including young people and their elders

The core collection of the data for this project includes:

- Three hours of video recordings of 17 Semai elders demonstrating cultural practices and traditions;
- 420 digital stills of cultural materials and traditional artefacts;
- Four hours of audio recordings of life story narratives from some 20 community members.

At present, translation from Semai to colloquial Malay has already been completed by a Semai research assistant. We are now working on the translation from colloquial Malay to Standard Malay and then to English. The edited volume will showcase all three languages. We have also been working on matching the collection of digital stills with appropriate narrated context obtained from the Semai elders. The drafting of chapters using this material is well underway.

The outcome of this work will enhance language revitalization, acquisition and use among the Semai and develop pride in their oral and cultural traditions. The completion of the edited volume will also enable the public to acknowledge and appreciate Semai language, culture and traditions.

4 Conclusion

Digital technology has provided opportunities for language and cultural maintenance and transmission for indigenous communities such as the Semai who are experiencing language and culture shift or loss. Thus, projects such as the ones highlighted in this paper show how the sustainability of indigenous traditional cultural knowledge is made possible with the help of modern technology. Both the projects presented in this paper highlight how through participation in meaningful community-based digital projects, non-literate communities such as the Semai are able to actively engage in projects that assist in the maintenance of their language and traditional cultural practices.

5 References

Aulakh, R. 2013. The Star.com. Dying languages: scientists fret as one disappears every 14 days. Retrieved from https://www.thestar.com/news/world/2013/04/15/dying_languages scientists_fret_as_one_disappears_every_14_days.html . Accessed on 5th June 2017.

Benjamin, G. (1976). Austroasiatic Sub-groupings and Prehistory in the Malay Peninsula. In P. N. Jenner, L.C. Thompson and S. Starosta, (Eds), *Astroasiatic Studies* (pp. 37-128). Hawaii: The University Press of Hawaii.

Benjamin, G. (2012). The Aslian languages of Malaysia and Thailand: An assessment. In Peter K. Austin & Stuart McGill (eds). *Language Documentation and Description*, 11, 136-230.

Karim, W. J. (2001). Minorities of the minority: Language death and patterns of cultural extinction. In Razha Rashid and Wazir Jahan Karim (Eds), *Minority cultures of Peninsular Malaysia: Survivals of Indigenous Heritage* (pp. 69 –74). Penang: Academy of Social Science (AKASS).

Kral I. & Renganathan S. (2018) Beyond School: Digital Cultural Practice as a Catalyst for Language and Literacy. In Wigglesworth G., Simpson J., Vaughan J. (Eds), Language *Practices of Indigenous Children and Youth: The transition from home to school* (pp. 365-386). London: Palgrave Studies in Minority Languages and Communities. Palgrave Macmillan.

Krauss, M. (1992). The world's languages in crisis. *Language 68*, 4-10.

Moseley, Christopher (Ed.). (2010). Atlas of the World's Languages in Danger, 3rd Edition. Paris, UNESCO Publishing. Online version: http://www.unesco.org/culture/en/endangeredlanguages/atlas

Nicholas, S. E. (2014). "How are you Hopi if you can't speak it? An ethnographic study of language as cultural practice among contemporary Hopi youth. In T.L. McCarty (Ed.), *Ethnography and language policy* (pp. 53-75). New York/ London: Routledge.

Renganathan, S. (2016). Educating the Orang Asli children: Exploring indigenous children's practices and experiences in schools. *The Journal of Educational Research, 109*(3), 275-285.

Renganathan, S. and Chong, S. L. (2009). Disparity in School's Literacy Practices and that of Home: Understanding Orang Asli Children's Educational Needs in Malaysia. Singapore: Paper published in the proceedings of the 2009 International Conference on Social Sciences and Humanities, ICSSH.

Renganathan, S. and Chong, S. L. (2010). *Exploring Multiliteracies and Social Practices of the Orang Asli Children in Perak.* Report submitted to the Educational Planning and Research Division, Ministry of Education

Renganathan, S. and Kral, I. (2017) Using Media as a catalyst for literacy, and language and culture maintenance, in an Orang Asli village in Peninsular Malaysia. Research talk presented in UPM, Malaysia

Simons, Gary F. and Charles D. Fennig (Eds.). (2017). *Ethnologue: Languages of the World, Twentieth edition.* Dallas, Texas: SIL International. Online version: http://www.ethnologue.com.

Schecter, S. R. and Bayley, R. (2002). *Language as cultural practice: Mexicanos en el Norte.* Mahwah: Lawrence Erlbaum Associates.

Theguardian. (2017). Endangered languages: the full list. From https://www.theguardian.com/news/datablog/2011/apr/15/language-extinct-endangered. Accessed on 5th June 2017.

UNESCO. (2018). Local and Indigenous knowledge system cited from http://www.unesco.org/new/en/natural-sciences/priority-areas/links/related-information/what-is-local-and-indigenous-knowledge/

UNESCO (2011). *Atlas of the World's Languages in Danger.* Paris: UNESCO.

UNESCO (2003). *Language Vitality and Endangerment.* UNESCO Ad Hoc Expert Group on Endangered Languages. Document submitted to the International Expert Meeting on UNESCO Programme Safeguarding of Endangered Languages. Paris, 10–12 March 2003.

Wallace, Lane (2010). What's Lost When A Language Dies The Atlantic. https://www.theatlantic.com/national/archive/2009/11/whats-lost-when-a-language-dies/29886/. Accessed 6th June 2017

Challenges of continuous technological disruption in the broadcasting industry and the search for 'job-fit' talents

Ramachandran Ponnan[1*] and *Azween* Bin Abdullah[2]

[1] School of Communications, Taylors University, Subang Jaya, Malaysia
[2] School of Computing and IT, Taylors University, Subang Jaya, Malaysia

Abstract: Job placement is becoming an uphill task for new graduates when they do not have the relevant knowledge to start up on their own. This happens when their education is either sub-standard, rendered obsolete upon graduation or cannot be translated into practice at their respective workplaces. Such a shortcoming in the broadcasting industry is not with the provision of training or the supply of skilled manpower. It is rather the continuous disruption of technology when graduates require more than operational knowledge. In the absence of 'job-fit graduates' the broadcasting industry looks out for the best talents outside of their traditional marketplace. To address this fallout, the industry is looking to recruit new talents with emerging computing knowledge instead of the traditional Mass Communication discipline. This empirical study explores the skills shortage in the broadcasting industry from the perspective of "Challenges of Industrial Revolution 4.0 in higher education (MyHE 4.0). Qualitative data was gathered from stakeholders, training providers and vendors. Two major findings: 1) The need for talents transformation to a modern converged setting in broadcasting; and 2) The establishment of pathways to 'Job-Fit' training in the broadcasting industry, have emerged from the study. The study proposes six recommendations in support of the transformation in the broadcasting industry.

Keywords: Technology disruption, technology fit theory, Job-fit interventions.

1 Introduction

In presenting the challenges of industrial revolution 4.0 in higher education (MyHE 4.0), Selamat (2017) argued in appreciation of a digitalization wave that will soon witness an increased use of artificial intelligence, robotics, computing and the Internet of Things (IOT) in most industries. He warned that Digital Disruption is occurring at production and dissemination of media content. As postulated by Manyika, Chui, Bughin and Dobbs (2013), the widespread emergence of new media technologies is disrupting traditional practices of conventional broadcasters compelling them to migrate to newer compatible broadcasting platforms. The rise in user-generated content is further disrupting traditional value chains of content production and delivery.

McKinsey Global Institute (2013) suggest that among 12 disruptive technologies, Internet of Things (IoT) has caused a 300% increase in machine to machine devices usage since 2008 leading to new ways of operation at the workplace. The introduction of such new technologies can affect society and therefore changes in required education must also take place in tandem. Meanwhile the Malaysian Education Blueprint 2015-2025 (Ahmad, 2016) proposes a curriculum that balances the development of ethical values and knowledge & skills as well. The blueprint (MyHE 4.0) strategies include open access to information with learners as constructivist of knowledge. The future of AI in classrooms, claims Kurshan (2016), is lagging behind. She cites the momentum of artificial intelligence in education as fading away again blaming poor strategies and the provision of interconnectedness and accessibility of internet in classrooms. Pointing to the same weakness, Grant (2017) observed that investments in emerging technologies and human connectivity, building digital resilience, as well as institutional capabilities in digital governance and accountability, are key strategies for moving on but was skeptical whether the higher education community is doing enough to adapt.

* Corresponding author: ramachandran.ponnan@taylors.edu.my

Among the major challenges facing the Fourth Industrial Revolution Challenges 4.0, as reported in the 2016 Global Industry 4.0 Survey (Report, 2016) is the lack of a clear digital culture and training policy among custodians of national education providers.

In an effort to redesign higher education, the Ministry of Higher Education (MOHE) via the Malaysia Education Blueprint 2015-2025 (Ahmad, 2016) has put in place follow-through and continuing programs designed to produce holistic graduates. These programs were expected to produce employable graduates with good academic results, experience in the industry and love for the nation. Along those lines, to better prepare students for a balanced studies-work orientation, eight universities initiated the "*Take a break*", a one-year program believed to inculcate ethical values and social skills. Here students took time off from studies to pursue their interests. In particular they were encouraged to volunteer with agencies such as the uniformed forces especially those under the National Blue Ocean Strategy.

MOHE believed it was important to harmonise public and private higher learning institutions as a single entity that adhered to a strict set of regulations that will ensure their quality (New Straits Times, 2017). Hence, ACT 555 was enacted in 1996 to regulate the operations of 487 Private Higher Education Institutions (PHEIs) in the country (Laws of Malaysia, 2006). With that students will benefit from a dedicated network for research and learning with its own separate and independent gateway to the Internet known as Myren-X. Students will afford an increases bandwidth capacity to a maximum of 100 gigabits per second (Gbps) and for other institutions under the MOHE, up to 10 Gbps. The provision of increase accessibility is expected to facilitate new ways and methods of teaching and learning such as collaborative online learning, virtual classrooms and Massive Open Online Courses.

1.2 Statement of Problem

The fourth industrial revolution 4.0 has rightly identified the lack of clear digital culture and training as the biggest problem inhibiting operational capabilities experienced by industries. This realization applies directly to the current technology disruption experienced by the broadcasting industry. The technology disruption is so overwhelming that the next generation of broadcasters is bound to experience a 'job-fit' challenge if stakeholders delay in ensuring their talents were 'Tech Savvy'. The speed of digital transformation and disruption has impacted human operational capacity at work requiring new critical skills for interaction with fellow workers and technology. Broadcasters who prefer to operate in 'silos', have now to merge their resources to meet new industry standards. To institute interventional strategies which are able to deliver these new skills, training providers have to better understand the future needs of their potential talents. In this regard, the study observes that the industry has failed to provide for some types of training needs that support continuous and supplementary training rather than relying on formal tertiary education. At the same time, collaboration between the universities and professional broadcast training providers have been nil or minimal at the best.

1.3 Objective

The study explores the state of skills shortage in the broadcasting industry at a time of continuous technology disruption in an exponentially growing multimedia environment.

1.4 Research Questions

RQ.1. Why is there challenges to 'Job-fit' talents in the broadcasting industry?
RQ. 2. How is the broadcasting industry responding to the challenges of the industrial revolution?
RQ.3. What recommendations are being offered to relief the skills shortage?

1.5 Significance of Research

An exploration of technology disruption related to the provision of education by the higher education services and training by the broadcasting industry is of great importance at this time. The areas of digital transformation and recommendations described in the study offers value for aspiring broadcasters and education providers from a practical perspective.

2 Literature Review

The model we use in this research is based on Task Technology fit theory (TTF) (Barki, Titah, and Boffo, 2007), (Baloh, 2007), (Cane, and McCarthy, 2009). This theory argues that information system use and performance benefits are attained when an information system is well-suited to the tasks that must be performed. Furneaux B. (2012) presented an overview of the diverse range of research contexts and methodologies that have been used to test and extend TTF theory. Key outcomes of interest to TTF researchers are also examined as are the various approaches that researchers have used to operationalize the notion of TTF.

In the developing nations, educators are clamoring to provide additional skills through new strategies. Agbo, Onaga, and Omeje (2018) raise their concern and are skeptical about the availability of opportunities in e-learning for effective technical and vocational education and training (TVET) programmes. In their absence they recommended that teachers be sent on capacity building programs to get acquainted with the skills to exploit the availability of e-learning opportunities in higher institutions.

Elsewhere Chimpololo (2017), points out the lack of teachers with critical skills in technical, entrepreneurial education and training (TEVET) skills citing the limited supply of quality and quantity of experts available in the job market. Well- equipped teachers would enable graduates to perform multiple-related functions once they are employed. To close the glaring mismatch in the demand and supply of technical, entrepreneurial, vocational education and training (TEVET) Skills, the study suggests the facilitation of academia-industry linkages to promote harmonious working relationships which would ensure that curricula are informed by industrial needs.

Zahraa, S., Sisodiab, R. and Mathernea, B. (1999) pose two perspectives to exploit links between competitive and technology strategies. They espouses that a company's competitive strategy and internal capabilities jointly determine its technological choices. And that technology is strategic to pursue its competitive goals. The interdependence between technology and competitive strategies emphasizes the importance of organizational learning and exploiting the knowledge required in this process. In the case of the broadcasting industry, it is new technology disruption requiring strategic educational interventions within the organization.

Media Planet (2018) in its latest report identifies three strategies to solve the skills mismatch existing in the media industry. First it is to solve the divide between the skills and credentials held by talents looking for work and the qualifications employers are saying they need. Next, jobs in the future will require post-secondary education, which are career-focused programs including apprenticeship training. The third strategy is the partnerships between job-seekers in the job market and skills training providers. With an uncertain job market, Students must be encouraged to explore the full range of post-secondary programs available. Job seekers ought to pursue programs that best fit their strengths and aptitudes to land jobs that best fit their skills and expertise

The idea of 'job-fit' is dependent on how individuals negotiate their tasks. Parkes (2012) experimented with the task–technology–individual fit model into three two-way interactions and ascertain how these interactions affect user attitude and performance.

The results demonstrated that user attitude (*behaviour*) is affected by the fit between individual (*broadcaster*) and technology (*machine processes*) whereas technology or the machine process in the case of broadcasting performance is affected by the fit between task and technology, and task and individual. Users of technology fitted to them as an individual can perceive it as more useful than it actually is, in terms of improving task performance. Finally, technology performance translates into task performance.

In suggesting innovations to broadcasting curricula to meet workplace expectations, Ponnan and Ambalavanan (2014) pointed to the need for academia to discuss the expectations of practitioners on developing evolving curricula that accommodate changes in the industry.

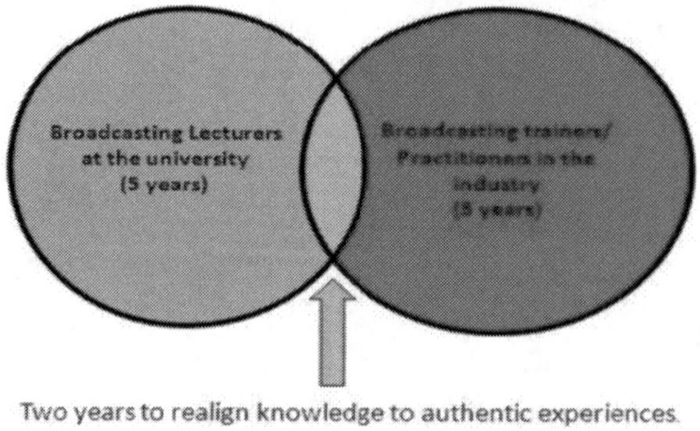

Figure 1: Aligning knowledge to authentic experiences

They put forward the idea that lecturers may be provided with sabbatical opportunity to work in the industry for two years after every five years of teaching and vice versa (see figure 1). Their study suggested classroom and lab work would effectively be current and challenging when they undertake project-based learning initiatives or collaborative projects involving the industry. Students and lecturers who return from internship and sabbatical should further identify worthwhile collaborative projects to continue with the same company, to create opportunities to revisit them to demonstrate new skills for hire.

A survey by the International Association for Broadcasting and Media (IABM, 2017) on staff employment in the Broadcasting Industry in the Asia-Pacific Region including Malaysia (2016-2017) informs the study briefly.

Only 20% of trained staff in the broadcast industry is due for retirement in five years. A shortage of staff has not reached crisis level yet as is the general assumption. That the average age of broadcasters on their feet now is 46, however, it is unclear if the succeeding cohort of broadcasters would be the ICT savvy generation of broadcasters.

Table 1: Nature of staff employment in the Asia-Pacific survey region (IABM 2017)

Selected Related Survey Items	Results
Well trained broadcasters on demand are working as expatriates	22%
A third of staff is self-employed	30%
Self-employed in broadcasting as primary occupation	14%
Self-employed in broadcasting as part-timers	17%
School leavers with shorter tertiary education	20%
Employees have Bachelor's degree	71%
Receive training from internal academies	51%
The average age of broadcasters on their feet now is 46 in the Asia-Pacific region. 45 years ⟷ 55 years	

3 Methodology

The study conducted face-to-face interviews with assistance from the HR department of two broadcasting organizations. As a result seven well informed experts were identified as informants from the broadcasting industry. Additionally, one broadcasting vendor, one independent training provider, one formal training provider and nine university students and one lecturer were also interviewed. A detailed breakdown is given below.

Table 2: Breakdown of Informants

Broadcasters	University
HR placement officer -1	Internship placement officer 1
Decision makers -1	Final year students 2
Executives interacting with interns -1	Students during internship 2
Engineers -2	Students after internship 2
Producers -2	Students after graduation 2
Vendors -1	Lecturers 1
Training providers – 2	

Literature reviewed informed the study on the current state of technological disruption affecting broadcasters in the developed and developing countries. Their action plans helped to inform the recommendations in this study. Content analysis of official government and university portals also provided perspectives on how they viewed their current needs and formulated action plans.

4 Findings

Changing broadcasting landscape as a result of convergence of technologies and digitalization has led to distinct technology disruptions. Two major findings have emerged from the study: 1) The need for talents transformation to a modern converged setting in broadcasting; and 2) The establishment of pathway to 'Job-Fit' training in the broadcasting industry.

4.1 Transformation to Modern Converged setting

Informants have dispelled the notion of critical shortage for skilled talents but confirmed the technology disruption experienced by the industry. Hence the need for talents transformation to a modern converged setting in broadcasting

As a senior engineer put it, "we remain 'siloed' in our respective job functions, particularly between creative, operational and engineering functions. We need to be more coordinated and collaborative, and have a broader focus on management of our core businesses".

The first illustration in figure 2 shows broadcasters performed tasks in their respective domains. Engineers in their engineering services, producers engaged in production and ICT talents providing complementary services. As convergence of technologies brought together the three entities of creative content producers, the traditional broadcast engineers and the advent of new ICT talents, they transformed and evolved into the second illustration working collaboratively. In the second illustration all three entities or services, creative production, telecommunication engineering and computing technology, converge to produce online compatible content for the internet enabled multimedia broadcasting. Broadcast technologies are constantly evolving to achieve online compatible content for broadcasting in the modern converged setting. It has now become inevitable for the new generation of broadcasters to be ICT savvy.

Figure 2: Talents Transformation to Modern Converged setting

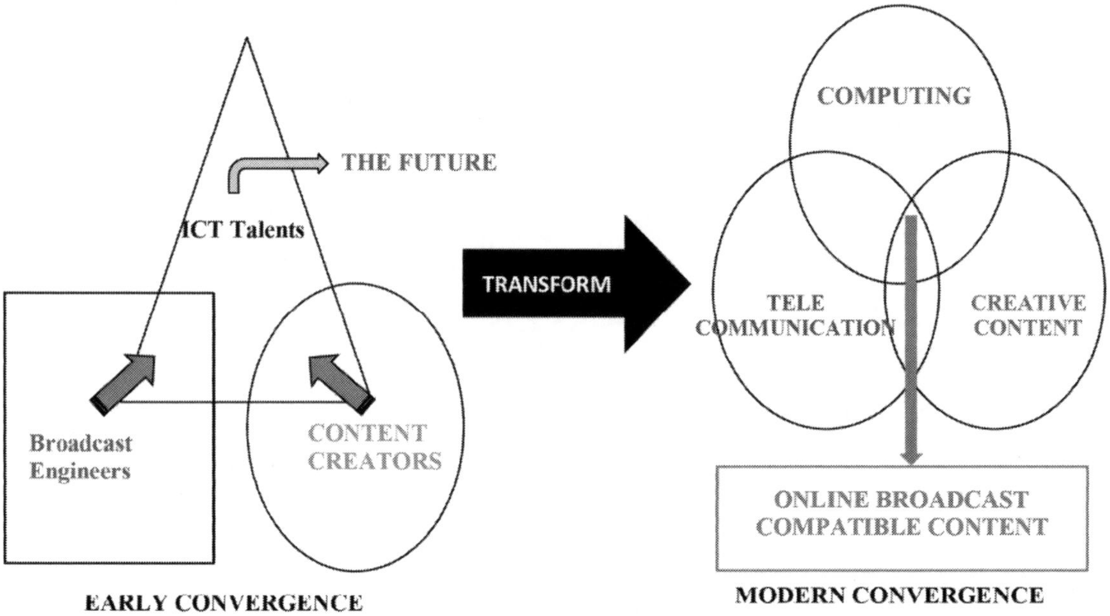

4.2 Pathway to 'Job-Fit' training in the broadcasting industry.

Stakeholders in the industry, who are cautious in decision making, expect the best from their staff but do not invest in the right human capital development. To get the right talents, our way of identifying the right talents have to change. Talents with passion may be better than the person with an on-paper qualification.

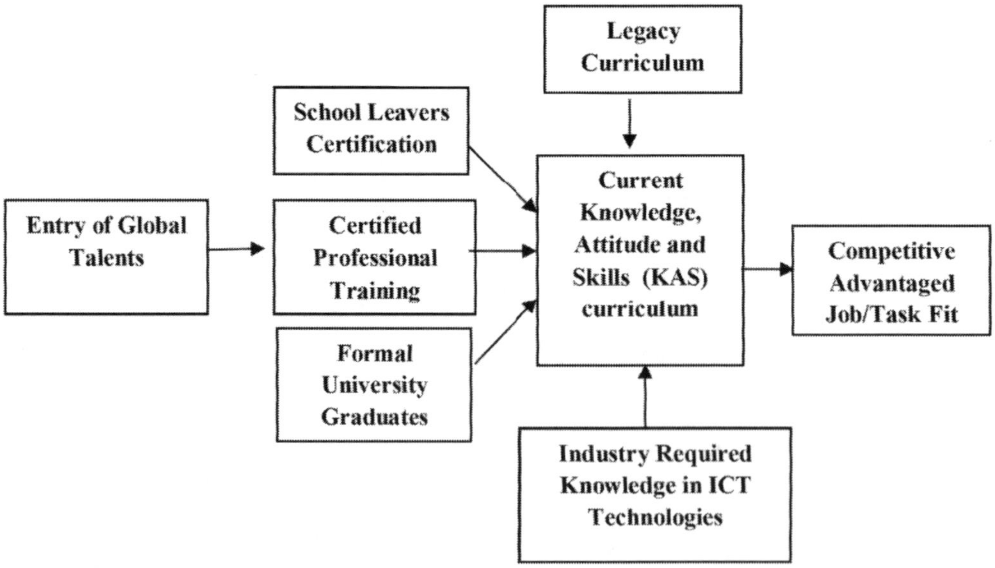

Figure 3: Entry Levels of Talents and pathway to 'Job-Fit' training in the broadcasting industry.

The study identified three entry levels of talents as pathway to 'Job-Fit' training beneficial for the broadcast industry. Namely, i) School Leavers Certification, ii) Certified Professional Training, iii) Formal university studies in broadcasting. For school leavers technical education by the government is in place (PADU, 2013).

PADU delivers and monitors the government's aspiration to develop world-class education that will better prepare young school leavers for the needs of the 21st century. Certified Professional Training is provided within the organization but is mostly self-funded training offered by vendors and commercial training providers. This is meant for up-skilling talents in the industry. There is great need to up-skill about 1,300 registered vendors in the creative content sector who rarely take up any form of training so much so their products face quality control rejection. Next, formal university broadcasting graduates land the higher level jobs in the industry. Nevertheless, they too need supplementary education to upskill themselves to fill in the gap where they lack optimal operational knowledge. All three levels of talents walk into the regular 'KAS' training curriculum but soon discover the need to relearn new tools and technologies as a result of technology disruption. Apart from local talents there is a flow of global experts in the industry now filling in the void created by technology disruption. These training initiatives have one challenge though. The burden of cost is gradually being shifted to individuals and broadcasters.

4.3 Tech savvy talents

Students returning from internship discovered that the broadcasting industry required usage of unfamiliar technology in areas they were less comfortable such as Cybersecurity and Cloud. The gap between past practices and current expectations in the industry is the technology disruption that is rendering interns incapacitated at their workplace. They ask that future students be better prepared with skills in these areas.

Decision makers in the broadcasting industry are wondering if broadcasting is beginning to become more like the IT industry. Traditional content production has now moved on to include many technical features, especially for live, liner and OTT multimedia consumption. The technology disruption is so overwhelming that the next generation of broadcasters is bound to experience a 'job-fit' challenge if it delays in equipping itself to be 'Tech savvy' with skills in:

- "IT & IP, especially in the area of coding"
- "IT Security & Networking"
- "Internet Streaming Formats - HEVC+HDR/WCG – DRM Technology"
- "DAM workflows"
- "Better Computing skills such as programming"

4.4 Learning needs in the technology disruption era

Emanating from the interviews of training providers, broadcasters, university students and lecturer, content analysis on strategies and action plans, several consistent themes on skills shortages, learning needs and interventions became visible to the study as a result of talents migration to a converged setting. Four areas of technology change were identified as being particularly disruptive:

Education and training needs' analysis of the industry is mainly situated in the workforce and the machine. The technology disruption is an urgent matter to be understood completely to understand the skills needs at the digital workflow, from acquisition to The move from an analogue to digital workflow have remained substantially the same, and so have the workforce in their silos remained the same too. Challenged by changes in technology and in work culture we will need to assess how new digital culture can evolve. Five technology change were identified as being particularly disruptive:

a) IT and IP for workflows: IT has change the face of broadcast technology. IP will further allow for different production and distribution models, as in the case of non-liner, VoD and OTT providers. The need for thorough technical knowledge is one of the skills-needs areas.

b) Cyber Security: there is a need for enhanced Cyber Security skills in the production process that prevents unauthorized manipulation of content, preventing malicious attacks on broadcast systems and ensuring security of data is completely maintained. This is a need that goes beyond an 'anti-virus' software and the need to the need for protection at every stages of the workflow.

c) Cloud: The use of cloud computing, storage and distribution is where there may be huge differences in the accessibility of cloud infrastructure. With good telecommunications infrastructure, the use of cloud is a viable option as part of an overall infrastructure strategy. The skills required for this type of infrastructure management are different from those required to manage an on premise broadcast technology operations.

d) Media Assets Management (MAM): Digital disruption has changed Media assets management even further with the advent of IT, IP and Cloud into broadcast technology. This too has raised the need for continuous training in media management for new forms of content management and deliverables.

e) The last area of learning needed is in the coordination of creative, operational and engineering functions in a more coherent way as in figure 2. The workforce here does not 'speak the same language'.

Broadcasters cannot afford this model to continue as it will impede them from getting new creative ideas, putting them at a competitive disadvantage. In the past, each department has tended to focus on workflows in their respective area, but in the future, management will need to consider what will work best across an end-to-end digital workflow. This can only happen through supporting internal collaboration and organizational learning.

5 Solutions as Recommendations

In developing potential solutions the study proposes six recommendations as viable approaches in support of the transformation in the broadcasting industry that will address the short supply of 'job-fit' graduates for the industry.

5.1 Recommendation 1

A skills Survey to assess training needs and gaps accurately in required skills should provide for the lack of quality data on skills levels in the industry. For the moment there is no single consistent dataset provided by the job market beyond company needs. A critical skills issue is having a good understanding of the base level of skills and where the gaps are. Such evidences are collated only periodically during summits. Support for regular data collection would provide good regional and national assessment of skills levels, provision, trends and gaps. This skills survey would provide vital data for broadcast industry stakeholders.

5.2 Recommendation 2:

Ministry of Higher Education (MOHE), *Institute Penerangan Penyiaran Tun Abdul Razak* (IPPTAR), the Asia-Pacific Institute for Broadcast Development (AIBD) should create an annualized skills survey in the region, which can be reviewed in a session during their respective summits such as the Asia Media Summit by AIBD. MOHE should extend an existing commitment to their official portal to include the curation of industry needs.

5.3 Recommendation 3

Master Skills Model for Broadcasting: A skills model that can be used to communicate with university students, government bodies, industry organizations, broadcasters and technology vendors is essential to address skills provision in the industry. Each organization makes its respective calls for training. No one addresses skills need as a master skills model for broadcasting, and none was particularly successful. Referring to figure 3, three levels of talents certification can be planned as the national broadcasting curriculum. The nature of intervention organized should be in consultation with the industry stakeholders and all education and training providers.

Local training providers currently provide training out of a commitment to the industry. IPPTAR, AIBD and elsewhere in Australia, Bangladesh, India, Pakistan, and Media Industry Technologist Certification have obtained the support of the government. The certification process is informally recognized for jobs application in the industry.

Despite the initial support from broadcasters and government, MOHE or rather the qualification Agency (MQA) does not equate such certifications for onward qualifications in universities. Such inhibition causes frustration among talents. These talents then seek self-funded but low-cost entry-level skills training, offering low-risk means of self-certification within specific areas. For example, the IT community used Cisco Certification process as a mandatory indicator of competence and capability in network technology. Create a 'Graduate Career Path' that would guide graduates' career progression. The industry needs to create an over-arching model for aspiring broadcasting talents that enables them to develop and learn with the industry.

5.4 Recommendation 4:

MOHE, IPPTAR to collaborate with qualified organizations globally such as the AIBD to design Master Skills Model for careers in broadcast and media technology that defines required training programs for the Industry. MOHE, IPPTAR and AIBD should consider how to support informal regional 'Workflow Innovation' style groups made up of broadcasters, academics, and training providers, to encourage the sharing of practical experience

5.5 Recommendation 5

Role of Higher Education: The study has identified the need for greater collaboration between the industry and educational institutions. Broadcasters and vendors would benefit from access to innovation from institutions and that educational staff would benefit from a practical understanding of broadcast operations as seen from figure 1. (Ponnan and Ambalavanan, 2014). Another proposition that could be considered was co-operation around commercial ventures or specific products from IT companies for university lecturers and students. A starting point for university lecturers could also be an R&D joint product development project, with broadcasters providing oversight. Technology disruption has created the need for new metadata indexing process and stringent technical quality control using machine learning focused on workflow innovation in the operations control room. This is an exploratory endeavor, and expectations should not be set too high. It is about starting on a simple but practical way with a view for stronger and more formal co-operation.

5.6 Recommendation 6

Universities, MOHE, IPPTAR and AIBD set up exploratory session to identify areas for co-operation between media industry, academic institutions and training providers.

6 Conclusions

In conclusion, globalization and the global movement of talents are bound to change broadcasting landscape especially in developing countries that are slow in implementing changes in tandem with the rest of the industry. For broadcasters to work collaboratively to make broadcasting a viable industry through engagement with non-traditional broadcasters, the intervention of educationists and training providers cannot be over-emphasized. The need for talents transformation at the modern converged setting in broadcasting; and the establishment of pathway to 'Job-Fit' training in the broadcasting industry are worthy of action considering that the successful digitalization and eminent analog shut down, will further cause technology disruptions. Six recommendations are proposed for transformation in the broadcasting industry that will address the 'job-fit' requirements among current broadcasters and fresh entrants. In that way, the study is hopeful that MyHE 4.0 will fulfil shortcomings by further technical disruptions.

References

Agbo, B. D., Onaga, P. O. and Omeje, H, O. (2018). Teachers' Opinion On Availability Of E-Learning Opportunities For Effective Technical And Vocational Education And Training (TVET) Programme In Tertiary Institutions In South East Nigeria. In *International Journal of Vocational and Technical Education*. Vol.10(1), pp. 1-5, January 2018 https://doi.org/10.5897/IJVTE2015.0176 [Article Number: D79FEFB55715].

Ahmad, B. (2016). *Ministry of Education Official Portal: Malaysian Education Blueprint 2015-2025* (Higher Education). https://www.mohe.gov.my/en/download/public/penerbitan/pppm-2015-2025-pt/5-malaysia-education-blueprint-2015-2025-higher-education. Retrieved on the 17th.2.2018.

Baloh, P. (2007). The role of fit in knowledge management systems: Tentative propositions of the KMS design. *Journal of Organizational and End User Computing,* 19(4), 22–41.

Barki, H., Titah, R., & Boffo, C. (2007). Information system use-related activity: An expanded behavioral conceptualization of individual-level information system use. *Information Systems Research,* 18(2), 173–192.

Cane, S., & McCarthy, R. (2009). Analyzing the factors that affect information systems use: A task-technology fit meta- analysis. *Journal of Computer Information Systems,* 50(1), 108–123.

Chimpololo, A. (2017). Glaring Mismatch In The Demand And Supply Of Technical, Entrepreneurial, Vocational Education And Training (TEVET) Skills In The Agricultural Sector: An Analysis Of The Oil Seeds And Sugar Industries In Malawi. Retrieved from: https://doi.org/10.5897/IJVTE2017.0237 [Article Number: 3319D2966147]

Furneaux B. (2012) Task-Technology Fit Theory: A Survey and Synopsis of the Literature. In: Dwivedi Y., Wade M., Schneberger S. (eds) *Information Systems Theory. Integrated Series in Information Systems*, vol 28. Springer, New York.

Grant, B., C. (2017) "Current Status and Readiness in Meeting the Fourth Industrial Revolution Challenges" In 9th University Presidents Forum2017. Retrieved from https://www.mohe.gov.my/muat-turun/awam/teks...dan.../2017/.../file; University+of+Southampton%2C+University+Presidential+forum+2017&oq=(Prof.+Dr.+Colin+B.+Grant %2C+Vice+President+(International)%2C+University+of+Southampton%2C+University+Presidential+forum+2017&aqs=chrome..69i57.3619j0j8&sourceid=chrome&ie=UTF-8

IABM Report (2017). "Engineering the future" In Tom McGann Memorial Summit-Tackling skills shortages in the broadcast & media industry. Kula Lumpur. Retrieved from: http://www.theiabm.org/training/Engineering-the-future

Kurshan, B. (2016), The 'evolution to social learning' and AI-teacher harmony. Retrieved from https://blog.acadly.com/the-evolution-to-social-learning-and-the-harmony-between-teachers-and-ai-5833918a61ea

LAWS OF MALAYSIA, Act 555 Private Higher Educational Institutions ACT 1996, incorporating all amendments up to 1 January 2006. Published by the Commissioner of Law Revision, Malaysia under the Authority of the Revision of Laws ACT 1968 in collaboration with Percetakan Nasional Malaysia Bhd. 2006.

Manjik, J., Chui, M., Bughin, J. and Dobbs, R. (2013) "Disruptive Technologies: Advances that will transform life, business and the global economy" In McKinsey Global Institute Report. May 2013. Retrieved from https://www.mckinsey.com/business-functions/digital-mckinsey/our-insights/disruptive-technologies

Media Planet (2018). Careers and Education: Solving the Skills Mismatch. http://www.careersandeducation.ca/industry-insight/solving-the-skills-mismatch. Retrieved on the 10th Feb 2018

New Straits Times (2017). "An education redesign in progress" in New Straits Times Dec 27, 2017. https://www.nst.com.my/education/2017/12/318734/education-redesign-progress. Retrieved on the 1st Feb 2018 ////

PADU- Education Performance and Delivery Unit (2013) - Ministry of Education Malaysia official portal. https://www.padu.edu.my/resources/meb-2013-2025/pppm/ Retrieved on the 2nd Feb 2018

Parkes, A. (2012). The effect of task–individual–technology fit on user attitude and performance: An experimental investigation. journal homepage: www.elsevier.com/locate/dss.

Ponnan, R. and Ambalavanan, B. (2014). Innovations to Broadcasting Curriculum to Meet Workplace Expectations. (2014). Journal Procedia - Social and Behavioral. 10.1016/j.sbspro.2014.01.1410

Report (2016). "Industry 4.0 Building the digital enterprise" In 2016 Global Industry 4.0 Survey - *www.pwc.com/industry40 https://www.pwc.com/gx/en/industries/industries-4.0/landing-page/industry-4.0-building-your-digital-enterprise-april-2016.pdf*

Selamat, A., Alias, R. A., Hikmi, S. N., Puteh, M., Tapsi, S. M., (2017). Higher Education 4.0: Current Status and Readiness in Meeting the Fourth Industrial Revolution Challenges. In Redesigning Higher Education towards Industry 4.0, Kuala Lumpur, Malaysia (23-24, August 2017).

Zahraa, S., Sisodiab, R. and Mathernea, B. (1999). "Exploiting the dynamic links between competitive and technology strategies". In European Management Journal Volume 17, Issue 2, April 1999, Pages 188-203

The representation of the Syrian conflict in the Syrian online news: A critical discourse analysis

Kais Amir Kadhim*

Department of Management and Humanities, Universiti Teknologi PETRONAS, 32610 Seri Iskandar, Perak, Malaysia

Abstract. This study investigates the usage of Syrian media tools in manipulating people's perception towards the uprising in Syria. The language of the Syrian media is analyzed using a critical discourse analysis. For this purpose, two frameworks are combined: the historical approach by Wodak (2002) and the socio-semantic network of social actors by Van Leeuwen (2008). The study particularly focuses on the way the Syrian government and the rebels are represented in the Syrian online news Cham press which is a pro-government news agency. The two research questions of this study are: (i) What are the referential and predication strategies used by Cham press in reporting the Syrian conflict? (ii) How do these strategies reflect the ideology that surrounds the representation of rebels in the Syrian news? Three hundred and ninety-seven articles were selected from Cham press based on the most relevant keywords. These articles were taken for a period of six months that is from June to October 2012 and May 2013. The study reveals that the Syrian uprising is negatively represented using terms with negative connotations like terrorism, foreign conspiracy, against the law, foreign fighters that are targeting people and responsible for the massacres that happened in the country.

1 Introduction: The Syrian Conflict

The uprising in Syria is a resemblance to the uprisings that erupted across a number of Arab countries like Algeria, Egypt, Libya, Yemen and Tunisia. It is a revolution against the long term dictatorship regime. The Syrian revolution started when 35 opposition representatives were arrested in March 16, 2011 after a promise from the government to solve the problems with the opposition with peaceful resolutions as stated in the Damascus Declaration in 2008. A protest started as a call to release the political prisoners. Yet the Syrian government did not hesitate to open fire on the demonstrators. 60 people were killed in the first 10 days. After that, the number of protesters rapidly increased. They were demanding to end the state of emergency that the country declared 35 years ago, since the time of the previous president, Hafedh Al-Assad, the father of the current president, Bashar Al-Assad. The state of emergency is a condition declared by the government in cases of civil unrest.

*Corresponding author: kais.amirkadhim@utp.edu.my

It gives authorities boundless powers of surveillance and arrest (Collin Dictionary; Syria ends emergency law). After a month of clashes between the rebels and the demonstrators, President Bashar Al-Assad declared an end to the emergency rule. Perthes (2004). Yet this declaration did not bring much of an action. Therefore, the demonstrators demanded Al-Assad to step down. Since then, the Syrian uprising was one of the bloodiest in the Middle East because the government used military power to stop the rebels. Tens of thousands of people were killed. The country was severely destroyed because of the clashes between the government and the rebels (*Global News*, April 19, 2011).

2 Discourse-Historical Approach

The discourse-historical approach which started in Vienna tries to incorporate much available knowledge about the historical sources and the background of the social and political fields in which discursive "events" are embedded. In addition, it analyzes the historical dimension of discursive actions by investigating the ways in which particular genres of discourse are subject to diachronic change. Lastly, it aims to integrate social theories to be able to explain text. On the methodical level, one of the most salient distinguishing features of the discourse-historical approach is its endeavour to work interdisciplinarily. It attempts to transcend the pure linguistic dimension and to include more or less systematically the historical and political dimensions in the analysis, theory and interpretation of a specific discursive occasion. Finally, the approach is problem-oriented and not focused on specific linguistic items. It aims to analyze the linguistic items in terms of their locations in the discourse with all the sociological and political ends embodied in it (Wodak, 2002; Meyer& Wodak, 2009). According to Wodak (2002), one of the main discursive strategies is referred to as "Meta strategy of us vs. them": the discursive construction goes in the following stages:

1. labelling social actors

2. generalization of negative attributes (Reisigl and Wodak, 2001)

3 Van Leeuwen's Socio-Semiotic Framework

This framework provides a useful analysis to newspaper articles that deal with marginalized and the silenced groups in the society who are represented in the media by the powerful agents in the way that serve the government's agenda leaving no room for the marginalized groups to represent themselves. According to Van Leeuwen's framework (2008), every social practice involves linguistic and non-linguistic actions. Certain criteria should be considered to critically understand the text: (1) Who are the participants? (2) What is the action being described and in what sequence? (3) In what "performance mode" was the action performed? (4) the eligibility of the participants, (5) "representation style" (6) "time"; (7) "location"; (8) eligibility of conditions of the location; (9) the tools and resources used in making the social practice; and (10) the eligibility of the resources.

4 Methodology

This paper will benefit from some concepts of Van Leeuwen's social actors' theory in addition to some aspects from Wodak's historical approach according to the availability of the concepts in the selected texts.

4.1 Data Collection

The data were collected from the *Cham press* online archives. *Cham press* is a Syrian online news agency that shows its support to the Syrian government. The reason behind selecting *Cham press* is because it is pro government agency; supporting Assad action against the demonstration. Furthermore, the news itself were written in Arabic then translated into English. originally in Arabic. Textual analysis can be used as a tool to systematically explain social contradiction and social change (Fairclough, 1995, p.15). The selection of articles is according to the most relevant keywords to the Syrian conflict like: terrorists, armed groups, gangs, government agents, gunmen, heroes, soldiers etc. 397 articles were selected in total; and the period from June-October, 2012 and May 2013 were considered the timeline of the Syrian Civil War. The articles were selected randomly as analyzing all the articles will be redundant repetition.

4.2 Research Design

This research will benefit from two strategies: referential, and predication strategies. The referential strategies include:

(i) **Inclusion and Exclusion**: these criteria refer to the way social actors and activists are presented in the text explicitly or implicitly, partially or fully depending on the ideologies aimed in the text.

(ii) **Activation and Passivation**: refer to the representing of the social action either with reference or without reference to the social actors depending on the ideologies aimed in the text.

(iii) **Genericisation and Spacification**: this is a clear criterion that distinguishes the language tools used for representing immigrants and citizens.

4.3 Data Analysis

The process of analysis will undergo the following procedures:

(i) Describing the theme addressed in the extraction
(ii) Explaining the sentence structure and detecting 'activation' and 'passivation'
(ii) Identifying the social actors with reference to vocabulary and grammatical components
(iv) Explaining the ideologies behind the representation

The following sections describe the findings and the discussion:

4.3.1 Representing the Government as victorious in the street battles:

A. *(<u>The army</u> cleaned al-Quneitra Street and the surrounding areas from <u>the mercenary terrorists</u> and killed large numbers of them."*

(Cham press, 18-09-2012, Terrorists killed in Aleppo and Damascus countryside...Infiltration attempts foiled in Homs)

The news article of the extraction (A) is describing a series of clashes between the government army and the rebels but the incidents are reported by one of the government's news agency "*Cham press*" and; therefore, from the government supporters' point of view. In addition to the clashes, the article informs that the army destroyed a number of trucks and cars loaded with weapons belonging to the rebels. According to the BBC Online (2013), during the month of September 2012, there wasn't any record of attempts to resolve the conflict between the rebels and the government. The only record is a number of confrontations between the army and the rebels. (Syrian profile, 2013).

In terms of sentence structure, the lexical word 'the army' is in the position of the subject of the sentence and the doer of the verb 'cleaned', while the rebels are referred to as 'mercenary terrorists' in the position of the goal of the action 'cleaned'. This is ***Metonymy***. ***Metonymy*** is a kind of substitution to the names used to refer to participates (participants) with names that imply different connotations to "objectivate" social actors (Reisigl & Wodak, 2001, p. 57). Therefore, 'the army' is given an active role in the sentence 'activated', while the 'mercenary terrorists' are given the passive role 'passivated'. The subsequent sentence phrase joined by the conjunction 'and' adds another verb 'killed' and treat the rebels as the recipient of the action of killing in '*and killed large numbers of them'*. This time the rebels are referred to as 'large numbers of them'. The reason behind this sentence structure is to show the power relation between the two rivals. 'The army', which is the government representative, is always activated to show that the government is in control of the situation, and is the most powerful rival.

In terms of the representation of actors in the context given, the government is referred to by its powerful force - 'the army'. It comes in a singular collective noun to show unity of the group. On the other hand, the rebels are referred to as 'mercenary terrorists'. Firstly, 'mercenary terrorists' are negative adjectives used very often by *Cham press* to refer to the rebels. The reason behind the very frequent reference of 'terrorist' is to legitimize any action taken by the government against the rebels. This is a form of Predication. According to Wodak & Meyer (2001), Predication is to give the social actors stereotypical or evaluative attributions with negative or positive traits in a linguistic form to give an implicit or explicit predicate. Secondly, the adjectives 'mercenary terrorists' come in a plural form to show that the rebels are not one in unity like the army. This is to give an indication that the rebels are distracted gunmen who are not united by an objective or action. Therefore, the negative attributes in that plural form 'mercenary terrorists' makes the act of rebelling look like a random form of violence rather than an uprising in the whole country. In addition, the rebels are referred again as the recipient of the action of killing in "*and killed large numbers of them"*.

4.3.2 Representing the Government as the Guardian of People.

(B) "The armed terrorist groups could not survive more than 6 hours behind the walls of Yalda town in the Damascus countryside as the brave soldiers of the Syrian Arab Army cleared the whole area of hundreds of terrorists who were provided with various advanced weapons."

(*Cham press*, 13/09/2013, *Security and calm back to Yalda town in Damascus countryside)*

The article is describing the ending and aftermath of the clashes between the army and the rebels in the town of Yalda which is located on the southern outskirts of Damascus. The article starts by informing that the clashes took six hours and hundreds of armed rebels were killed. Then, the article refers to the families of the town who went back to their houses when the clashes were over. By reference to extraction (B), there are two social actors identified: 'the army' who was referred to as 'the brave soldiers of the Syrian Arab Army'; Then, the rebels who were referred to as 'the armed terrorist groups', and 'hundreds of terrorists'. In this regard, the rebels –as in extraction (A) above- are described in association with terrorism in 'the armed terrorist groups' which is according to Wodak an example of predication in her self-other-schema, because the rebels are labelled in this adjective that bare a negative trait and they are referred to accordingly. In addition, according to the data, the word 'terrorist' refers to any armed force other than the government army. There is no mention of lexical words like 'resistance, rebels, activists, etc.' which might give the act of rebelling a positive or neutral trait. This type of reference as explained in extraction (A) is used to make legitimate any kind of action taken by the government against the rebels. Also, they are referred to in a plural form in 'the armed terrorist groups' to indicate that there is no unity among the rebels. The important social actor's representation is the army itself. The army in this extract is referred to as '*the brave soldiers of the Syrian Arab Army*" the adjective of bravery is used with the army men. The use of the adjective "brave" is to represent the army men in a heroic manner. The writer wanted to add the spirit of patriotism through the use of the adjective '*brave*'

4.3.3Representing Rebels as Foreign Fighters Rather than Local Rebels.

(E)"Hazim M. described ...blasted the ongoing killings and destruction by the armed terrorist groups, which consist of gangs and killers, Calling on Libyan and Arab fighters not to enter to Syria because of the ongoing... there is never a popular revolt."

(A Libyan terrorist: Arabs and Africans fighting in Syria gangs and killers, Champress, 16 July 2012)

The article in general is reporting of an interview in *Libya Now* agency with the Libyan fighter, Hazim M., who participated in the Syrian revolt. In terms of representation of social actors, the social actors in this text are: *Hazim M.* and *the armed terrorist groups, which consist of gangs and killers.* Hazim M. is one of the foreign fighters belonging to the Syrian rebel group. In line with the previous articles, the Syrian rebels are referred to as, *"armed terrorist groups*" which is a form of predication that involves negative connotations as explained earlier. Similar to the previous articles analyzed, the rebels are given an active voice in association of violent acts like "*killing and destruction*". This time, the rebels are further identified with "*gangs and killers*". Therefore, only the rebels are included in the above text.

Yet, the statement "*there is never a popular revolt*" supports the government's point of view. Although the *sayer* of the statement as a "verbiage", according to Haliday's functional grammar, is one of the rebels, yet the statement belongs to the Syrian government. The statement is used as a justification to support the government, and not the rebels because it says that the revolution is not popular, since it refers to the popularity among the Syrian citizens. The statement that the Syrian citizens do not welcome the revolution makes the revolution illegitimate. Yet the reality is that the foreign fighters do not constitute more than 20% of the rebels which means that 80% of the rebels are from the Syrian citizens. In addition, the statement here is not quoted from Hazim M., but it is rephrased by the writer.

4.3.4Representing Revolution as a Foreign Conspiracy

In *Cham press*, there are many evidences in *Cham press* where the government blames rebels for receiving foreign military aid, like in the following example:

(F)"DAMASCUS– Information Minister Omran al- Zoubi said that what is happening in Syria is a massive conspiracy and an aggression using barbaric tools whose elements are

obvious, stressing that Syria is facing a terrorist, Wahabi and Takfiri mentality and an extremist culture in various guises and under various names."

(Cham press, 03-09-2012, Information Minister: Syria facing massive conspiracy and terrorist, extremist mentality... solution is purely Syrian)

The interview of the Information minister in the extraction above reveals Al-Assad's point of view of the foreign conspiracy in Syria. The writer named some of the terrorist organizations as tools for the foreign conspiracy against Syria. The use of the term 'terrorist' to describe the groups mentioned in the extracts (*Wahabi and Takfiri*) is to make the readers think that the rebels are only the *Wahabi* and *Takferi* groups. There is no mention either of the total number of those groups' members or on the source of the information. It is meant to make a random generalization to make the whole revolution look like a terrorist action. Therefore, the information is not accurate.

4.3.5 Rebels are Portrayed as Responsible for the Massacres in the Country.

(G) "PROVINCES- Al-Qaeda-linked terrorist groups committed two horrible massacres against a number of families in the towns of al-Shumariyeh and Taldo in the countryside of Homs province, in escalation of their crimes that comes usually before the Security Council sessions on Syria and this time in coincidence with the announced visit of the UN envoy to Syria, Kofi Annan."

(Cham press, 26-05-2012, New massacres by al-Qaeda-linked terrorist groups against families in al-Shumariyeh and Taldo in Homs countryside

In this extract, the government is naming the rebels as, "*Al-Qaeda-linked terrorist groups*". There is no evidence in the text to verify whether this group is linked to Al Qaeda or not, but linking the group to Al-Qaeda gives the group a negative image and labelling them as 'groups' implies distraction and less unity. In this context, those groups are said to commit a number of massacres against civilians. The victims here are called 'families' rather than civilians to gain the sympathy of the readers. In order to convince the readers that the rebels are responsible for the massacres, the writer further added that there is a rhythm detected in the timing of actions of the rebels which is "*usually before the Security Council sessions on Syria and this time in coincidence with the announced visit of the UN envoy to Syria, Kofi Annan.*". This statement is used as the only evidence to claim that the rebels are responsible for the massacres.

4.3.6. Representing Rebels as Against the Law.

(H) "The Assembly also approved the draft law on demobilizing state workers or employees and depriving them of their wages, salaries and pension rights if charged with committing terrorist acts, joining terrorist groups or providing support to them.

(Chm press, 28-06-2012, *People's Assembly approves draft laws on combating terrorism)*

On the other hand, *Cham press* repeatedly called the government's men as the force of law to give them more credit. The government's men have been called as 'law enforcement' in more than thirty articles in our data. Yet, the definition of terrorism is not clear. Terrorism is defined as any help or support given to the groups against the government. Therefore, it makes no distinction of peaceful or militant, internal or external, organized or scattered groups. There is also no list of the groups that are considered terrorists and the groups that are not. This is because, to the government, all groups fall in one basket leaving no space of any opposition to the government.

4.3.7 Representing Revolution as Business

In a few articles of our data, *Champress* keeps referring to the presence of smugglers and external sources of money being paid to the civilians to help the rebels. For example:

(I) "He said that they agreed with Amoun to make each delivery of weapons in exchange for SYP 35,000, adding that they also smuggled journalists and militants into Syria in exchange for various sums of money, with them transporting weapons and individuals to Amoun's house in Jusiyeh or to a school occupied by Amoun and his group."

(Champress, Lebanese terrorist confesses to participating in smuggling weapons and militants to Syria and committing acts of terrorism, 10-09-2012)

The use of verbs like 'agreed', 'make delivery', 'smuggled' are all associated with illegal business making the revolution look like a form of mafia business. This is because the verb 'smuggling' and 'agreed' indicate a form of secret, dirty business. The irony lies in the idea of 'smuggling journalists'. These journalists are seen in this contexts as part of the weapons against the government. Therefore, the writer used the word 'smuggled' with various objects that the country does not allow, including journalists. Yet the government does not admit that it is preventing freedom of speech; yet, it lists journalists as part of the illegal objects like 'militants'. Yet, the writer is referring to the source of information '*Lebanese terrorist Mohammad Hussein Fares*' who is referred to as 'he' in the abstract above in order to give more credibility to the information. On the other hand, the government also receives militant support from Russia and Iran but it is not referred to as smuggling but as 'support or help' and not mentioned anywhere in the data. This is because the government allows itself to import weapons from the outside yet prohibits rebels to do so.

4.3.8 . Rebels Force People to Support Them

The government keeps referring to people as the passive entity which falls between the rebels and the government's men. This is supported by the following extract:

(J) "Fares said that they opened fire on citizens to force them to join them against the government or leave the area, with the terrorists continuing to open fire for nearly 90 minutes."

(Champress, Lebanese terrorist confesses to participating in smuggling weapons and militants to Syria and committing acts of terrorism, 10-09-2012)

In extract (J), 'they' refers to the rebels. The rebels are associated with a material process, '*opened fire*'. The object is 'citizens'. This statement is made to make the rebels look like terrorists to the readers and to show that without the influence of guns, people would not have supported the rebels. Similar to the previous examples, *Cham press* gives negative connotations to the rebels and represents them negatively. In contrast, the government punishes anyone who helps or even supports the rebels as the previous example (E) shows yet it is not considered as a force to support the government. These contradictions of actions and statements have become the signature of the government.

4.3.9. Rebels Target People

Cham press again portrays the rebels as violent as in the following extract:

(K) "An armed terrorist group attacked the town of al-Ghassanyeh in al-Qseir area in Homs countryside. Five citizens were martyred in the attack. A source in the province told SANA reporter that the terrorists also kidnapped 17 citizens from the town, adding that the relatives of the kidnapped persons fear that the terrorists are planning to kill them."

(Champress, Crackdown on terrorists continues in several provinces,05-09-2012)

In this extract, an unspecified source of information, 'a source', has given the news of the rebels attacking the town of *al-Ghassanyeh* . The rebels are referred to as 'armed terrorist groups' to employ violence, danger and evil adjectives to the rebels. Furthermore, the use of the material process 'attacked' is used because the writer wants to reiterate that the rebels are the culprits who fire innocent citizens. There was no mention of any kind of resistance from people because, as in most of the extracts, the people are seen as passive recipients of the rebels' actions. Thus, the use of verbs like 'attacked', 'kidnapped', and 'planning to kill' are all violent actions associated with the rebels. Yet, the government's role is absent and the people's action is absent. Citizens were killed by the rebels according to the writer, but they gain the support of the readers as the writer used the term 'martyred'. Yet, earlier in the article, the government's men were clearing several areas from rebels "*The armed forces targeted the terrorists in the areas of al-Halk, Sheikh Said, al-Nayrab, Bustan al-Qasr and Hanano in Aleppo, killing scores of them and destroying their DShK-equipped cars."* There is no evidence as to who did the attack on *al-Ghassanyeh* because according to this scene, both the rebels and the government's men were in action.

4.3.10. Government Portrays Itself as a Peaceful Government

As *Champress* portrays rebels as violent, evil, unlawful and terrorists, it portrays the government with a positive image like the safeguards, heroes and being peaceful as in the following extract:

(L) "Title: Al-Halqi: Crisis is in the final stage, working on reaching peaceful solution"
"Premier al-Halqi added that the state is working on solving the crisis through field work and political, economic and social methods with the aim of stopping violence and ensuring the security and stability of the country."
(Champress, 17-09-2012, Al-Halqi: Crisis is in final stage, working on reaching peaceful solution)

In this extract the government is represented by its prime mister, Al-Halaqi, that the government is trying a peaceful way to solve the problem. This statement is supported by a sense of security and peace like 'ensuring security', and 'stopping violence'. The government is here portrayed as the protector with peaceful intentions. Yet, in reality the government did not show its peaceful side because the statement above is contradicting with another statement where the government declares its rejection of dialogue with the rebels as shown in the following extract:

(M) '<u>They</u> also <u>stressed rejection of dialogue</u> with <u>those who got involved in killings</u>, underlining necessity of enhancing national reconciliation."

(Champress, 28-06-2012, People's Assembly approves draft laws on combating terrorism)

There is no clear definition on how the peacefulness goes in line with the government's actions and how it goes in line with the rejection of dialogue. The term 'involved in killing' is a broad term but from the evidences above, readers can assume that it refers to any force against the government because unfortunately it is difficult to know who is involved in the killing and who is not. Thus, leaving the term open and unspecified gives the government more powers and more excuses to use violence against the rebels.

5 Conclusion

This study shows that ideology has a very crucial role in manipulating the language of the media to frame the characters involved in the discourse in a particular way. In the case of political conflicts, the media becomes one of the tools to get local and international support by reporting events from a particular point of view to make one as the villain and the other as the victim. In the Syrian scene, the pro-government news agency *Cham press* has used various linguistic techniques to represent the revolution as illegitimate, to direct the audience's attention to certain events rather than others, and to view the rebels as terrorists and to gather the audience's support against the revolution. *Cham press* used terms and arguments earlier used by Bashar Al-Assad in his speeches and interviews to support the government's actions. In terms of predication strategies, the government is represented positively by the use of positive predicates to frame its position as the victorious, the protector, the saviour and the caring government. However, the rebels are viewed as criminals, terrorists armed groups, street gangs who create fear, violence, and are accused of being foreign fighters from different countries who kill innocent people. We find no reference to the local fighters, the people who support or join the rebels' army. We find no reference to the unity between the rebel's parties or their achievements against the government's oppressive treatment. The ideology is served in supporting the government for leaving no room for the other opinions. Any opposing opinion falls, in the *Cham press*'s point of view, under the umbrella of foreign conspiracy that has an unclear definition or boundary. Having the media in favour of the government and banning the international media from reporting in Syria is a way of manipulating people's minds and brainwashing the Syrian people to reduce their support of the revolution.

6 References

Fairclough, N. , (1995) *Discourse and Social Change*. United Kingdom .Cambridge.

Meyer, M., Wodak, R. (2009). The Discourse Historical-Approach (DHA). In Meyer, M., Wodak, R. (Eds.), *Methods for Critical Discourse Analysis* (87-121). London: SAGE.

Perthes, V. (2001) Politics and Élite Change in the Arab World. In Perthes, V. (eds.), *Arab Élites: Negotiating the Politics of Change*. Boulder: Lynne Rienner Publishers.

Reisigl M. & Wodak R., (2001) *Discourse and discrimination: Rhetorics of racism and anti-Semitism.* London: Routledge.

Van Leeuwen, T. (2008) Discourse and Practice : New Tools for Critical Analysis. New York: Oxford University Press

Wodak, R. (2002). *Does Sociolinguistics need Social Theory? New Perspectives in Critical Discourse Analysis.* [Notes] Retrieved from http://www.univie.ac.at/linguistics/forschung/wittgenstein/unemploy/handoutbristol11.htm

Character Pleasantness in Malaysian Animated Cartoon Characters

Mohd Rosli Arshad[1*], Kim Hae Yoon[2] and Ahmad Azaini Manaf[3]

[1]Graduate School of Design, Dongseo University, Busan, South Korea
[2]Graduate School of Design, Dongseo University, Busan, South Korea
[3]Faculty of Applied and Creative Arts, University Malaysia Sarawak. Malaysia

Abstract. Current global trends have proved the creative industry to be one of the important sources of economic growth among developed countries. Creativity and its importance for Malaysia have made it imperative for any business organization to use creativity in a range of ways including multimedia content and animation. Malaysian animation viewers are rapidly influenced by digital media entertainment. The rise of such entertainment tends to drive them away from understanding what lies behind it that affect their emotion and thoughts. Therefore, the focus of this paper is to look into the experiences of "pleasantness" in viewer's emotions that stimulate the perception of pleasure when watching Malaysian animated cartoon characters. A descriptive and One-Way Anova will be implemented in this study to examine the design aesthetics and perception from the animation viewers that affects the psychological experiences in emotions that determines the pleasantness feeling. Overall, the results indicate that perceived pleasantness on Malaysian animated cartoon characters did not differ between age and gender. We believe this finding will benefit the creative content creators and help them to understand more about local animation viewers.

1 Introduction

Animation films, as a part of entertainment has long been recognized and accepted as a media content that influences people in any society. The importance of character design in animation and other media entertainment may have an indirect impact on the emotion of the viewers. According to Hodges's 2011 (as cited in Adi Bunyamin, Dahlan Ghani & Amir Asrieff Amir Hussin, 2017), the importance of graphically rendered characters has been paramount in this new form of entertainment. He suggested that a successful animated character is comparable to a successful actor whose popularity can be used to sell products, concepts, information etc. A simple graphic representation becomes instantly recognizable, appealing, emotionally accessible and engaging. Animated feature films are extremely expensive to create, costing some studios hundreds of millions of dollars. On a large scale

* Corresponding author: mrosli@unikl.edu.my

production, it is critical that every detail is perfect. In order to avoid failure, it is important to understand human visual attention when designing characters of animated films.

In the Malaysian animation industry, we have seen many products failed in the market. One prime example is the big budget local animated feature film called Oh La La in 2015, that spent almost RM 4 million and has failed to attract local audiences. Even though it is considered as the first Malaysian musical animated film, it failed miserably. Aesthetically, the entire film was visually not up to industry standards and one of the reasons behind it is the lack of quality in the design aesthetics. However, there are also other animated feature films which are very successful in the local market. For examples, films such as Upin and Ipin (2007), Boboiboy (2011) and Ejen Ali (2016) have successfully implemented the aesthetic of designs that helped to flourish Malaysian animation visual contents. Thus, this study explores the phenomenological experiences of "pleasantness" in viewers' emotion that stimulate the sensory perception of feeling pleasure and excitement when watching Malaysian animated cartoon characters in terms of design aesthetics. The aim is merely to understand the difference between gender and age that influence the sense of pleasure on character design aesthetics in animation film and to identify whether gender and age preferences gave a significant impact to the overall animation films.

2 Research background

2.1 The Malaysian animated films

Malaysian animated film is maturing and its contents has been accepted globally. The progress of technology and multimedia has tremendously assisted in the development of animation contents in Malaysia. It all started with the initiative of Kamn Ismail in 1995 where he produced the first animated local series called Usop Santorian. Since then, the demand of local animation kept increasing due to the effort taken by Multimedia Digital Economy Corporation (MDEC)[1] to boost and promote local animation to the international market. Over time, Malaysian animation products have successfully attracted many audiences not just locally but globally as well. In the beginning, local animation content was created based on Malaysian culture as well as to promote harmony among the different society in Malaysia (Mohd Nazri & Hassan, 2013). Later, the Malaysian animation content began to change to a more universal content in order to attract international viewers. Bola Kampung (2006) animated series was a starting point for the local animation titles to have been accepted globally. The series became the first local animation to break into the international market broadcast network. In November 2007, it aired in 16 Asian countries over the Disney Channel (Hassan, 2016, p.153). It was then followed by Upin and Ipin (2007), which became a huge phenomenon especially in South East Asian countries that shared similar cultures and language. The popularity of Upin and Ipin has lead the commercial success of Malaysia's first CGI-animated film, Geng: The Adventure Begin (2009), which features Upin and Ipin that successfully earned RM 6.31 million and being considered as one of the most successful computer 3D animated television series in Malaysia (Abdul Ghani, 2015). As a result, Upin and Ipin has opened a way, to many other local animation content such as Super Tots (2010), Boboiboy (2011), Dunia Eicak (2012) and recently Ejen Ali (2016).

[1] Malaysia Digital Economy Corporation (MDEC) is an agency under Ministry of Communication and Multimedia Malaysia that responsible to build, organizing and promoting digital economy and communications technology (ICT) in Malaysia. For more information, please refer to www.mdec.my.

2.2 Malaysian animated cartoon characters

Anime (Japanese manga comic characters) has very large following in Malaysia. The popularity of Japanese animation has influenced the local designers and animators. In early 1990's where local animation industries is just started, local animation studios had brought in the anime-style to the productions they worked on. This is because early animators had no proper training in art and design nor were exposed to the various styles of animation of other countries (Hassan, 2016). As time goes by, many local animators started to realize the importance of design aesthetics in their creative works and started to explore their own root.

Today, more local animation company has developed their own unique animation stories and appealing animation characters. For animators, when designing the animation characters, they have almost the same art direction. Visual style, color and demographic factors has to be considered before coming up with a finishing design for the characters. From Usop Sontorian animation in 1996 until today, the Malaysian cartoon characters have changed drastically from anime-like design into local influences visual style designs that capture the imaginations of animation enthusiasts and are widely appreciated by Malaysian animation viewers.

2.3 Design aesthetics in animated films characters

The importance of design and color in animated films is very significant in order to portray a specific mood or emotion. Sajjad (2012) stated that the combination of basic shapes and colors are relied heavily on character design. The artist with the intention to create a greater emotional response in the viewer emotion uses the appropriate design and color. Appealing designs contribute to a sense of emotional engagement between the visuals and the ultimate goal depicted to the audience. The great character designer visualizes inspiration from history, culture and everyday life. Culture plays a significant role in defining a good and aesthetic design. According to Welsh-Asante (1994), an aesthetic draws upon the history, mythology, motif and creative ethos of a group of people in which is something obtained by fully understanding the ways of the certain community. By definition, character means the mental and moral qualities distinctive to an individual ("character"). It is also to determine how the person looks, acts, speaks, and even interacts with others. Su and Zhao (2011) defines it as the "design of human or humanlike characters of distinctive uniqueness and rich features for all kinds of visual media" (p.12). The perfect combination of colors in design creates a wealth of information that supports the idea that graphics enhance the emotional experience. Ionescu, Lambert, Coquin, & Buzuloiu (2007) put forward the claim that color is a major feature of animation movies: almost each movie uses a particular color palette, which is related to its artistic content, and the colors are selected and mixed by the artist according to art concepts to express particular feelings or to induce particular sensations in emotions. In fact, in animation the aesthetics of design and colors is not only applicable for the characters but is also essential to the visual background and animation environment to establish the mood or scenario of a scene that will integrate with the characters involved.

It is apparent from the previous studies that design and color aesthetic plays an important role in revealing human emotion towards animation contents. Recent evidence suggests that online gaming contents positively give hedonic or pleasantness experience towards the players and it significantly affects all dimensions of emotions according to PAD Model developed by Mehrabian and Russell in 1974 (Huang, Ali, & Liao, 2017). Online gaming is almost similar representation of animated films where all the visual is created using the same process where the design of characters and environment is developed at early stage. Thus, the characteristics of design and color aesthetics is significantly important and serve a promising effect to the viewer's emotion due to the fact that the content can be created as imaginatively as possible (Napier,2001, p.157).

2.4 Pleasantness (emotion)

Influence on emotion is a key factor to satisfy the movie viewers. One way to understand emotion is through positive and negative feeling. In theory, the psychology effects when people are asked the emotional factors of watching film, the common sense answer is to be entertained, to escape boredom, to get respite from their problems, to get new experience or information and, sometimes, to get an artistic thrill (Tan, 1996, pp. 17–18). From this finding, we can see that human beings are looking for relief, comfort or escape in their lives and watching movies is one way to reconnect or reawaken the sense of pleasure that effects their emotion and cognitive abilities.

Emotion knowledge is recognized as being split into two or three-dimensional structure, with positive-negative valence, activity or arousal, and potency or dominance (Dormann, 2006). The dimension of pleasantness is defined as the positive-negative valence of emotion and ranges from unpleasant to pleasant (Clore, 1994). Many psychology researchers have demonstrated that induced pleasure encourages the viewer to explore the stimuli in more detail. For example, Isen, Daubman, & Nowicki, (1987) purported that positive emotions increase an individual's sensibility to the stimuli in an environment. The word pleasantness is referring to pleasure or happiness. By the term positive effect, it means a conscious feeling of pleasure, a quintessentially subjective phenomenon. Conscious pleasure is the only form of pleasure of which many people can conceive (Berridge, 2003). Mehrabian and Russell (1974) developed environmental psychology model called PAD (Pleasure, Arousal and Dominance) or emotional state model to describe human psychology that conceived three basic dimensions of emotion responses that indicate people's state of feeling. They suggested that pleasure as a continuum range from extreme pain or unhappiness to extreme happiness and used adjective such as happy-unhappy, pleased-annoyed, and satisfied-unsatisfied to define a person's level of pleasure (Huang, Ali, & Liao, 2017).

3 Research methodology

The groundwork of this paper is to explore the experience of pleasantness in animated cartoon characters among Malaysian animation viewers. In order to investigate whether the character design aesthetics contribute to people's perception and emotion, we hypothesized that there is a significant difference between gender and age among the respondents. A set of questionnaire were used which consists of 9 items with categorical and five-point Likert scale sections ranging from 1 (Strongly Disagree) to 5 (Strongly Agree). Statistical analysis was performed in order to examine the difference between characteristics in emotion and design aesthetics in cartoon characters. A questionnaire's survey has been conducted to N=143 respondents consists of Malaysian youth age from 17 to 27 years old. The questionnaire employed in this research contained four parts, which in the first part is related to the demographic variables of the respondents such as gender and age. The second part contains nine selected titles of Malaysian animated films with the character lineup to give an emotional responsiveness to the respondents. These animation films are Upin and Ipin (2007), Bola Kampung (2007), Super Tots (2010), Boboiboy (2011), Dunia Eicak (2012), The Amazing Awang Kehnit (2014), Puteri (2014), Rimba Racer (2015) and Ejen Ali (2016). The last part was divided into three groups of questions that asked respondents' awareness and perception about local animated characters, character design aesthetics and psychological effects towards pleasantness in animated cartoon characters. In order to estimate the result, first we will use a frequency table for three items regarding the respondent awareness and perception about selected case studies for this research purposes.

Second, the Analysis of Variance (One-way ANOVA on rank) is used to determine the significance between independent variables groups in this paper. The rank statistical tests, is conducted to examine two categories variables, which is design aesthetics and feeling pleasantness between respondent's gender and age.

4 Research analysis and results

The goal of the current study was to examine the design aesthetics and perception from the animation viewers that affects the psychological experiences in encouraging positive emotions that determine the pleasantness feeling. The data in Table 1 were analyzed in frequency table to show the differences between gender and age to obtain numbers of participation in this research. Likewise, descriptive statistics concerning like to watch and like the appearance is presented in table 2.

Table 1: By frequency (N=143) for Gender and Age

Gender		Age	
	Frequency & (%)		Frequency & (%)
Male	78 (54.5%)	less than 20	34 (23.8%)
Female	65(45.5%)	20-23	43 (30.1%)
Total	143 (100%)	24-27	66 (46.2%)
		Total	143 (100%)

Table 2: By frequency (N=143) for Awareness and Perception

Like to watch (Awareness)		Like the appearance (Perception)	
	Frequency & (%)		Frequency & (%)
Yes	94 (65.7%)	Yes	92 (64.3%)
No	49 (34.5%)	No	51 (35.7%)
Total	143 (100%)	Total	143 (100%)
Factor's why I like the characters (Perception)			Frequency & (%)
Character's role			29 (20.3%)
Character's appearance			32 (22.4%)
Character's dialogue			18 (12.6%)
Character's theme			57 (39.9%)
None			7 (4.9%)
Total			143 (100%)

Table 1, indicates the total sample of gender is N=143 (M=1.45, SD=0.50) Malaysian youth consists of male is 54.5% (n=78) compared to female 45.5% (n=65). Next table, summarizes the respondents aged (M=2.22, SD=0.80) less than 20 years old accounted for 23.8% (n=34), 20-23 years old is 30.1% (n= 43) and 24-27 years old for about 46.2% (n=66) which brings the highest participation in this study.

In Table 2, the percentage of awareness among sample shows that like to watch local animation (M=1.34, SD=0.47) indicates that 65.7% (n=94) of the total respondents really love to watch Malaysian animation compared to 34.3% (n=49) dislike and preferred to like non-local or foreign animation. In general, the results clearly show that most of the respondents who took part in this survey really love to watch local animation. Obviously, it will benefit the purposes of this study and might influence on the results of this study. The findings from Table 2 also indicate that the perception of like the character appearance in local animation (M=1.36, SD=0.48) have showed the percentage of 64.3% (n=92) from the respondents which give an impression that the character's appearance gave a significance impact into their emotion. Further analysis is to identify the main factor that contributes to their personal preferences of why they like the characters in the animation (M=2.87, SD=1.27). The data revealed that character's theme shows the highest percentage 39.9% (n=57) compared to others. While data from character's appearance and character's role appears to have a 22.4% (n=32) and 20.3% (n=29) that have not so much different percentage which strongly suggested that character appearance and its role indeed play an important factor in animation films. Besides, the results can be connected back with the like the character appearance results which majority of the respondents strongly believe that characters in local animation have the positive psychological effect into their emotion. The relationship between character's theme and character's appearance is very important in the early stage of the animation production. The theme and concept of the animation are carefully planned to ensure the visual looks of the final products is according to the storyline.

In this test, the exact significance level was generated and we use the exact $p < .05$ to determine our results, and not to reject the null hypothesis of the p-value if greater than 0.5 (i.e., $p > .05$). A Mann-Whitney test was conducted to determine if there are any significant differences in character design aesthetics scores between males and females. Mean scores result reveals that males (mean rank = 73.35) and females (mean rank = 70.38) were not statistically significantly different, U = 2.430, z = -.4.55, p = .649 ($p > .05$). Age scores was conducted in Kruskal-Wallis H test reveals that distribution scores were dissimilar for all groups range from less than 20 years old to 27 years old. Median scores were not statistically significantly different between age groups, $X^2(2) = 4.007$, p = .135 ($p > .05$). While for the role of colors also brings a non-significant results, $X^2(2) = 1.666$, p = .435 ($p > .05$). As for the pleasantness feeling, the distribution of mean as assessed by visual inspection of frequency, indicates that gender distributions were not similar between males (mean rank = 73.83) and females (mean rank = 69.80), U = 2.392, z = -.619, p = .536 ($p > .05$) and were not significantly different. Finally, test for age groups showed that the distribution of visual inspection in boxplot were highly dissimilar for all age groups. Median scores for age groups and pleasantness were not statistically significant where $X^2(2) = 4.351$, p = .114 ($p > .05$). These results strongly suggest to keep the null hypothesis and shows that there is no significant difference between gender and age groups when comparing with pleasantness feeling towards character designs aesthetics in local animated cartoon. Based on our observations from the results above, we found that the character design plays an important role in determining the successful factor in local animation film. We had studied N=143 from the population consists of Malaysian youth age between 17 to 27 years old. We conclude that there is no difference between males and females when associating the feeling of pleasantness and character design aesthetics. However, the findings can be argued because there are certain conditions that can be questioned. The possible explanation for this might be:

i. Focusing on local animation products only, might affect the overall feeling of the respondents.
ii. The selected animation characters from the case studies might not interest them.
iii. The respondents might be confused and find it difficult to distinguish the aesthetics of each characters because the design is different.

Nevertheless, we found that the results have shown a strong evidence that the character animated cartoon in Malaysian animation film have successfully given a significant impact to the emotion of the animation viewers.

5 Research finding and discussions

The aim of the present study is to examine how the participants perceived pleasantness that may affect their feelings when watching Malaysian animated cartoons. The strength of this paper is to highlight the importance of understanding human emotion towards local animated cartoon characters. By examining the feeling of "pleasantness", we tried to understand what lies behind their feelings that stimulate the sensational feeling of pleasure when watching local animated cartoon characters. However, some weaknesses could be identify in this study. The clearest point of view in this paper is only to discuss the findings on the positive valence of emotion that described the conscious pleasure of watching animated characters but not the negative valence that leads to the feeling of unhappiness and annoyance. The PAD (Pleasure, Arousal and Dominance) psychology model is not fully utilized to examine the emotional response since the model conceived three basic dimension. Here, we only looked at one dimension, which is 'Pleasure' as a basic dimension to comprehend the meaning of pleasantness to describe level of pleasure of human emotion.

At the beginning, we hypothesized that there are higher chances of differences between genders and age. The results indicated that there were no significant differences between gender and age when pleasantness is perceived towards animated cartoon characters. As mentioned in the literature review, character appearance in animation was considered as one of the important element in animation. The results of this study has shown a positive result which supports previous research. Another important finding is the character's theme in the story is the main factor that contributes to the sense of pleasantness in animation. This outcome reveals that the theme for the overall looks of the animation including the character role and theme in the animation gave a significant impact to the emotional states that gives pleasurable experience to the viewers. Creating a dominant and attractive character design in animation is one of the key features to enhance the mood of the animation viewers.

This study may be important for the exploration of relations between emotion and the animation content. By defining the importance of understanding animation viewers, it symbolizes an extended meaning to the animation makers and the animation industries. From the business and marketing perspective, the results of this study provide significant insight for the purposes of producing and selling animation products in Malaysia. According to Hassan (2000), animation production must be viewed as a business that promotes folklore heroes and stories to the local audience. The local society shall become the consumer of local culture. In fact, one of the advantages of having a good character design is the translation into successful IP (Intellectual Property). The important aspect of creative work is ownership of the intellectual property. As an advantage, popular cartoon character has the value to position itself in promoting local brands. Hence, despite the limits, the data from the present study appear important for research concerning the relationship between animation viewer's emotion and the animation contents. It is beneficial for the creative content creators who produce animation for the Malaysian viewers.

6 Conclusion

Therefore, we hope that this study will benefit the creative content creators to understand how Malaysian animation viewers perceived pleasantness. This is to develop successful commercialized animation content, especially on the creation of animated characters. Finally, we would like to highlight the results that suggest the character's theme is the main factor that contributed to the psychological effects to the viewers and may be a worthwhile direction for future work. We recommend that further study should explore the perception and emotion that stimulate the sense of enjoyment among animation viewers especially local audiences. Additional studies need to be done to understand the psychological and phenomenological experiences from the eyes of the local animation viewers.

7 Acknowledgement

This paper is part of author's doctoral research sponsored by the Majlis Amanah Rakyat (MARA) and supported by the Universiti Kuala Lumpur, Malaysia.

8 References

Adi Bunyamin, Dahlan Ghani, Amir Asrieff Amir Hussin, M. Z. K. (2017). Technological advancement through character design: Darth vader as case study. *Annales Series Historia et Sociologia*, *1*, 1–8. https://doi.org/10.19233/ASHS.2017.01

Abdul Ghani, D. Bin. (2015). Upin & Ipin: Promoting Malaysian culture values through animation. *Historia Y Comunicacion Social*, *20*(1), 241–258. https://doi.org/10.5209/rev-HICS.2015.v20.n1.49558

Berridge, K. C. (2003). Pleasures of the brain. *Brain and Cognition*, *52*(1), 106–128. https://doi.org/10.1016/S0278-2626(03)00014-9

Ionescu, B., Lambert, P., Coquin, D., & Buzuloiu, V. (2007). Color-based content retrieval of animation movies: A study. *CBMI'2007 - 2007 International Workshop on Content-Based Multimedia Indexing, Proceedings*, 295–302. https://doi.org/10.1109/CBMI.2007.385425

Isen, A. M., Daubman, K. A., & Nowicki, G. P. (1987). Positive affect facilitates creative problem solving. *Journal of Personality and Social Psychology*, *52*(6), 1122–1131. https://doi.org/10.1037/0022-3514.52.6.1122

Hassan Abd. Muthalib. (2000). Sang Kancil – Ke Mana Kau? Filem animasi di era globalisasi – masa depan dan cabaran. Working paper presented in November 2000 at *Minggu Sastera Kebangsaan*, Dewan Bahasa dan Pustaka, Kuala Lumpur.

Hassan Abd. Mutalib. (2016). From mouse deer to mouse: 70 years of Malaysian animation (1st edition), Kuala Lumpur: Akademi Seni Budaya dan Warisan Kebangsaan (ASWARA).

Huang, M., Ali, R., & Liao, J. (2017). The effect of user experience in online games on word of mouth: A pleasure-arousal-dominance (PAD) model perspective. *Computers in Human Behavior*, *75*, 329–338. https://doi.org/10.1016/j.chb.2017.05.015

Napier, S.J. (2001) Anime from Akira to Princess Monokoke: Experiencing Contemporary Japanese Animation (1st Ed). New York: Palgrave Macmillan.

Sajjad, S., Mohsin, S., Riaz, S. & A. Abdullah. (2012): Enemy Character Design for Graphical Imagery Therapy (GIT) Game. International Journal of Information and Education Technology, 2, 4 (August 2012), 345–347.

Tan, E. S. (1996). Emotion and the structure of narrative film. Mahwah, NJ: Lawrence Erlbaum Associates, Publishers.

Social media as political hatred mode in Malaysia's 2018 General Election

Sara Chinnasamy[a,*] and *Norain* Abdul Manaf[b,*]
Universiti Teknologi MARA, 40450 Shah Alam, Malaysia

Abstract. In the era of technological advancement, with smartphones and digital devices, almost everybody can have access to the Internet. We move at very fast pace along with the information that we carry and every each of it will rapidly processed. Especially on social media platforms, where false news or scandals that revolve around pundits or the higher up often create chaos and triggers public interest. Malaysia had its 14th General Election (GE14) dated 9th May, 2018. It was believed to be the most nerve-wracking, fiercest general election to ever happen where after 61 years, Barisan Nasional (BN) had to bow down to the voice of Malaysians where social media is part of the factor, utilised ultimately to influence the people and powerful enough to change people's perception, thus, create political hatred. A textual analysis was conducted on 187 news on six online news stations' social media platform during the 11 days of campaigning. In-depth interviews were also conducted to discover the political hatred issues on social media. The findings suggested issues like Goods and Service Tax (GST), candidates' reputations and 1 Malaysia Development Berhad (1MDB) scandal are the main causes of political hatred expressed on social media platforms.

Keywords: Malaysia, social media, political hatred, politics.

1 Introduction

Nowadays, Internet is an essential part of our daily life. With the emergence of new technology and evolution of the smartphones, people are more connected. As cited in Lenhart, Purcell, Smith and Zickuhr (2010), the usage of social networking sites (SNSs) among American young adults and teenagers has risen significantly since 2006 as it has become a social phenomenon among their peers (Yusop & Sumari, 2013). Instagram, Twitter and Facebook are essential applications today as many of us have these apps for photo sharing and live update event. Every netizen is eagerly posting photos of their daily schedule, pictures of their pets, videos of attending events, course to show off luxuries and to express opinions on political topics happening in the country especially circa GE14. Social networking sites are built with micro blogging technology that supports blogging activity with smaller sizes of word posts and their users-friendly features invite wide usage among millennial (Yusop & Sumari (2013), as cited in Chin and Yusop (2010). The current generation is so technology-savvy where in their research findings state that 90 percent of the respondents own various types of handheld telecommunication devices such as cellphones and i-Pad, and almost everyone own computers and utilise them for communication purposes. This statement is tally with previous facts that the root cause of the phenomenon is the smartphone penetration that drives more users to access the Internet than ever (Amirudin Abdul Wahab, 2016).

Social media is a place like the actual world which people share ideas, cooperating and collaborating to create art, encourage thinking and build commerce, start a debate or discourse, finding new friends, allies or even lovers which people have built several civilisations on (Mayfield, 2008). He added that, due to these reasons, it is spreading so quickly, not because of anything complicated but solely because of 'us', the society. This is because the revolution of technology that continue to create new ideas, services, business models and many more.

Millennials in Malaysia actively participates in social media activities, from entertainment to argument and even political discussion. Politicians or even non-politicians actively talk about the political topic. According to Malaysian Communications and Multimedia Commission (MCMC), the number of active users in Malaysia crossed the 20.1 million mark in 2015, on the back of a 72.2% broadband penetration rate, and an estimated 16.8 million are also active social media users (Amirudin Abdul Wahab, 2016) whilst 18.3% of the internet users shared their political view online.

1.1 Social Networking Sites as a Political Platform

As cited in Bennett (2014), Instagram is a platform for photo sharing and is increasingly popular and it doubles the users within a year as of March 2014 (Abel, Buff, & Burr, 2016), as there were 200 million users on this social networking sites. A research done by the University of Missouri showed that majority of Instagram users are less likely to engage with political or controversial images (University of Missouri-Columbia, 2018). The research stated that many people view Instagram as an oasis to escape troubles and concerns of their lives, thus, they feel bad liking a photo of tragedy and prefers other sources for serious news.

* Corresponding author: sarachinnasamy@outlook.com

Twitter on the other hand is a useful platform for political communication, as the content of political tweet from politicians must be expressive than factual. Twitter has been the battleground for political actors and the place where scandal surrounding the politicians or the current environment can set in motion. It enables global communication immediate and unfiltered. Twitter has been quite the catalyst as it has become the way to connect directly with constituents and other interested audiences, dodging the traditional media (Sullivan, 2013).

Whilst, Facebook is actively used by political institutions for the purpose of entering into direct dialogs with citizens and encouraging political discussion in groups or pages in Facebook (Stieglitz & Dang-Xuan, 2014). They added the use of Facebook is the most logical location for a discussion on political issues to an extent the discussion appears to have succeeded in the over-coming polarisation of online discussion that previously hard to be done (as cited in Kushin & Kitchener, 2009). A research by Vitak et al. (2011) showed that undergraduate students tend to engage in lightweight political participation both on Facebook and also other platforms (Stieglitz & Dang-Xuan, 2014). When people gather online, they tend to discuss issues of public interest as people share in such inter-connected society and if the issue is discussed by many, more users tend to join the bandwagon to keep up to date of what is going on. The topics usually get heated up especially when the general election is near. Malaysia's political arena has undoubtedly moved along with technology, thus, the usage of social media in political campaign become very crucial especially when the campaign season approaches.

1.2 Problem Statement

In the era of explosive growth of technology, social succumb to media influences teenagers' decisions to use drugs, no privacy, competition and peer pressure. As social media is so influential, there is no doubt that hot debated topics such politics is discussed through social media platforms which tally with Memon (2017). His article states that social media channels are now the main sources of news, opinions and debates among millennials. According to MCMC (2016), 18.3% of the Malaysian Internet users share their political views online.

Political hatred is a common feeling among party supporters as they tend to believe that their party is superior to the other. Brooks (2015) stated that through anonymous hate, people could just post anything in this Internet age and scales up the ability to express political hate with astounding efficacy, as political hate is a demand-driven phenomenon and we are the one creating the platform for it.

Internet users cannot help but be bias on what they see on the social network due to political preferences. Thus, political hatred will move along with the ideology of the political parties or the politicians themselves. Political hatred may sound normal to many as people are joining the discussion, but the power that political hatred holds may sway the political parties as Malaysia is a multiracial country. Among the issue that has sparked political hatred is the 1MDB issue, implementation of GST, price hike of gas and many more. Netizens sure would not want to miss out on the issues that have been discussed internationally, as every Internet users have their 'two cents' to share. Thus, this study will explore the social media as the medium for political hatred in Malaysia and the exposure have led to findings that the social media is a platform for political hatred.

1.3 Research Objectives

The purpose of this qualitative research is to look at social media as the platform of political hatred in Malaysia. The objectives of this research are:

1. To explore the political hatred posts in Instagram, Twitter and Facebook.
2. To find out the Malaysian citizens response about political hatred on Instagram, Twitter and Facebook.

2 Methodology

This study utilises two approaches in qualitative research; (a) textual analysis, and (b) in-depth interviews. The interviews were conducted using an interview protocol among five (5) selected participants from different fields of expertise. For textual analysis approach, this study analysed 187 selected news on six (6) news agencies' Facebook, Instagram and Twitter during 11 days of campaign, namely *Malaysiakini.com, Free Malaysia Today, Berita Harian, The Star* and two international online newspapers, *Bloomberg* and *The Straits Times Singapore*. The news were selected based on the comments received on the articles, likes, re-tweets and shares. Comments related to issues, have sense of anger, sarcasm towards issues of the news were classified according to exploratory perspective in data analysis to code and theming by researchers. This study did not require a big sample because it is highly structured and the participants are consistently participating in social media activities and have to vote in GE14.

The in-depth interviews were conducted using semi-structured questions that were opinion-based and related to their experience and knowledge. The questions structured for these interviews revolve around their opinions about social media and political hatred in Malaysia and during the general election period, where the participants were asked about their direct and indirect involvement in sharing opinions online and seeing political hatred in social media platforms. Questions regarding the social media usage as other postings in the social media relating to political hatred and issues regarding Malaysian politics were discussed in detail during the interviews.

The understanding of these participants in political hatred would help lead to accurate findings. The participants were free to express their thoughts and view through open ended questions. Thus, that is why in-depth interview approach was adopted. The researchers believe this approach is important in determining the political hatred that have arisen from social media in Malaysia during this period. The researchers deem this could identify the causes of political hatred especially during the GE14 period, because each individual online or offline had their own beliefs, perspectives and viewpoints. Thus, the findings on political hatred especially on social media have found a big reversal for GE14. Both approaches used a thematic analysis to generate key themes.

This research has revealed the type of political hatred occurred in social media and getting the opinion from the people why political hatred happen during the GE14. The political parties have used multiple approach mediums in order to win the hearts of the people. It provides the explanation for winning and losing factor during the election and may become the source to overcome the factor for the upcoming election for both government and also opposition.

3 Findings and Discussion

GE14 saw the power of the people with the help of power of social media. Various issues and opinions were pointed out during the campaign and also through their manifestos. Foucault Theory suggested that power is not exactly a tool that were used by people but rather are actually the tools of power or simply power is a relation and embedded in the web of power relation (Wolf, 2013). Pastoral Power viewed as the power of an individual or simply productive power that produces subjects (English, 2004). She further added, "concept of pastoral power is an extension of disciplinary power and an alternative perspective to sovereign power that focuses on repression or having power over" as cited in Dreyfus and Rabinow (1982). Thus, Foucault's work helps to understand the rise of opposition power againts the previous government over the people. Hence, issues that highly mentioned both in textual and also during the interviews are explained next.

1. Goods and Services Tax (GST)

This issue was among the most debated topic from the moment it was informed to the public until today. It was supported at its initial stage, but also rejected by members of the society due to the concern of Malaysian economic capability. The concern by society was it will impact every level of society member as the amount charged by six percent (6%). There is more middle income and below average income earners than people who can afford a hefty lifestyle. One of Pakatan Harapan (PH) manifesto for GE14 is to abolish GST and replace it with the old taxation system, the Sale and Services Tax (SST). The initiative of abolishing GST has brought many responses from social media during on the 11 days of campaigning, as many have hoped that PH to walk the talk as it was levied on most transactions since implementation in April 2015. It has caused a major debate among BN divergent as the idea was of Dato' Seri Najib Razak and the opposition party at that time held an assembly to protest against the implementation.

For Malaysians, GST as seen as a burden by according to Bakar and Sawandi (2018), *'isu ketidakpuasan GST masih mendapat tempat dalam kalangan rakyat yang cukup terasa kenaikan kos sara hidup'* (dissatisfaction towards GST issue still flames among the society member who were affected by the significant increase of the cost of living). Thus, this issue continues to spark interest of Malaysian and highly debated among social media user and also recognised by the participants of this study. Bakar and Sawandi (2018) also added *'sokongan rakyat terhadap sistem GST telah semakin merosot selepas GST dilaksanakan, berbanding sebelum perlaksanaannya. Ini ditambah dengan sokongan terhadap kerajaan juga merosot dengan sangat banyak selepas perlaksanaan'* (the society support for the GST system declined after it implementation compared to before the implementation. The support to the government also decreased drastically after GST was implemented). It cannot be denied that with the implementation of GST, it has jacked up the cost of living and are one of the major factors for political hatred on social media.

2. Candidates' Reputation and Promises

Globally, politics and scandals cannot be separated; it moves in a circle following one another. Politicians are always associated with political scandal as it is a way to put the politicians at loggerheads with others. Usually, they offer sweet promises to change voters' belief. According to Aragons, Palfrey and Postlewaite (2007), politicians make promises to secure themselves in the administration and assumed that candidates' reputation at every election can be either good or bad. Therefore, candidates with good reputation are the ones that never reneged in past elections, whilst the candidates with bad reputations are the people that break promises that they made.

Politicians in Malaysia actively take part in online community to reach out their supporters and the community. They gave opinions, share political belief, criticising other politicians and be part of the community in offline arena too. Online political participation by the society is when people look into how the political parties and the political leaders debate and also engage each other on political discourse; the public give opinions based on what has been circulated from the debated topic online. Thus, it is important for political parties and politician to be involved in debates and engagements as these will impact their reputation towards society's impression.

Therefore, the analysis focused on comparing the content of the news presented both in headlines and comment sections/replied tweets. Applied thematic analysis helped researchers to identify the presence of political hatred whilst helping readers to understand the issues being highlighted by netizens that led to political hatred. All six media platforms highlighted headlines about candidate's registration for polls on nomination day, which represented 21 out of 33 selected news (based on the words nominations, candidates and register). This is expected as it was the day stipulated by Election Commission (EC) for candidates to send their nomination forms. However, there were candidates who failed to comply with the nomination regulations. Generally, this issue can affect the credibility of the candidates, for example in Tian Chua's case which he was disqualified by EC for not fulfilling the requirement needed in order to contest as one of the qualified members of parliament (Suruhanjaya Pilihan Raya Malaysia, 2018). The Star Online Facebook had a page that shows the netizen backlashed him in Facebook commentary section (see Fig. 1).

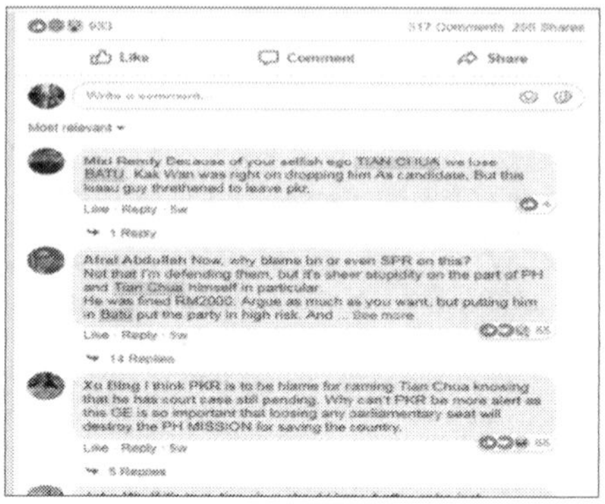

Fig. 1. Netizens' comments on Tian Chua disqualification.

The early campaigning stage saw that the selected news agencies reported mostly on what the political parties' leaders have to offer. For example, offering minimum wage to be raised if BN wins the election, continued development if BN were chosen, Malaysia stands to lose internationally if people vote for the opposition and few more. The chosen news mostly reported on BN's promises to the people. The responses given by the netizens were not as positive as the title proposed.

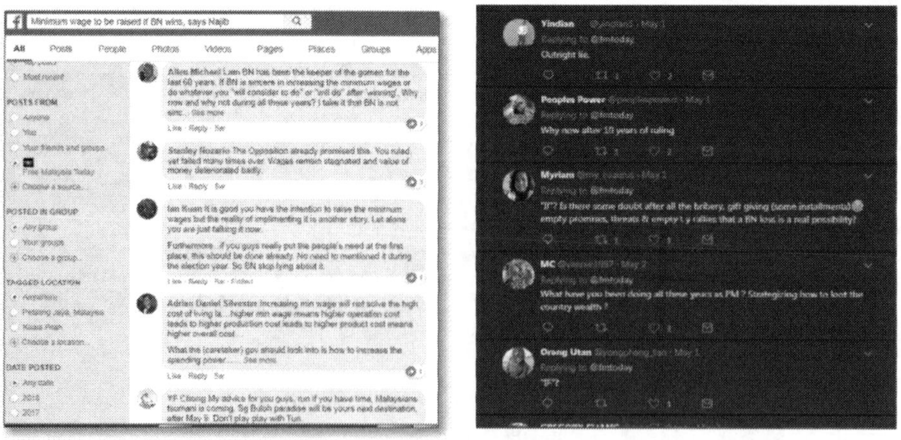

Fig. 2. Netizens' comments on minimum wages offered by the BN on Facebook (left) and Twitter (right).

The minimum wages offered by BN received unfavoured response from the social media users. This is aligned with the issue raised by Lo (2017) that the rising living cost does not match with the issue of minimum wages. The middle and low income consumers' were declining in terms of goods and services purchases that put a limit on Malaysia to become developed or high-income country and the wage inequality has widened and negatively impacting the national growth. The local talents had to compete with floods of foreign workers in order to secure a job. With minimum wages offered to foreign workers, employers prefer the latter as they had to pay more to them because of minimum wages regulations by the government. Thus, the issue of raised minimum wages sparks questions among the society, as the step should have been taken ages ago.

Promises of developmental projects were part of political agendas of the parties. To win over the people at the area, they usually make promises to the society of the area needed, such as education, access to any basic necessity such as marts, safe surrounding, etc. As for example, Free Malaysia Today (FMT) dated 29th April 2018 mentioned, "Amanah man promises to give Kamunting people a university" where there are already two universities in Taiping; Universiti Teknologi MARA (UiTM) and Universiti Teknologi Petronas (UTP). Thus, with the existence of a new university was said to raise interest among society to further their education. But promises on the developmental projects were not fulfilled, it merely just the candidate's way of gaining and piling trust to gain votes from society. Another example from FMT dated 3rd May 2018, "Choose BN for continuous development, Salleh tells Semporna". Felda in Segamat area also received a convincing offer where Subramaniam (MIC) showered Segamat Felda with RM 1 million replanting funds by polling day. Thus, making promises are sure one of the way to obtain votes from the people but also will become the main reason for the party to be criticised back if the promises failed to be fulfilled by the winning government.

Candidates' personal attacks were one of the popular counter attack modes in order to sway the people's attention and make themselves convincing. Personal attack may either bring a fruitful result to the political party or boomerang effect to the party. During the 11 days of campaigning, the technique was seen used multiple times by the candidates in order to sway people is trust and loyalty from the other candidates. The most popular candidates worth mentioning are Dato' Seri Najib Razak and Tun Mahathir Mohamad as they both contended for the prime minister position. A lot of attempts to downgrade the persona in order to influence people's perceptions, which not only happened to the two parties, but also to PAS candidates. They called out the opponents with many negative remarks or brought up scandals revolving around them, such as corruption (Najib's 1MDB scandal), spendthrift family member lifestyle (Tun Mahathir's grandchildren's fancy lifestyle, Rosmah Mansor's influence in Najib Razak decision making/massive spending), being obsequious and few more. Other example mentioned in Sarawak Report in 2016, "How Jho Low used 1MDB Cash to Fund Rosmah's Diamond Shopping Sprees". However, Sarawak Report was banned from reporting on 1MDB in 2015. The ban was lifted after PH won the election.

3. The Emerging Influence of Opposition Parties

The opposition party role in politics is important to ensure a balanced political environment. It gives chances to the society to pick the party of their choice to govern the nation and also at the same time the public can give constructive criticism to the current ruling government. The power of the opposition can be seen in reports headlines such as, "Tok Batin: Semai Tribe Orang asli are completely rejecting BN for GE14", "*Calon BN gigih pikat pengundi bandar*", "Angry urban Malaysian chip away at PM Najib's dominance" and also "Ex-finance minister, Daim, campaigns for opposition". These headlines indicated that the preference towards opposition party was showing. The figure below shows the reactions and comments received from news headlines "BN loss will see bad future" which score amongst the highest interaction amongst netizen on selected news reported on May 2nd.

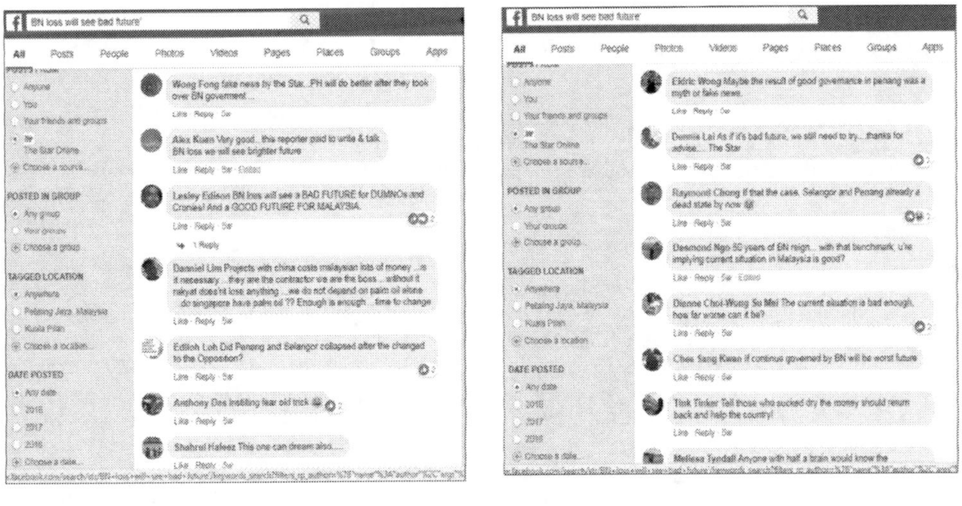

Fig. 3. Netizens' disagreement on The Star's article.

Amongst the issues mentioned in the comment section by Facebook users were the influence of China's project in Malaysia and the state of Selangor and Penang (administration by DAP and PKR). Users also did not hesitate to criticise BN as the headlines indicated that Malaysia's future will be in uncertainty if BN not in charge. The influence of China's large infrastructure project in Malaysia began ever since Najib Razak became Prime Minister in 2009 where Malaysia has tilted decisively towards Beijing's orbit that saw 14 memorandums of understanding (MOUs) worth $34 billion were signed under his watch (Capri, 2018). This statement is supported by Jennings (2018) where he stated that the opposition alliance suggest that Najib Razak has allowed too much of China's projects in Malaysia which usually relied on exports. These were the reasons that netizen were brave enough to point out their opinions.

At the final phase of campaign, the news showed the negative depiction of candidates, where they attack each other. They dissed each other's political movement that could be seen through news headlines such, "Sabah no longer 'fixed deposit' for Malaysian PM Najib, says party rebel Shafie Apdal" (The Straits Times, 6th May 2018), "Rafidah: Sacked from UMNO? I'm not a member of the kleptocratic organization" (Malaysiakini.com, 5th May 2018), "They're bussing people in, claims Najib of large Harapan turnouts" (Malaysiakini.com, 5th May 2018), "*Gambar beribu hadir ceramah pembangkang palsu - Zambry*" (Berita Harian, 5th May 2018) and many more.

The news headlines in the newspaper depicted no issue or concern of the root of why political hatred happened. Apart from that, in the midst of political hectic, former Deputy Minister of Home Affairs, Nur Jazlan Mohamed was badly criticised by netizens due to his comment on "Overseas voters do not affect the GE14 result" in Malaysiakini.com. The article received many responses from netizen as shown as below:

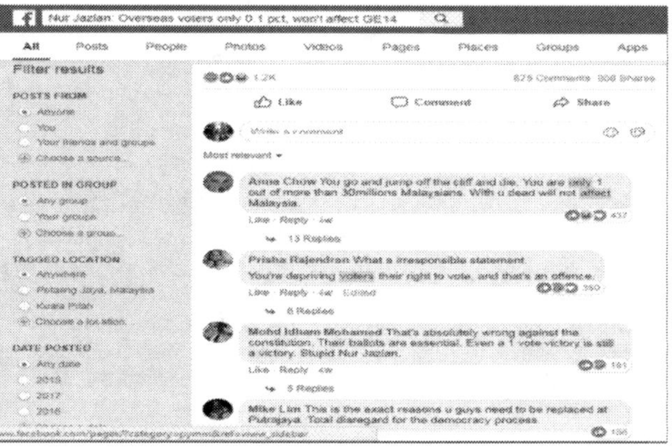

Fig. 4. Netizens' responses on Nur Jazlan's statement on Malaysiakini.com.

It is undeniable; his statement has sparked anger and hatred among netizens towards himself and also the political party that he represents. Thus, hatred comments on social media platforms can happen if it triggers the society with such situation, they questioned Nur Jazlan's statement of him denying the votes from the overseas voters.

4. Mahathirism

Mahathirism is an ideology that derived from thoughts and logical categories in achieving intellectual satisfaction and understanding. Mahathirism is believed to influence a lot of Malaysians preferably the younger voters as they feel only he can 'save Malaysia', considering a lot of outside influence in Malaysia's administrations. Mahathirism issue was mentioned in Malaysiakini.com (30th April 2018) (Najib: Here's Mahathirism for you–'his way or highway'), The Straits Times (30th April 2018) (PM: Dr. M's grab remark is typical Mahathrism, Malaysia election: BN communications chief says Mahathir trying to get sympathy votes) and Berita Harian (1st May 2018) (Kita pilih Mahathir untuk tarik sokongan Melayu). Mahathirism is believed one of the thriving factors influencing the process GE14's 11 days of campaign. The Straits Times reiterated Mahathirism factor as it seen on international media portraying Tun Mahathir as one of the successful political icons. Tun Mahathir was keen in convincing Malaysians to vote for PH to save Malaysia. Ever since, the 'Save Malaysia' slogan was used expressively by social media user in Malaysia. The sign of strong authority and prioritising the community by becoming an ally with other opposition parties have made the party stronger in terms of 'walk the talk', which can be retrieved from their "10 Janji 100 Hari PH" manifesto. The coalition managed to grab the attention of young voters and also his prime-time supporters. In Mahathirism ideology, it is over the individual, and a strong family-based society and it draws upon the experience of the western world in order to assess state and society considering modernity. For other political parties, Mahathirism is a mere autocratic administration and will ruin the country if they were to win. It is not only capable of swaying people's thoughts and mind but also can be a good tool for PH in winning the heart of young voters and existing Tun Mahathir supporters.

5. Conflicts within and inter-political parties

Political conflicts happen to every political party and it can happen either within the party or inter party. It happens because of differences in ideology and approaches that they want for the party. Some may focus on the prominence of the party and some parties prefer to have a 'save face/individual reputation'. During the 11 days of campaigning, a lot of conflicts happened, namely Nik Omar (Nik Aziz's eldest son who was known as a PAS leader) contending on behalf of PH's Parti Amanah Negara candidate that has caused a stir among PAS member and supporters, whilst his brother Nik Abduh was contesting as PAS candidate as reported by The Star (30th April 2018). According to Habibu (2018), "PAS led a bruising campaign against so-called "moderates" who were also called "Anwaristas" and "jebon (weasels)". These leaders then left PAS to form Amanah". Conflicts involving ex-member of UMNO/BN also mentioned as "UMNO defector shows up at Sungai Besar nomination centre in PH colours". A report by Straits Times (2nd May 2018), 'Ex-Finance Minister Daim campaigns for opposition" foresaw that Malaysian are taking sides on the opposition on social media as all the former ministers are turning their back from BN and joining the PH's force. The conflicts have brought consequences towards the party as the supporters come and go, and sustaining the trust of members of the party take the effort of the whole committee members as everyone has different opinion and views for the sake of the party.

6. Preserved Malay Special Rights

The Malays special rights in Malaysian constitution (Article 153) were also debated during the campaigning period, where Universiti Teknologi MARA (UiTM), known for 'Bumiputera only university' was mentioned by Dato' Seri Najib Razak. "Najib: Opposition has no agenda to safeguard UiTM interests" was mentioned in various newspapers including the researcher's selected newspaper. Due to the special privilege that was given to Bumiputera's, many supporters and political actor politicise the idea and make an argument out of it. Social media users have a different thought on the idea of opening UiTM to non-Bumiputera but some also opposed the idea as to preserve the Malay's special rights during the campaigning period.

The following interview transcriptions are the answers to Research Objective 2. The researcher had conducted an in-depth interview with participants to explore their perspectives on the topic of political hatred on selected social media platforms. The transcription begins with descriptions of all the candidates namely Candidate N1 to N5. All informants are Malaysian based in Alor Setar and Kulim in Kedah, Shah Alam and Klang in Selangor and Kuala Lumpur, respectively. The participants' responses were tabulated in Table 1 as shown as below:

Table 1. Responses from the participants.

Name	Responses
N1	*It's challenging but competitive between Najib and Tun Mahathir.*
N2	*I think the political situation of the GE14 campaign is so nerve-wracking.*
N3	*It's a fierce competition, but we are already expecting the winning party.*
N4	*It's a drastic political changes showing political maturity among Malaysians especially 40-43% young voters. Social media was highly utilised for the purpose of electoral gates and electoral political discourses.*
N5	*The election's campaign totally changed with the usage of Twitter, Instagram, Facebook, and Youtube.*

The participants shared many commonalities in giving out rational opinion. The categories and themes derived from the in-depth interviews on the response on political hatred on social networking sites namely Facebook, Twitter and Instagram are listed.

3.1 Political Strategies on Social Media

Participation in social networking sites increases from day to day. Advancement of technology and digital communication has enabled society to share their life, and opinions and it has become a social norm in our society to share everything online. The political strategy on social media is gathered in three themes: (i) crucial political tools; (ii) the emergence influence of opposition; and (iii) opinion about politic on social media. These themes are derived from the N responses towards political strategy on social media.

1. Crucial Political Tools

Conformity of social media as a political tool is one of the supporting factors in this study. The responses from the N1 in Malaysia were observed when they expressed on political hatred in selected social media platform during the GE14 campaign. Shim and Oh (2018) stated that as voters, people not only voice their opinion on social media but later establish political movements via political engagement and participation. All the N stated that social media is indeed crucial political role and their responses were restated below:

"Yes, it is." (N1)

"Of course social media plays a crucial role in political campaigns." (N2)

"Yes, social media is a very crucial platform for either government or opposition to reach out to the voters itself to find the latest current issues." (N3)

"I claim this year is a Facebook election year." (N4)

"As journalist, we follow the leaders and candidates. We opened the candidates' Twitter, the Instagram or Facebook on their activities. It's a new kind of campaigning." (N5)

Thus, it can be concluded from the participants feedback that social media is indeed a crucial tool for political campaign as it can help to the success of the campaign as people are very dependant of technology to reach out to others. It also helps journalists to monitor current activities of the political actors and parties without having to follow them around.

2. The Emergence Influence of Opposition

The opposition plays a very important role in order to maintain a healthy competition in politics. Below are the statements from the participants seeing the emergence influence of opposition in social media and the responses are stated below:

"People from PH were quite dominating compared to the BN. After the election, I followed some PH politicians through social media. Although I follow Tun M's accounts, PH politicians were sensational in social media." (N1)

"I spend most of my time on social media in Twitter especially Muar's MP Syed Saddiq and Lembah Pantai MP, Fahmi Fadzil (both are PH representatives). They really dominated Twitter to tackle the younger voters." (N2)

"Opposition's Nurul Izzah, one of the new faces." (N3)

"PKR, they are good bloggers. For examples, Rafizi Ramli and Tony Pua. They got the power to attract people." (N5)

Based on the responses given by the participants, during the GE14 campaign, the opposition parties from PKR, Parti Pribumi Bersatu (PPBM), Parti Amanah Negara (PAN) and others, unite to form a coalition named Pakatan Harapan (PH), dominating the social media platforms as the candidates are very popular offline as they have been vocal towards the previous ruling government. Nurul Izzah known for her more than decade-long involvement in politics since at a very young age. She has been actively involved in any march against the previous government along with Rafizi Ramli, who is a political influencer and other opposition politicians.

3. Opinion about Politic on Social Media

According to Shim and Oh (2018), people who share a high level of opinion congruence would feel more fear of isolation. Therefore, they tend to be anonymous online in expressing their opinion on political issues by using pseudonyms. They also added that the anonymity offered online enables individuals to do what they usually do not in reality (as cited in Suler, 2005). Yet, there are also groups of people that bluntly give their opinions online or joining the topic that attract their interests. The statements from the participants seeing the opinion about politic on social media and the responses are stated as below:

"It's better to preserve your vote from online discussion, and most important thing is, it's not healthy in social media and keyboard warriors." (N1)

"I do share my opinions, as I can hear and get other's feedback as well. But, I always have to keep in mind that I need to be neutral and not biased to any political parties." (N2)

"Yes, I participate in voicing out my opinion in social media which I feel the issue is breaking issue and there's something I must say about it." (N3)

Participants had mixed feelings on social media as they have a preference on how they would share their opinion online. Sharing opinions online might not be in a serious issue like politics, given the strict law and political situation in Malaysia as the wall to freedom of speech.

3.2 Political Hatred on Social Media

The responses were observed and for the researchers to understand the sense of political hatred that happens on social media among Malaysian through online news portal. Political hatred is common in political topics as one has a different opinion from another and stand to disagree on it. It could be mild or can get so excessive which sometimes leads to an unexpected situation. The statements from the participants seeing the opinion about political hatred on social media and the responses are stated as below:

1. Political Hatred among Malaysian

"Hatred is more to the arguments between political parties and supporters. Their online manners ridiculously to accept." (N1)

"Political hatred happens when you are too obsessed with one political party or individuals that you can't accept other's views. There many political hatred messages on social media and people even get paid by certain parties or individual to spread hate messages online." (N2)

"Political hatred is common in any nations it just that the constituencies need to be smart in justifying what issue were raise by the political parties as is it for the sake of putting politicians at loggerheads with one another. Political hatred issue during election week prone to cyber troopers who is driven by the government towards the opposition where the attack by the cyber troopers towards the opposition was very intense." (N3)

Participant N1, N2 and N3 believed that political hatred among social media users could go to the extent that cyber troopers and use pseudonyms to attack the opponent take the opposition or government. On the other hand, N4 believe that:

"Political hatred previously was raise by the PH where there was defamation factor. They used to attack opponent leaders also political parties. Social media played an important role in displaying the political events indirectly raised online political participation and debates on political discourses".

Hence, political hatred comes from disagreements on political beliefs and supporters on social networking sites, either the group supporting the opposition or the government. As supported by N1, N2 and N3 statements, it can be said that party attached supporters were likely to spark the political hatred itself.

2. Harsh and Hatred Opinion

Malaysians are known for their humility. Yet on social media, it seems that Malaysians are also very bold and harsh at expressing opinions particularly on political topics. This statement is supported by Shim and Oh (2018) where they stated, "Even political underdogs or minorities can cause their voices to be heard on social media." and it could continue offline. The statements from the participants seeing the harsh responses of Malaysian on social media and the responses are stated below:

"I go through one or two on the above replies under one post. It's so hard for me to scroll down to read one by one. They are more than daring due to the freedom of expression, the accessible social media, and public manners beyond the line." (N1)

"Social media is their space freedom of speech." (N2)

"Due to social media, the whole Malaysia system is different now. It is also people's power." (N5)

Thus, it can be concluded that Malaysians are very prone to hatred comments on social networking sites as Shim and Oh (2018) stated that anonymity setting of individual online predicted that the use of social media is for political engagement. Chinnasamy (2017) claimed that social media had increased the online political participation among Malaysians through her two cases of Malaysian election previously.

3. Issues that Triggered Political Hatred

The main aim of this research is to look issues that triggered political hatred among Internet users namely in Twitter, Facebook and Instagram. The participants were asked what issues had triggered Malaysians the most and issues that also raise political hatred among netizens. The statements from the participants and the responses stated as below:

"The most political hatred issue is Najib's scandals and his wife. It will invite political hatred easily, and politicians also do spread the hatred including 1MDB issues." (N1)

"1MDB and the rise cost of living are the most important issues during GE14. There are thousands of tweets and post by Malaysians!" (N2)

"The issue that highly debated on social media would be GST and 1MDB because these issues have been prolonging until the Election Day without solution." (N3)

"One of the issues is GST. The GST issue somehow been played by the opposition to gain voters attention in their 100 days of promises. It attracted the people to vote for the opposition." (N4)

"People are angry at Dato' Seri Najib. 1MDB and GST were created by the opposition now the current government to gain people attention." (N5)

It can be concluded that the political hatred was derived mainly from GST and 1MDB issues. Other matters such as candidates' reputation, Mahathirism, cost of living (which leads back to GST) are among the main topic that triggered the netizen as stated by all the participants. The participants stated that the issues triggered social media users because:

"It was considered as sensitive issues for the people." (N1)

"The youth are the largest group of voters that been effected by GST and the cost of living. Most of the middle-income group, living in the city, doesn't benefit most of the past government initiatives to help the people. They are really desperate for their voices to be heard. By talking about it on social media, it will at least get the opposition attention." (N2)

"It involves public fund, 1MDB where the government misuse the fund to afford projects that involved corruption, the GST implementation itself basically not fair and only benefitted the government." (N3)

"The politicians should become the role model to the society. They have to respond very fast to certain issue it's very hard to control people perception." (N4)

The 1MDB issue is regarded as a factor for the rise of political hatred among social media users and also participants of the study. As 1MDB involves a public fund and people were informed on it, thus, when the scandal of money laundering by the organisation was exposed by PH it triggered public scrutiny both on social media and offline.

4 Conclusion

This research has successfully reached its objectives of exploring the political hatred on social media during Malaysian 14th General Election. Through the textual analysis of 187 news and in-depth interview with participants, it looked into the opinions and feedback from students, executives, government servants and political analysts. The interviewees have expressed opinions on political hatred on social media. In general, they all agreed that social media is an important tool during election campaigning. The emerging influence of the opposition on social media is seen as a supporting factor for political hatred among supporters.

Social media is such a powerful tool that even experts in any field cannot deny it capability whether it brings bad or good impact, as it powerful enough to bring you down and cogent enough to bring you to the state of euphoria. It is a very essential tool for sharing and disseminating information. It eases the information to flow faster than traditional medium such as print and broadcast media. In the state of technological advancement, it is also being used in political activities where politicians or political parties all over the world use this medium to reach out to their constituents. It is also a platform where political hatred spreads very fast as social networking sites developer always improves their applications features for their users to share information, pictures, videos and many more. It is a good move to provide easier and handier functions to the applications but at the same time, they cannot control the content posted by users unless reported by other users. As information travels fast, constituents or social media users should be smart enough to judge the issues rose from political hatred matters.

Political hatred not only happens because of political disagreement, but the nature of not being able to accept other people's points of view and also the unheard voices of the society members. Fear of isolation from political participation also could be the contributing factor to political hatred where the minority joining the bandwagon for the sake of a dominant influence. Political disagreement should be discussed to create a healthier political environment. The differences in opinions can help one another to achieve a better development for the nation and also create a better society. Thus, through power theory, it has shown that the opposition of the government over the people where the struggles of question status of the people where they assert the right to be different whether in having opinion or freedom in expressing their thoughts. There have been a lot of discussions by the participants on the political hatred and why the issues spark political hatred towards the politician or parties. There lots of knowledge and thoughts shared that can help to give more perspectives on political hatred that occurs on social networking sites. This study has given a clearer idea and perspectives that has helped researchers to point out the surfaced themes based on the analysed comments on selected social media platforms. Researcher was overwhelmed by the thorough explanation by participants and their willingness to give opinion.

5 References

Abel, J. P., Buff, C. L. & Burr, S. A. (2016). Social Media and the Fear of Missing Out: Scale Development and Assessment. *Journal of Business & Economic Research, 14*(1), 33-44.

Aragonès, E., Palfrey, T. & Postlewaite, A. (2007). Political Reputations and Campaign Promises. *Journal of the European Economic Association, 5*(4), 846-884.

Bakar, A. S. & Sawandi, N. (2018). Impak Sebelum dan Selepas Perlaksanaan GST di Negeri Kedah. *4th Annual ECoFI Symposium 2017*, 1-10. Alor Setar: Pusat Pengajian Ekonomi, Kewangan dan Perbankan.

Brooks, A. C. (2015). *The Thrill of Political Hating*. Retrieved from The New Yorks Times: https://www.nytimes.com/2015/06/08/opinion/the-thrill-of-political-hating.html

Capri, A. (2018, May 10). *Why Malaysia's Suprise Election Result Should Be a Wake-up Call for Global Leader*. Retrieved from Forbes: https://www.forbes.com/sites/alexcapri/2018/05/10/why-malaysias-suprie-election-result-should-be-wake-up-call-for-global-leaders/

Chinnasamy, S (2017). New Media Political Engagement and Participation in Malaysia. Routledge.

Amirudin Abdul Wahab (2016). *Creating a safer cyberspace for future digital natives*. Retrieved from The Star Online: http://www.thestar.com.my/tech/tech-opinion/2016/06/24/creating-a-safer-cyberspace-for-future-digital-natives/

English, L. M. (2004). *Foucauldian pastoral power and feminist organizations: A research direction for adult education*. Sheffield, England: SCUTREA 34th Annual Conference, University of Sheffield.

Habibu, S. (2018, April). *Nik Aziz sons to contest in opposing camps*. Retrieved from The Star Online: https://www.thestar.com.my/news/nation/2018/04/28/nik-aziz-sons-to-contest-in-opposing-camps/

Harun, N. M., Hamid, N. A., Yahya, M., Ahmad, B. & Ismail, M. I. (2017). The Effects of GST Knowledge Towards Students' Spending Pattern. *e-Academia Journal, 6*(1), 116-126.

Jennings, R. (2018). How Malaysia became a 'model' for China's Overses Economic Expansion. The Forbes.

Lo, A. (2017). *Salary increment cannot match rise in cost of living*. Retrieved from The Star Online: https://www.thestar.com.my/metro/views/2017/03/31/salary-increment-cannot-match-rise-in-cost-of-living-locals-continue-to-get-low-wages-due-to-presenc/

Malaysian Communications and Multimedia Comission (2016). *Internet Users Survey 2016*. Retrieved from Malaysian Communications and Multimedia Comission: https://www.mcmc.gov.my/skmmgovmy/media/General/pdf/IUS2016.pdf

Mayfield, A. (2008). *What is Social Media?* Retrieved from iCrossing: www.icrossing.com/insight_pdf_files

Memon, S. (2017). *How Millennials and Social Media Changed The World*. Retrieved from HuffPost: https://huffingtonpost.co.uk/shaz-memon/how-millennials-and-socia_b_15537484.html

Palil, M. R. & Ibrahim, M. A. (2011). The Impacts of Goods and Services Tax (GST) on Middle Income Earners in Malaysia. *World Review of Business Research, 1*(3), 192-206.

Shim, K. & Oh, S.-K. (2018). Who creates the bandwagon? The dynamics of far of isolation, opinion congruency and anonymity-preference on social media in the 2017 South Korean presidential election. *Computers in Human Behavior, 86*, 181-189.

Stieglitz, S. & Dang-Xuan, L. (2014). Social Media and Political Communication - A Social Media Analytics Framework. *Springer-Verlag*.

https://doi.org/10.1051/shsconf/20185302005

Sullivan, S. (2013). *What Twitter has meant for politics (and what it hasn't)*. Retrieved from The Washington Post: https://www.washingtonpost.com/news/the-fix/wp/2013/03/21/what-twitter-has-meant-for-politics-and-what-it-hasnt/?utm_term=.3061f030ba47

Wolf, S. (2013). Climate Politics as Investment: From Reducing Emissions to Building Low-carbon Economies. *Springer Fachmedien Wiesbaden*, 1-11.

Yusop, F. D. & Sumari, M. (2013). The Use of Social Media Technologies among Malaysian Youth. *Procedia Social and Behavioural Sciences, 103*, 1204-1209.

The Consciousness of History in Malaysia: The Ideas on *Merdeka* Day in Sarawak

Muhaimin Sulam [1*], Norena Abd Karim Zamri[2], Ahmad Murad Merican[3]

[1]Department of Management and Humanities, Universiti Teknologi PETRONAS, 32610, Bandar Seri Iskandar, Malaysia

[2]Department of Management and Humanities, Universiti Teknologi PETRONAS, 32610, Bandar Seri Iskandar, Malaysia

[3]Centre of Policy Research and International Studies, Universiti Sains Malaysia, 11800, Pulau Pinang, Malaysia

Abstract. For decades, the Malaysian national history is in a stable condition and not being challenged by anyone. The enhancement of national history is constantly progress through educational system starting from primary school to university level. The consciousness of national history is the foundation for the unity of the people and country development. Though, from time to time, the status quo starts to face the challenges and disputes which began with small scale and got bigger. This could trigger a national issue. Thus, this study focus on the date of Independence Day amongst Sarawak citizens. As one of the states that form the Federation of Malaysia in 1963, the history of Sarawak as a separate nation with the history of Malaysia has been detected. The consciousness of Merdeka Day for Sarawakian is more to refer to the establishment of the federation rather than the date of August 31, 1957 as the day of independent. To add, this research is to identify the level of consciousness in revising the national history amongst Sarawakian. This study also delves the understanding of "Merdeka" from Sarawakians' point of view and discuss whether the nation history is proportionate for the Sarawakian. For the methodology, this study applies quantitative and qualitative approaches, a concurrent triangulation mixed method design. The phases of data collection are two including surveys and focus group discussions. Generally, the result shows that the Sarawakian have a new consciousness of the national history.

Type of Paper: Empirical

Keywords: Consciousness; National History; Merdeka; Sarawak; Malaysia

1 Introduction

The idea of Malaysian history is now facing a new volume of consciousness, giving a huge challenge to the interpretation of history especially in the school textbook. As being claimed by Hein & Selden (2000), "history and civics textbooks in most societies present an 'official' story, highlighting narratives that shape contemporary patriotism". The contestation over textbooks contents is due to the system of education, which is authoritative by the nation, delimit proper behavior of citizens, and sketch the imagine nation. "The nature of the textbook is carried directly or indirectly, they carry the imprimatur of the state, and they have enormous authority" (Hein & Selden, 2000).

National history through educational system has been questioned. Thus, this paper will focus on the event of independence to view the consciousness of Sarawakian, which covered the study area. National independence can be described as one of the vital event in a country's formation. The development of a new media helps to promote different ideas and interpretation about events that significance for the country, as such Independence Day or 'Hari Merdeka' has become one of the contestation.

* Corresponding author: muhaiminsulam@utp.edu.my

The root word of 'independence' is independent. Independent means free from colonialism, confinement, protectorate, not relying on others, stand alone and sovereign (Dewan, 2002). Commemorate the independence of a country or nation that is free from any form of colonialism described as a country's greatness day. Independent also is one of the liberty to develop a country or nation according to own rules without the intervention of others.

Independence Day is a day to commemorate the birth of a country. Independence is a climax of a crusade of the locals and a symbol of nation country's sovereign. To add, independence is a yield of struggle and unity of citizens which requires high sacrifices including a soul. Independence Day is the past, however in Malaysia, since it was celebrated on 31 August 1957 each year, the date and day of independence was not being questioned until there is a wakeup call by some the Sarawak community. Today, the factual errors and misleading assumptions occurred among Sarawak community especially the on the topic of Independence Day.

For the majority of Sarawakian, the Independence Day is no longer to be reconciled on 31 August 1957 but believed it should be on 22 July 1963. This date reconciles the date when Governor Colonial Sir Alexander Waddell appointed Stephen Kalong Ningkan to be the first Sarawak Chief Minister and the formation of High Council Members as the first Sarawak's cabinet (Sarawak, 2016). To the Sarawakian, the event is the declaration that Sarawak is free from British by forming a sovereign state.

2 The Idea of '*Merdeka*'

Starting from the year of 2010, government declared 16th September as Malaysia Day (Hari Malaysia) as to commemorate the establishment of the Malaysian Federation on the same date in 1963 (Razak, 2011). It marked the joining together of Malaya, Singapore, Sarawak and North Borneo (Sabah) (Stockwell, 2008). The formation of the new federation was planned to occur on 1st of June 1963, but was later postponed to 31st of August 1963, in order to coincide with the sixth Malayan Independence Day. However, due to several issues related to the Indonesian and the Filipino objection to the formation of Malaysia delayed the declaration to September 16 of the same year. The postponement was also done to give the United Nations' team sufficient time to conduct referendums in North Borneo (now Sabah) and Sarawak regarding the two states participation in the new federation (Stockwell, 2008).

Generally, in August and September, Malaysia will celebrate its national day on two dates, the Independence Day (*Hari Kemerdekaan*) and Malaysia Day (*Hari Malaysia*). Both of these date play a significant role in the heart of Malaysian. However, the idea having Independence Day celebration and Malaysia Day was not meaningful for Sarawakian compared to celebrating 22nd July. It was claimed to be Malaya-centric and caused heated debate among Sarawakian (Fernandez, 2013; Kaung, 2012).

It took 47 years since Malaysia was formed for us to start celebrating September 16 as Malaysia day and a public holiday beginning 2010. Thus, it also took 50 years since Malaysia's birth for Sarawak to start celebrating its own independence anniversary on July 22 beginning 2013. Apart from that, the idea of "Sarawak Merdeka" is more dominant in social media especially amongst youngsters. Historically, Sarawak gained its independence on July,22 in 1963.

Through "google search", the earliest blog sparked the questions about Sarawak Independent on 22 July 1963 is Radin's Blog Discussion from his writes posted on 21 July 2012 entitled "*22 Julai: Hari Kemerdekaan Sarawak*" (Radin, 2012). The climax debate on this issue and the effect with Malaysia's Independence Day on 31 August 1963 is when Sarawak Chief Minister, Adenan Satim announced 22 July is a Sarawak Day throughout his pre-election tour "*Sejiwa Sanada*" in Kuching on 23 April 2016. (Yu Ji, 2016). This is the official statement from the highest rank leader Sarawak state. This statement is the climate amongst the public debate since 2012. Later, the statement is supported with the events organized by Sarawak government such as "*Upacara Memperingati Hari Sarawak Merdeka Daripada Penjajah*" on 22 Julai 2013 by Yang di-Pertua Negeri Sarawak, Tun Datuk Patinggi Abang Muhamad Salahuddin along Sarawak's Riverbank.

Before that, during press conference on 20 July 2013 in "*Wisma Bapa Kemerdekaan*", Kuching, Sarawak's Minister of Resource Planning and Environment II, Datuk Amar Awang Tengah Ali Hasan stated that their side are thrive to include Sarawak Independent event in Malaysian education syllabus. To add, Datuk Seri Tiong King Sing, Bintulu's member of parliament, is one of the politician who fighting for the event to be included into the national education syllabus (Awani, 2015).

In fact, there is one article wrote by Mohamad Razif bin Awang Yusuf (UKOM Sarawak) entitled "*22 Julai Hari Sarawak Merdeka Daripada Penjajah: 50 Tahun Sarawak Dahulu, Kini dan Akan Datang*" published in website of Jabatan Hal Ehwal Khas (JASA), Malaysian Communications and Multimedia Ministry on 4th July 2014. The article gives a view on the establishment of self-government headed by local citizens. There is one statement in that article claimed that Sarawak Independence Day is to be on 22 July 1993. Thus this created different views from the main version of a nation's history. Ironically, this controversial statement also comes from the ministry.

The opposition also support the statement from Sarawak government. One of the *'Gerakan Pemisah'* so called "*Sabah Sarawak Keluar Malaysia – Sabah Sarawak Merdeka*" used online internet as a platform to spread their influence in order to support the effort making the date as Sarawak Independence Day (William, 2015). Other than that, Jofri Jaraiee from PAS Sarawak, demand the state government to announce the date as a Sarawak Day (Jaraiee, 2013)

3 Research Method

This study employed a quantitative and qualitative approach as the methodology, a concurrent, triangulation mixed method design. This study utilized a mixed methods study design, which Cresswell and Plano Clark (Cresswell & Plano Clark, 2007) describe as a type of research in which qualitative and quantitative data are collected, analyzed and integrated to better understand a research problem. By using these methods, it allows researchers to employ a variety of strategies to answer questions that cannot be addressed by qualitative or quantitative methods exclusively. Specifically, this research employed an exploratory sequential design. In employing a transformative exploratory sequential instrument design, this study began with the collection of quantitative data which then guided the development of a qualitative instrument- focus group discussion.

3.1 Phase 1: Surveys

In this study, the Sarawak citizens view from these four sections have been collected. Under school section, the set of questionnaires have been distributed to five national schools with each of the school comprises of one History teacher and two form 4 students. In addition, the other eight and seven set of questionnaires have been distributed to two national Chinese schools with each of the school consists of two history teacher and the balance is the form 4 students. The balance of 10 questionnaires have been distributed to Independent Chinese School. For the others sections, the criteria as stated below. From the 61 questionnaires distributed, the collaborator managed to return 314 answered questionnaires, and the total of 314 questionnaires have been used for this study. The other 6 questionnaires have not been returned. Collected data were exported to Statistical Package of the Social Sciences (SPSS) for analyses. Data were analyzed and descriptive statistics were used to describe findings.

Table 1. Division for the survey distribution

Section	Criteria	Number
School	National School	40
	National Chinese School	
	Independent Chinese School	
University	Academic and Non- Academic Staff	40
Public Sector	According to Ethnicity	40
NGO	According to Ethnicity	40
	TOTAL	**320**

3.2 Phase 2: Focus group

Two focus groups with 25 participants were conducted at Miri and Kuching, Sarawak. The participants in focus group discussions were selected along the basis of purposiveness in a given subject rather than representativeness. This is because; they are able to provide relevant information on the topic was given and provide insights that are personally important to them. Research collaborator was appointed in selecting and recruiting the focus group participants. The participants were chosen on the basis of socio-demographic variations, in terms of age, gender, ethnicity and profession. This was desirable to reflect a wide scope of experience. An email outlining the objectives of the research was sent out to the participants before the discussion.

During the discussion, participants were informed about anonymity and confidentiality. Written informed consents were obtained from all participants. The focus group questions were developed based on a review of the literature on historical consciousness; studies related to history trajectories and the constitution were designed to look how communities in Sarawak viewed national narrative (mainstream history), promptly how they see themselves as part of it. The questions and statements were used to guide the discussion, but probes were also used to further explore certain comments or ideas. The data were audio-recorded with the permission. Each group discussion lasted approximately two hours.

4 Findings

		Frequency	Percent	Valid Percent	Cumulative Percent
Valid	*Malay*	74	23.6	23.6	23.6
	Chinese	63	20.1	20.1	43.6
	Indian	8	2.5	2.5	46.2
	Iban	62	19.7	19.7	65.9
	Bidayuh	25	8.0	8.0	73.9
	Melanau	14	4.5	4.5	78.3
	Dayak	5	1.6	1.6	79.9
	Melayu Sarawak	3	1.0	1.0	80.9
	Kenyah	4	1.3	1.3	82.2
	Orang Ulu	5	1.6	1.6	83.8
	Penan	3	1.0	1.0	84.7
	Lun Bawang	3	1.0	1.0	85.7
	Kadazan	2	.6	.6	86.3
	Kedayan	3	1.0	1.0	87.3
	Kayan	7	2.2	2.2	89.5
	Seluko	1	.3	.3	89.8
	Bisaya	2	.6	.6	90.4
	Chinese Hakka	1	.3	.3	90.8
	Berawan	1	.3	.3	91.1
	Jatti Miriek	1	.3	.3	91.4
	No Respond	27	8.6	8.6	100.0
	Total	314	100.0	100.0	

Table 2. Ethnicity

4.1 Phase 1: Surveys

4.1.1 Ethnicity

The results of the type of ethnicity of the respondents in Sarawak is presented in the Table 2. The total respondents are 314 and from the data obtain, the number of ethnicity involved in this study are 20. From the survey collected, most of the data are gathered from Malay ethnic which make up of 23.6% of total respondents (74 respondents). Then, follow by Chinese that with 20.1% from the total respondents (63 respondents), *Iban* that with 19.7% from the total respondents (62 respondents) and *Bidayuh* that with 8.0% from the total respondents (25 respondents). To add, 14 (4.5%) respondents are from *Melanau* ethnic group, 8 respondents (2.5%) are from India ethnic group, 7 respondents (2.2%) are from *Kayan* ethnic group, 5 respondents (1.6%) are from *Dayak* and *Orang Ulu* ethnic group, 4 respondents (1.3%) are from *Kenyah* ethnic group, 3 respondents (1.0%) are from *Kedayan* and *Lun Bawang* and *Penan* ethnic group and 2 respondents (0.6%) are from *Kadazan* and *Bisaya* ethnic group. Also, *Seluko, Chinese Hakka, Berawan* and *Jatti Miriek* share the same number of respondents which are 1 respondent (0.3%). The other 8.6% of the respondents (27 respondents) are not respond to this question.

In response to ethnicity and online media usage, the descriptive results reveal that majority of the respondents participated in this study use online media between 1 to 2 hours which left 53 out of 314 or about 16.9% use online media in less than 1 hour. The study also found out that 53.2% of the respondents' views are being influenced by online media, while 12.1% of the respondents do not being influenced by the online media.

4.1.2 Level of Consciousness

Respondents were then asked to rate their agreement and responses were then tabulated according to the ethnicity (refer Figure 1.0). From the survey, it records majority of the respondents agree that Malaysian history needs to be revised by counting the "strongly agree" and "agree" and giving a result 66.24% (208 respondents). This make up of *Malay, Chinese, Indian, Iban, Bidayuh, Melanau, Dayak, Melayu Sarawak, Kenyah, Orang Ulu, Lun Bawang, Kadazan, Kedayan, Kayan, Seluko, Bisaya, Chinese Hakka, Berawan, Jatti Miriek* and 'No Respond'. Then, there is 26.75% (84 respondents) at the state neither agree nor disagree about this matter. From the table also, it shows 7.01% (22 respondents) representing *Malay, Chinese, Iban, Bidayuh, Penan, Kedayan, Kayan* and 'No Respond' disagree that the History of Malaysia should be revised. This is by counting the "disagree" and "strongly disagree" as indicated below.

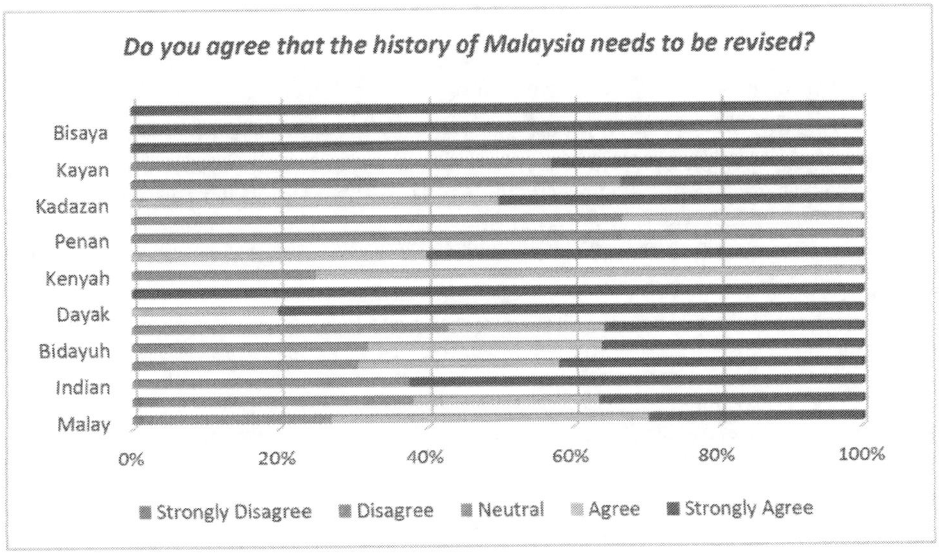

Figure 1. Level of consciousness according to ethnicity

 https://doi.org/10.1051/shsconf/20185302006

In addition of historical consciousness, respondents were also asked to rate their agreements. From the table below, it shows that, the highest mean score will represent the most agree issue by respondents whereas the least mean score will represent the least agree issue by the respondents in term of consciousness about Malaysian history for the Sarawak citizens. From the survey and analysis, the most respondents agree is about the History of Malaysia need to be revised with the mean score of 3.952. Whereas the least agree issue is the issue of online sources where they agree with the online sources related to Sarawak perspectives of Malaysian history with a mean score of 3.086 (refer Figure 2.0). This highlights that most Sarawak respondents agree that the History of Malaysia needs to be revised thus new insights, fair and true Malaysian history can be generated especially from the Sarawak perspectives.

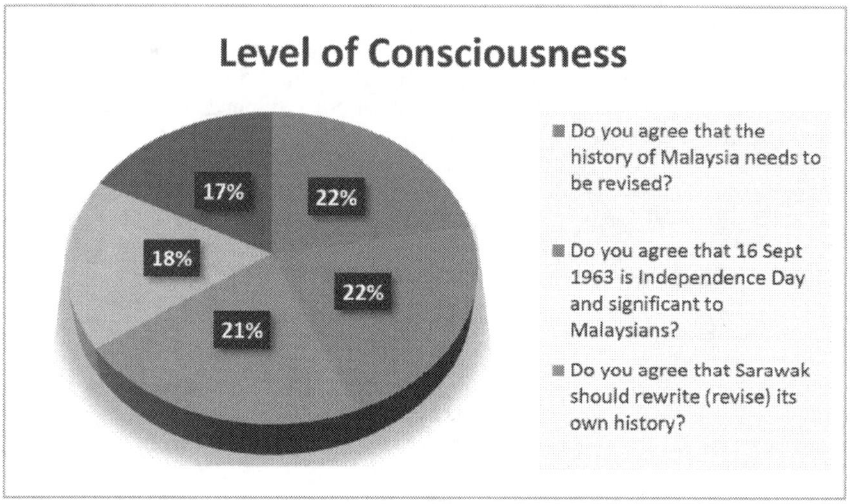

Figure 2. Level of consciousness

4.2 Phase 2: Focus Group Discussion

This study adopts focus group discussion to collect data in order to have an in-depth data as compared to survey. Two focus groups were conducted in Kuching and Miri and the participants came from different ethnic with a different education, gender, politic, age and career. There were 13 participants in Miri and 12 participants in Kuching. Each session was conducted within two hours and covered the topic that have been prepared by the team.

First, based on the transcript, there is a strong connection in both places between history consciousness and new media. Generally, the participants stated that national history have been controlled by elite group who have the power to write a text book that become the source of nation history. Majority of the participants have a view that national history is based on text book. Currently, the expansion of new media such blog, online news, WhatsApp and other social media have changed their views. The major topic highlighted throughout the discussion was mainly on Sarawak Independence Day.

For Sarawakian, the term '*Sarawak Merdeka*' has significant impact to their lives. For them, Sarawak was granted self-government on July 22, 1963. It was the day that the flag of Sarawak hoisted. Thus, the joy of celebrating 31st August as Independence Day does not significant to the heart of Sarawakian. The implication of the indictment of '*Sarawak Merdeka*' and recognition Sarawak as '*Negara*' (nation)-proclaimed by some of the Sarawakian, seems contradict with the act in Malaysia constitution amendment 1976. However, this interpretation is dominant amongst Sarawak government leader and public.

The major issues highlighted from the focus group discussion was most of them wanted the national history to be revised. They wanted the fact of 'Sarawak Merdeka' to be included into the syllabus. However, some of the participants whom worked as government servant represented themselves as neutral, although deep inside their heart they have their own stand on the issue. The reason behind of this attitude shown is because they refuse to involve in sensitive issues.

Majority of the respondent showed negative responds as they could not accept 31st August 1957 as Malaysia's Independence Day. For them, Sarawak is still colonized by the British in 1957 and only gained their self-governance in 1963. On the other hand, most of the participants agreed that their views were strongly being influenced by the new media. Through the new media, the new insight of the importance of 22 July, 1963 were elevated.

According to Prof Michael Leigh, Sarawak was granted to govern their own state since 22 July, 1963 and lasted until 16 September 1963. It is prior to the condition that it would join to form the Federation of Malaysia on 16 September the same year (Welman, 2011). Apparently, it is not an independence but more to a debut to a process of transition towards the independence.

5 Conclusion

Different version of history in one nation is not something new or as a hitch to be accepted by society. It has been accepted in European countries such United Kingdom, Sweden and Germany (Körber, 2015; Luisa, 2011). For them, it is a process of thinking skills, providing looking at certain phenomena from multi-angle or perspective.

The issues of national history have become political issue due to the nature of the history itself as political history and less focusing on the country's social and economic development (Ting, 2014). When the history is more prone to highlights political history, it will be influenced by the changes that happens in the country's political issue. Those who have interest will use the new media as a medium to spread their propaganda.

Contrastingly, the new media work as a tool to evaluate the history. Hence, the sense of history consciousness is increasing from time to time. One of the ways to surpass the dumping of history information in new media is to have an individual strainer who have a credibility and open mind towards new findings and views. This is also a contribution to the national development.

The current national history is incomplete and always in a process of amendments, improvements and reinterpretation from time to time. Thus public historian who has the integrity plays a crucial role in developing national history. Although from the reality, we are going to deal and jungle with different opinions and views but in the end, through the maturity process, the result will be benefited for the sake of the country.

The data collected in Kuching and Miri showed that the national history consciousness for the Sarawakian is 'multiple'. Affirmations from the people to revise and rewrite national history is an alarm on history consciousness. An effort to revise national history should be no issue as the process to reproduce new syllabus will include all parties. The process to revise textbook can be inclusive and fair to the Sarawakian who believes their history has been marginalized. An open and wide involvement is a condition towards the process to revise and rewrite national history.

However, the effort to include all historical events into the syllabus is a tough process due to the time and space limitation. One of the best way to overcome this issue is Malaysia historiography must have one character that can is significance by all parties and it is not a just a matter of propaganda. For instance, "*1 Malaysia*" concept is a concept that is significance by all parties or only for part of government politics? Is our forefather- Tunku Abdul Rahman, former first Malaysia Prime Minister being owned by all races and parties?

As what have been explained before, this study discovered that a new consciousness towards Sarawak Independence date have been identified. From the gathered data, Sarawakian are mostly incline to celebrate Independence Day (*Hari Kemerdekaan*) 31 August, 1957. The new date, 22 July, 1963 as being claimed as *Sarawak Independence Day* is more to have a political value rather than history value. Generally, historian reject the event which they claimed as a symbol of *Sarawak Independence Day*. The event is during the inauguration of Chief Minister and formation of Sarawak cabinet to rule Sarawak. The event is actually a process towards independence.

The history on Malaysia's Independence Day (*Hari Kemerdekaan*) should remain on 31 August, 1957. The celebration to commemorate the inauguration of Chief Minister and formation of the first Sarawak cabinet can be claimed as a process for the Sarawak to get independence. This ritual has been practiced in United States which is in federal country that unites 50 states. United States have been declared as an independency country on 4 July 1776 through the merger of 13 states. Then, a few states joined the country on different dates. Apparently, all states in United State is maintained with 4 July 1776 as Independence Day (Hole, 2001).

National history consciousness will be expressive with the consciousness towards the nature of history and the history documentation to be more open and comprehensive. In the end, this consciousness will form more solid understanding in a multi ethnic society and separated geographically. The consciousness is the main priority for the society to move forward in the challenging future.

Acknowledgements

This research has been supported by research grant from the Malaysian Communications and Multimedia Commission Networked Media Research Collaborative Programme with Universiti Teknologi PETRONAS (Project Number: 0153AB-G16). The authors are also thankful to the members of this research grant for providing support and commitment.

6 References

Awani, A. (2015). Semak buku teks sejarah, masukkan Hari Kemerdekaan Sarawak. Retrieved October 9, 2016, from http://www.astroawani.com/berita-malaysia/semak-buku-teks-sejarah-masukkan-hari-kemerdekaan-sarawak-66654

Dewan, K. (2002). *Kamus Dewan*. Kuala Lumpur: Dewan Bahasa dan Pustaka.

Fernandez, J. (2013). Sabah, Sarawak: 50 Years in Malaysia plagued by bad politics — Joe Fernadez. *Malay Mail Online*.

Hein, L., & Selden, M. (2000). Censoring History: Perspectives on Nationalism and War in the Twentieth Century

Hole, R. (2001). The American Declaration of Independence of July 4th , 1776 Why Declare Independence ? Retrieved October 15, 2016, from http://www.historytoday.com/robert-hole/american-declaration-independence-july-4th-1776

Jaraiee, J. (2013). Hari Kemerdekaan Sarawak wajar diwartakan. *Sinar Online*. Retrieved from http://www.sinarharian.com.my/karya/pendapat/hari-kemerdekaan-sarawak-wajar-diwartakan-1.185546

Kaung, M. (2012). Merdeka " no relevance " to Sabah , Sarawak. *Free Malaysia Today*. Retrieved from http://www.freemalaysiatoday.com/category/nation/2012/08/27/merdeka-no-relevance-to-sabah-sarawak/

Körber, A. (2015). Historical consciousness , historical competencies – and beyond ? Some conceptual development within German history didactics, 1–56.

Luisa, B. M. (2011). *History teaching today: approaches and method*. Retrieved from http://www.theewc.org/uploads/content/archive/History_teaching_today_manual_1.pdf

Radin. (2012). 22 Julai; Hari Kemerdekaan Sarawak. Retrieved October 5, 2016, from http://radin87discussion.blogspot.my/2012/07/22-julai-hari-kemerdekaan-sarawak.html

Razak, D. S. M. N. B. T. H. A. (2011). Perutusan Hari Malaysia. Kuala Lumpur Retrieved from http://www.pmo.gov.my/home.php?menu=speech&page=1908&news_id=525&speech_cat=2.

Sarawak, G. of. (2016). Self - Governing Sarawak 1963. Retrieved September 29, 2016, from http://www.sarawak.gov.my/web/home/article_view/238/249/

Stockwell, A. J. (2008). The Journal of Imperial and Malaysia : The making of a neo - colony ?, (October 2014), 37–41. http://doi.org/10.1080/03086539808583029

Ting, H. (2014). The battle over the memory of the nation: Whose national history? In L. K. S. Mark Baildon, G. İ. and Ivy Maria Lim, & J. Jaffar (Eds.), *Controversial History Education in Asian Contexts* (pp. 41–57). Oxon: Routledge. Retrieved from http://samples.sainsburysebooks.co.uk/9781135014216_sample_717252.pdf

Welman, F. (2011). *Borneo Trilogy Sarawak: Volume 2*. Booksmango.

William, L. (2015). Sarawak Was Granted Self-Government on 22 July, 1963. Retrieved from http://www.sabahsarawakmerdeka.com/2015/07/sarawak-was-granted-self-government-on.html

Yu Ji. (2016). Adenan : July 22 is now Sarawak Day. *The Star Online*. Retrieved from http://www.thestar.com.my/news/nation/2016/04/23/adenan-sarawak-day/

Awareness of Ethics and Integrity at a Malaysian Government District Office

Khalidah Khalid Ali *
Department of Management & Humanities, Universiti Teknologi PETRONAS, Perak, 31750 Malaysia.

Abstract

The public sector, operated through local/ state governments is the most important instrument in management and administration of a nation. Apart from delivering services and sustaining national development, it is also the social institution entrusted to restore law and order of a nation. These functions and responsibilities indicate direct relationships with morality, ethics and social justice. The district office is the closest level of the government to the people. Hence, issues of ethics and integrity are clearly seen as the people whom they serve directly and who are affected by the government. This paper reports on a pilot study conducted at a district office in Malaysia to investigate employees' responses on awareness initiatives to imbue ethics and integrity at the workplace through a questionnaire with mixed methods. This study has positively found that the district office employees are highly aware of ethics and integrity. They perceive that all initiatives taken at the workplace to create ethical awareness are highly effective and their understanding of ethics and integrity is high (Mean: 4.34 and 4.25 respectively). Posters are viewed to be the most effective medium to create awareness, followed by websites and signages. Superiors, as role models have taken a committed responsibility to remind subordinates through various channels of communication on ethical values and integrity. However, there are still employees who do not seem to be aware of the information provided on ethics. In addition, less than half of the respondents (46.3%) are directly involved in the launched ethics campaigns in office. The study findings will draw some insights on effectiveness of initiatives taken at ground level to create ethical awareness among the public-sector employees as Malaysia transforms her socio-economic landscape in the globalization era.

Key words: Public sector, Malaysia, ethics, integrity, awareness initiatives

1 Introduction

Human values are undergoing change in the past few decades (Khalidah *et al.*, 2012a). The challenges posed towards human values, humanity and human civilization have been even greater since the beginning of the 21st Century, featured by globalization, information and communication technology (ICT), trade liberalization and the free market system. The public sector of a nation forms the government machinery that develops and implements socio-economic policies for the society's prosperity and general well-being.

In fact, the public sector, operated through local/state governments is the most important instrument in management and administration of a nation. Apart from delivering services and sustaining national development, it is also the social institution entrusted to restore law and order of a nation which directly relates to ethics, a study of morality (i.e. right and wrong acts) and social justice. The public sector/local government therefore shoulders a high social responsibility and accountability towards the community at large in the course of providing its services. As any social institution, it is constantly evolving through new leadership, environmental influences and socio-political environment (Article entitled Malaysia: Ethics and Excellent Public Servants. Source: Bernama.com, 18 October, 2014).

The district office, a decentralized set-up within the Malaysian state-government is the closest level of the government to the people. Issues of ethics and integrity are clearly seen as the people whom they serve directly and who are affected by the government. In view of the significance of public service and the pertinent role of the local/state government to imbue ethics and integrity within the civil service employees themselves, an exploratory research has been initiated at a district office in Malaysia to investigate employees' responses on awareness initiatives to imbue ethics and integrity at the workplace.

This study will contribute to the limited value-based studies within the Malaysian context; not to mention the public sector as a whole. It will investigate on public sector employees' awareness on ethics and integrity as well as provide some empirical evidence and insights for the government to assess/ take constructive steps towards instilling ethics and integrity among its employees while balancing economic considerations. At the decentralized level, it will be indicative of the state public service after all the ethics and integrity campaigns at local/ national levels. In addition, it will support

* Corresponding author: khalidah_kh@utp.edu.my

Perak State government's call to promote an ethical culture in the process of transforming the state's socio-economic landscape under Perak Amanjaya plan.

2 Literatures and Discussions

2.1 Ethics and Integrity in Concepts

Ethics is derived from the Greek word, *ethos* which means the character, spirit and attitudes of a group of people or culture (Khalidah *et al.*, 2014; Rahman, 2013). As a concept, it can be described as a set of principles that contains behavioural codes to determine what is right or wrong. Hence, in the study of ethics, one is trying to rationalize what actions are right or wrong and assess whether they are just or fair from different perspectives - individual, organizational and/or societal. To summarize from the writings of Hanafi and Hamid Salam in a book, Islamic Principles of Business Organization and Management, ethics may be defined as a model of behaviour that should be followed in order to harmonize human relationship and minimize abuse and work for the welfare of society (Faridi, 1997).

Integrity, on the other hand, is a virtue (good value) derived from the Latin word *integri* (Stanwick and Stanwick, 2009). Referring to Oxford Dictionary, integrity relates to honesty and goodness, wholeness and unity. It is an ethical value among many other universally accepted virtues. According to Mohd Tap Salleh (2007), former President of Malaysian Institute of Integrity (Integriti Malaysia), "*Integrity is the quality of being honest and upright. It is about individual and organizational characteristics which are perfect based on noble values such as honest, truthful, trustworthy, accountable, transparent, efficient and wise. Individual integrity is the harmony between what he/she says and does. Actions are in accordance with moral and ethical principles as well as laws and regulations and do not go against public interest. They also reflect promptness, correctness and quality.*"

Ethics relates to character building and development that indeed places on virtues to be emulated in life. While upbringing, socialization process, culture and self- reflections are factors influencing ethics (Khalidah *et al.*, 2012a; Khalidah *et al.*, 2014a; Shaw, 2011), moral objectivists view that religion has a strong influence on individual ethics as it preaches one to embrace fundamental principles and virtues to lead rewarding lives (Weber, 1958; Khalidah *et al.*, 2014b). Malaysia's *Rukun Negara*/ National Principles directly states the belief in God as a doctrine to be followed by all Malaysians irrespective of religious and cultural differences, implying the importance of religion in Malaysian society despite cultural diversities (Khalidah *et al.*, 2014a). Interestingly, the West, through relativistic thoughts has incorporated religion as a cultural component of a society when the East most often views these two as separate components in practice although we do not deny that religious beliefs do influence a society's cultural landscape (Note: Culture is defined as a set of norms, values and beliefs of a particular group or community and shared by its members (Hofstede,1980). Culture covers a wide spectrum in one's life; from religion to customs and rituals as well as language and membership in ethnic groups (S Awang, M Maros and N. Ibrahim, 2012; Mahathir, 1970; Syed Husin, 2008).

Recognizing the role of religion in shaping ethics while at the same time noting that modernism through industrialization has transformed human civilization since the 19th Century, measured in terms of material well–being, the 1980s and past few decades saw changes in human values and societal norms (Khalidah *et al.*, 2012a; Sharifah Hayaati 2007). Aptly, corruption, ethics and integrity have become contemporary issues in the practice and theory of politics, public administration, law, economics and society nowadays, not only from international contexts but within local contexts. Interestingly, these trends seem to coincide with globalization, introduced in the 1980s and facilitated by information and communication technology (ICT). The ICT/cyber age that we are going through today to some extent influences the way we conduct our lives, not to mention human interactions and values, be it on individual or societal levels. Within the public services environment, "*more than ever, citizens are calling for state institutions that are democratic, efficient in the use of public funds, effective in delivering public goods while in the meantime remain strong players in the increasingly competitive global system (*Capacity Development Action Briefs by United Nations Development Programme (UNDP) May 2007, Number 2). In another phrase, "*the need to build confidence in governments in light of the increased concern on corruption and unethical behavior of their civil servants prompted governments to review their approaches to ethical conduct. There are two broad approaches by which governments attempt to improve public service ethics, i.e. compliance-based ethics management and integrity-based ethics management*".

2.2 Government Initiatives to Instill Ethics and Integrity in Malaysian Public Service from 1980 onwards

Malaysia, under Prime Minister Tun Mahathir Mohamad, embarked on a rigorous industrialization policy in the beginning of 1980s to generate economic growth and social well-being. Alongside industrialization is the greater attention given to ethical/productivity-based concepts. Leading all the efforts to improve the Malaysian Public Service is the Panel for Improvement of the Public Service (PANEL), established in 1986 and headed by the Chief Secretary to the Government. Its role is to "investigate and offer solutions to service problems of common concern. The investigations by PANEL have resulted in several reformed ideas and innovations, not only seen as physical changes but behavioral and attitudinal changes (Sharifah Hayaati, 2011) to accelerate the emulation of virtues and ethics in public service. To name a few, they include enforcement of code of ethics at the workplace, Client Charter, Quality Assurance Standard such as ISO9000, Quality Control Circle and Manual of Work Procedures. The government has also listed ethical values to promote work ethics in the public service. They are quality, productivity, innovativeness, discipline, integration, accountability and professionalism. These values have been introduced through policy formulation such as Look East Policy (1982), Leadership by Example (1983), Name Tags (1985), The Inculcation of Islamic Values Policy (1985), Excellent Work Culture (1989), Clean Efficient and Trustworthy Government (1989) and Vision 2020 (1991) (Noore Alam Siddiquee, 2010; Sharifah Hayaati, 2011).

In fact, Vision 2020, introduced during the Mahathir era sets a new agenda for the future direction of civil servants, i.e. to take the lead role and bring changes not only to the nation's economy but to the community's ethical values. Six core areas have been identified as the major framework/focus for government initiatives on work ethics and values: (1) basic personal values to be imbued i.e. sincerity, trust, self-discipline, commitment, cooperation and moderation; (2) customer-related values i.e. courtesy, humanity, politeness, efficiency and carrying responsibilities expeditiously to fulfill customers' needs; (3) leadership-related values; (4) professional values; (5) productivity and quality values; (6) religious values (Muhammad Rais B Abdul Karim, 1995; Nor 'Azzah Kamri, 2004; Sharifah Hayaati, 2011). These government policies, mostly implemented through awareness campaigns have also been consistently communicated to employees through the chain of commands within the Malaysian public service itself, be it at national and state levels.

Following these, the institutionalization and awareness of ethical values within the society (including public service) continued with the establishment of Malaysian Institute of Integrity (IIM) on 23 April 2004 when Prime Minister Tun Abdullah Ahmad Badawi took office. Its prime objective is to educate and encourage Malaysians at all levels of the society to uphold integrity and embrace noble values as envisaged under the National Integrity Plan (NIP). The NIP is a master plan to develop a fully moral and ethical society whose citizens are strong in religious and spiritual values and imbued with the highest ethical standards. Under the Target 2008, the following missions were identified: effectively reduce corruption, malpractices and abuse of power; increase efficiency in the public delivery system and overcome bureaucratic red tape; enhance corporate governance and business ethics and corporate social responsibility; strengthen the family institution and community and improve the quality of life and people's well-being. The development of National Integrity Plan (NIP) and establishment of IIM have certainly justified the importance of integrity, a virtue to be enculturated and internalized among the Malaysian community as the nation aspires to be developed, in line with Vision 2020.

In April 2009, Prime Minister Datuk Seri Najib Tun Razak introduced the Government Transformation Programme (GTP) and National Key Result Areas (NKRAs) with two main objectives, first, to transform the government to be more effective in its service delivery and be accountable for outcomes that matter most to the *rakyat*; and second, to help Malaysia move forward to become an advanced, united and just society with high standards of living for all. In fact, eradication of corruption through effective enforcement has been listed as one of the six NKRAs to sustain public sector integrity. The Integrity Pledge was also developed and launched by IIM in 2014 to be internalized by government employees as well as the private sector.

2.3 Local Studies on Malaysian Public Service

While government's initiatives to imbue ethics within the public service have been a long way, few local studies have been conducted to link the Malaysian public service with ethics before the millennium. Several were only seen in the last five years (Amalina Abdullah *et.al* 2014; Baharom Mohamad *et al.* 2014; Sharifah Hayaati Syed Ismail 2011). To mention some, a study by Noore Alam Siddiquee (2010) on combatting corruption and managing integrity in Malaysia has critically outlined the anti-corruption strategies being followed by the Malaysian civil service and explores the problems and limitations of the current approach to fighting corruption and managing integrity in the society. Tracing from the establishment of Anti-Corruption Agency (ACA) in 1967, the setup of Public Complaints Bureau (PCB) in 1971 until the introduction of National Integrity Plan in 2004 by IIM, the study concluded that corruption has become a central issue in Malaysia's political discourse. Evidence shows that despite governmental campaigns and initiatives, corruption remained acute, widespread and worsened in recent years.

Next, a study by Sharifah Hayaati (2011), in attempting to introduce the Effective Governance (EG) as a concept from contemporary and Islamic perspectives have concluded that conceptualizing EG from both dimensions i.e. Islamic and contemporary theories may assist the government in continuously improving public services. Her study concentrated on the role of public service and the process of transition in the move towards EG. It has recognized that with globalization, managing a nation by a virtuous leader is not adequate; it takes the state, private sector and the citizens to ensure effective governance in nation building.

Another quantitative study was conducted by Baharom Mohamad *et al.* (2014) on 390 Islamic education teachers (IET) to identify the influence of work satisfaction within the perspective of Islam (WSI) through varied dimensions i.e. Spiritual WSI, Intellectual WSI and Material WSI towards work integrity and performance. The study provided an alternative answer on how the integrity of IETs based on work satisfaction from Islamic perspective can determine high work performance. Integrity of IETs is formed through spiritual satisfaction, reasoning and social relationship. This is developed within Islamic education management which balances intellectual, emotional, spiritual and physical aspects of life. While Noore Alam Sidiqque (2010) has walked though the government initiatives to combat corruption as a matter of public ethics and integrity, studies by Sharifah Hayaati (2011) and Baharom *et al.* (2014) have respectively linked governance and work satisfaction within an Islamic ethical system. Interestingly, a local study by Azharsyah Ibrahim and Nor 'Azzah Kamri (2013) has developed a measurement for Islamic World Ethics (IWE) and, an alternative approach to an available measurement, Protestant Work Ethics (PWE). Earlier academic study in IWE began in early 1980s. The first of such studies was from Naqvi (1981) and Nasr (1984).

3 The Research Setting and Methodology

This pilot study was conducted at a branch of a district office in Perak State, Malaysia. Although it reports to the main district office, it is a separate location and given the empowerment to manage its daily activities. Headed by an assistant district officer, this branch has been set up to serve the rural community more efficiently due to the sizable district. It has a total number of 73 employees.

Prior to this study, library research was carried out especially through texts and on-line resources on the Malaysian public service. A survey questionnaire was developed with two sections: Section A to compile demographic data and Section B to gather information on employees' awareness on ethics and integrity at the workplace. Structured interview questions have also been developed and selective interviews followed during the pilot study after gathering questionnaire data from the respondents. However, this paper will only report the findings and analysis from the gathered demographic data in addition to responses on questions related to awareness of ethics and integrity from the questionnaire. The developed questionnaire contains demographic details of respondents, two self- assessed questions on awareness and understanding of workplace ethics by using mean score analysis and 7 added questions to relate the role of media in creating ethical awareness. The 7 added questions are listed below;

1. Have you seen any type of media information on the values of ethics and integrity in the workplace?

2. If you do, what types of the media information?

3. Does your superior remind, through various channels, the values of ethics and integrity while working?

4. What are the examples of behaviours shown by superiors to remind subordinates on the importance of ethics and integrity?

5. Do you know that there are campaigns conducted on importance of ethics and integrity in the workplace?

6. Are you involved in one of the campaigns conducted?

7. Do you recite the Integrity Pledge every month? Write down the Integrity Pledge.

To analyze the demographic data and questions 1, 2, 5, 6, and 7 above, SPSS Version 2.0 was utilized. Mean scores and frequency tests were used for data analysis. Referring to the seven questions above in Section B, except for question 4, all questions have been analyzed by applying frequency tests. For Question 4, the researchers have qualitatively analyzed the written comments of respondents through conventional content analysis by counting the most number of responses belonging to categories/themes derived while going through the responses. For question 7 Part 2, (i.e. Write down the Integrity Pledge), the researchers have analyzed and assessed based on the writings of the respondents respectively.

4 Findings, Analysis and Discussions

4.1 Demographic Data

TABLE 1 below shows the respondents' demographic details.

TABLE 1: Demographic Details of Respondents
(N:60)

Gender	a) Male: 55. 2 % b) Female : 44.8 %
Age	a) 17 – 20 : 1.7 % f) 41 – 45 : 10.0 % b) 21 – 25 : 8.3 % g) 46 – 50 : 8.3 % c) 26 – 30 : 25.0 % h) 51 – 55 : 11.7 % d) 31 – 35 : 20.0 % i) 56 – 60 : 6.7 % e) 36 – 40 : 6.7 % j) 60 and above : 1.7 %
Religion	a) Islam : 100 % b) Others : 0 %
State of origin	a) Perak : 98.3% b) Other States : 1.7 %
Place of upbringing	a) Rural: 41.7 % b) Small Town : 50.0 % c) City : 8.3 %
Position	a) Management : 5.0 % b) Professional : 18.3 % c) Middle management : 28.3 % d) Support : 48.3 %
Qualification	a) PMR/SRP/LCE : 6.7 % b) SPM/MCE : 41.7 % c) STPM : 3.3 % d) Certificate : 18.3 % e) Diploma : 20.0 % f) Degree : 10.0 %
Average family/household income	a) Below RM 1500 : 13.3 % b) More than RM 1500 but less than RM 3000 : 53.3 % c) More than RM 3000 but less than RM 5000 : 25.0 % d) Above RM 5000 : 8.3 %

Note:
1. Total number of staff at the research setting: 73
2. Total number of respondents: 60
3. Professional category applies to all levels that have technical degrees and certificates.
4. Participation rate for pilot study: 82.2%

Participation rate for this pilot study is encouraging with 82.2% respondents. **Table 1** shows that although there are more males compared to female employees, the participation rate of women in the workforce is substantial (44.8%). 55% of the employees belong to age categories of between 17-35 years of age, implying that majority of the workforce belong to the Generation Y group (i.e. born between 1982-2000). All respondents are Malays and they embrace Islam as a religion. 98.3% originate from the same state they are serving. More than 90% of the respondents have been brought up in rural and small towns. An analysis on item (6) above suggests that the organization structure of this branch office is pyramidal/bureaucratic with a majority of support staff (48.3%), followed by the middle management and upper management at the top, overseeing the tasks and responsibilities of the lower ranks. Only 18.3% hold professional qualifications which come with the positions held at the district branch office. In terms of educational background, majority of the employees (41.7%) hold SPM/MCE qualifications. Only 30% of the workforce possess tertiary level

 https://doi.org/10.1051/shsconf/20185302007

qualifications (i.e. diploma and degree). Assessing their socio-economic status, majority (53.3%) of the workforce earn an average family/household income of between RM1500-3000. On the other hand, 33.3% of the respondents earn a household income of RM3000 and above. This implies that more than half of the respondents came from low income groups. (Note: In Malaysia, the definition of urban poverty is a household with a monthly income of RM3000 and below while hardcore poor family is defined as one with a household monthly income of less than RM540 per month. In Peninsular Malaysia, hardcore poverty is further classified; urban- RM663, rural-RM657 and overall- RM657). Demographic data has captured the average household income of respondents since it will be interesting to see their responses on ethical awareness as low income may drive one to compromise on ethics for survival "to make ends meet."

The following section shows the findings on employees' awareness on ethics and integrity.

4.2 Employees' Awareness of Ethics and Integrity

TABLE 2: Assessment on Employees' Awareness and Understanding of Workplace Ethics (N=59)

	ITEMS	Mean	SD
1	All initiatives and actions taken at my workplace to create awareness on the importance of ethics and integrity are effective and impactful	4.34	0.685
2	I fully understand the concept and meaning of ethics and integrity	4.25	0.685

Employees perceive that all initiatives taken at the workplace to create awareness of ethics and integrity are highly effective and their understanding of ethics and integrity is high (Mean: 4.34 and 4.25 respectively). This is a positive finding on the quality of public sector employees at district level.

The following reveals the findings from the responses on the 7 questions in the questionnaire to investigate the role of media in creating ethical awareness among respondents.

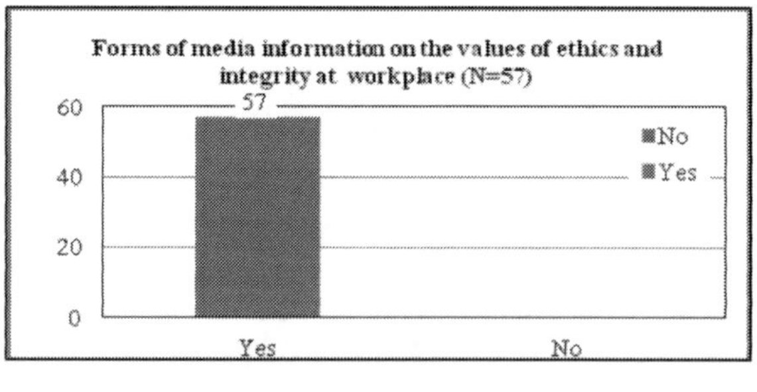

Figure 1: QI- Have you seen any type of media information on the values of ethics and integrity in the workplace?

Most respondents are aware and all have seen media information on ethical values and integrity at their workplace. Nevertheless, although majority have seen them, there are employees who do not seem to be aware of the information provided, indicating ignorance or perhaps indifference towards ethics.

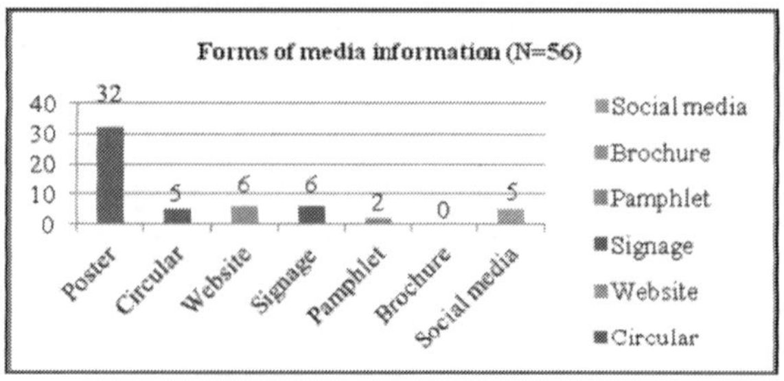

Figure 2: Q2 - If you do, what types of the media information?

57.1% of the respondents view that posters are the most visible form followed by websites (10.7%) and signages (10.7%). This implies that to create ethical awareness, the three mediums may be the most effective at the workplace.

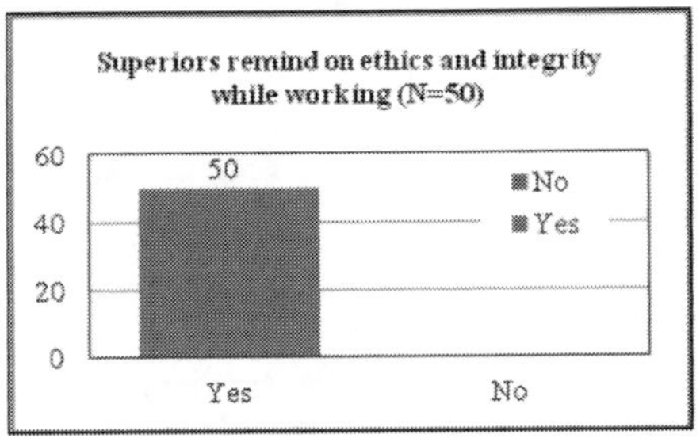

Figure 3: Q3- Does your superior remind, through various channels, the values of ethics and integrity while working?

Respondents also view that superiors have effectively played their leadership roles to remind subordinates through various channels of communication on the importance of applying ethical values and integrity at work.

The following have been found through content analysis on Question 4 (Q4) responses (i.e. What are the examples of behaviours shown by superiors to remind subordinates on the importance of ethics and integrity?);

1) Compliance to punctuality at work and dressing. Leadership by Example.
2) Through face to face communication and reminders to subordinates. This is done through meetings, workshops and addressed speeches.
3) Writings and postings in social media.
4) Subordinates express that superiors display honesty as a value.

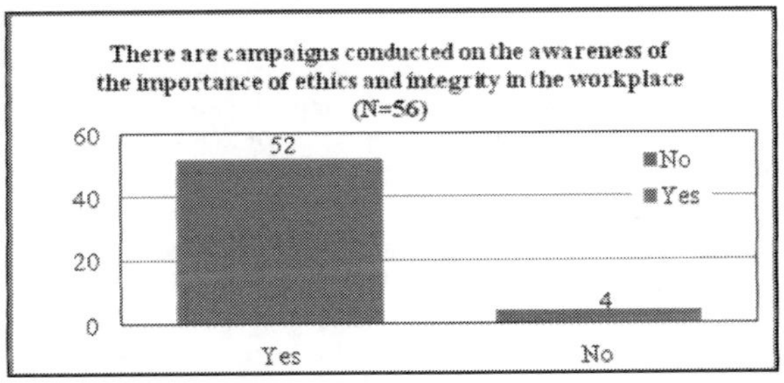

Fig. 5: Q5- Do you know that there are campaigns conducted on the awareness of the importance of ethics and integrity in the workplace?

Most (92.9%) of the employees are aware of the campaigns on ethics and integrity at the workplace.

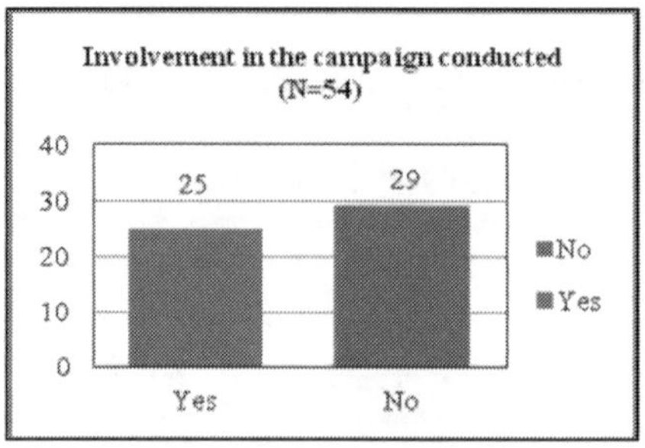

Fig. 6: Q6 - Are you involved in one of the campaigns conducted?

Although employees are aware on the campaigns conducted, only 46.3% are directly involved in it. This seems to question on their commitment to the campaigns conducted in office.

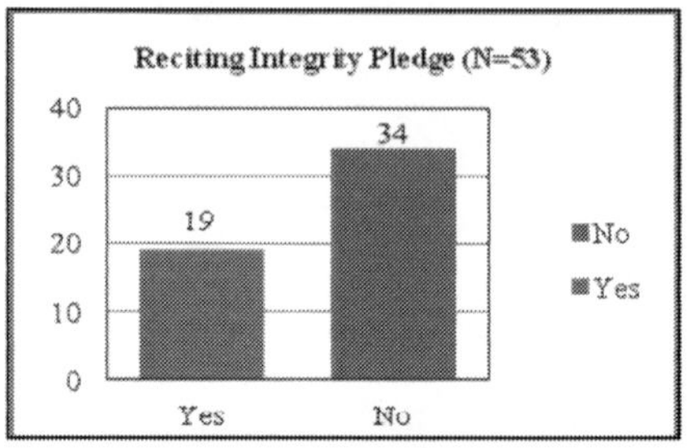

Fig. 7: Q7- Do you recite the Integrity Pledge every month?

Although employees are aware of the Integrity Pledge developed by Institut Integriti Malaysia (IIM) and the launching of a monthly programme to gather employees to recite the Integrity Pledge at the district office, only 35.8% has participated and recited it at work. In addition, only 5 out of the total sample of 60 respondents (8%) wrote down the Integrity Pledge as required in the questionnaire. Analysis seems to show employees are not familiar with the Integrity Pledge although they have heard about it, what more to internalize Integrity Pledge at work.

To elaborate further, interview result findings of selected respondents further reveal that employees nowadays do not have sense of urgency and commitment to serve the public. Plausibly, it may imply that initiatives such as the launching of Integrity Pledge is just one of those initiatives that may be regarded as formal but not necessarily fully bought by employees for practice and internalization. Deeper analysis has found that there is clash of views and opinions between upper management and support staff, leading to teamwork problems. Perhaps, these "clashing" environments may lead to indifferent attitudes among employees while the top management is trying to implement the Integrity Pledge at the district office. Furthermore, it is a challenge to make employees internalize good moral values as it is ultimately an individual's responsibility to be good beings or otherwise (Refer to the full study report as listed in No. 4 of the reference list of this paper, page 73). It is also interesting to note that although ethics awareness initiatives (such as Integrity Pledge recitation, reminders by superiors, etc) have been undertaken at the district office level, there are not many courses that are specially organized to seriously address ethics as a discipline to further create ethical conscience among public sector employees.

5 Conclusion

This study has positively found that the district office civil service employees in the research setting are highly aware of ethics and integrity initiatives taken at their office. They also perceive that all initiatives at the workplace to create awareness of ethics and integrity are highly effective and their understanding of ethics and integrity is high (Mean: 4.34 and 4.25 respectively). Posters are viewed to be the most effective form of creating awareness, followed by websites and signages. Superiors, as role models seem to take a committed responsibility to remind subordinates through various channels of communication on ethical values and integrity at the workplace. Despite these favourable findings, and recognizing that a majority of respondents have seen the various types of media information on ethics and integrity, there are still employees who do not seem to be aware of these information in office. In addition, despite the high awareness among the majority of employees through ethics-related campaigns and other initiatives at the workplace, less than half of the respondents (46.3%) are directly involved in the launched campaigns. The Integrity Pledge, introduced by IIM has yet to be made fully aware to employees, what more to be internalized. As ethics relates to moral behaviour that may not be consistently sustained over time, awareness initiatives including talks, forums/dialogues, courses and workshops must be rigorously organized as reminders to employees on the importance of work ethics and ethical conscience in general. There is indeed room for the district office to continue enhancing employees' awareness on ethics and integrity to promote good work values as well as further develop employees' ethical consciousness while serving the community as a social responsibility.

The research team would like to express their heartiest gratitude to Institut Darul Ridzuan Perak (IDR) for the generous allocation of funds to complete this investigation. Appreciation also goes to the district officer and his staff who have given their utmost cooperation and support throughout this research. We are also grateful to Universiti Teknologi PETRONAS management, our family members and colleagues who have given constructive comments, encouragement and continuous moral support throughout this research and publication. The author would also like to acknowledge the contribution of Prof. Dato' Dr. Ahmad Murad bin Noor Merican of Universiti Sains Malaysia for his invaluable advice and guidance in the conduct of this research.

6 References

A.A. Hanafi and Hamid Salam (1997), Business Ethics: An Islamic Perspective. Book Chapter in *Islamic Principles of Business Organizations and Management.* Edited by Dr. F.R. Faridi. Percetakan Sooriya, Kuala Lumpur Malaysia ISBN 883-899-808-7 pp3

Al-Quran; King Fahd Complex: Madinah Munawawarah, K.S.A. For the Printing of the Holy Quran; www.qurancomplex.org

Ahmad Mahdzan Ayob (2004). *Malay images in economic affairs: Viewing through the lenses of a Malay.* Invited paper presented at the Second International Conference on Malay Civilization held at the Legend Hotel Kuala Lumpur, Malaysia, on 20-28 February 2004 (unpublished) Rahman, A.R.A. (2003), Ethics in Accounting Education, Contribution of the Islamic principle of Maelaiah. *IIUM Journal of Economics and Management*, 11(1), 1-18.

Amalina Abdullah, Zunaidah Sulong and Ridzwana Mohd Said (2014), An Analysis on Ethical Climate and Ethical Judgement among Public Sector Employees in Malaysia, *Journal of Applied Business and Economics,* Vol. 16 (2).

Articles entitled the Malaysia: Ethics and Excellent Public Servants retrieved from http://www.unpan.org/PublicAdministrationNews/tabid/115/mctl/ArticleView/ModuleID/1467/articleId/23064/Default.aspx on October 2014

Articles entitled the NIP- Enculturating ethics, noble values and integrity in Malaysian Society retrieved from http://www.iim.org.my/documents/10157/df845017-98b0-4591-8570-98c574e7b515 on 20 August 2014.pp. 3

Awareness of Ethics and Integrity at Perak Tengah District Office: A Discourse Study. Universiti Teknologi PETRONAS. Final Report, 30 January, 2015. Research Funded by Institut Darul Ridzuan (IDR), Perak. Azharsyah Ibrahim and Nor 'Azzah Kamri (2013), Measuring the Islamic Work Ethics: An Alternative Approach, *International Convention on Islamic Management,* Kuala Lumpur, Malaysia.

Baharom Mohamad, Halimatus Saadiah Mat Saad and Sharifah Hayaati Syed Ismail (2014). The Role of Integrity as A Mediator between Work Satisfaction and Work Performance in the Perspective of Islam: An Empirical Approach Using SEM/AMOS Model, *International Journal of Research in Applied, Natural and Social Sciences,* Vol. 2, pp. 71-84, Jan 2014.

Case Evidence on "Ethics and Values in Civil Service Reforms", United Nations Development Programme (UNDP), May 2007

Dr. Danial Zainal Abidin, Public Lecture at Universiti Teknologi PETRONAS delivered on 29 July, 2008 in conjunction with the 9th PETRONAS Tilawah Al Quran Assembly (28-29 July, 2008).

Faridi, F.R. (1997), Islamic Principles of Business Organization and Management (Selected Papers of the Seminar of Organizational Behaviour held at Herdon, Virginia, USA, September, 1988), Kuala Lumpur, Malaysia, ISBN 983-899-808-7, S. Abdul Majeed & Co, p.3).

G. Hofstede (1980), *Culture's Consequences: International Differences in Work Related Values,* Beverly Hills: Sage Publications.

Green, K (2008). What is Taqwa? Retrieved August 11 2010 from http://islamicbeliefs.suite101.com /article.cfm/ what_is_taqwa.

H. Mutalib (1993). *Islam in Malaysia- From Revivalism to Islamic State.* Kent Ridge, Singapore, Singapore University Press, pp17-19.

H. Singh (2001), Ethnic Conflict in Malaysia Revisited. *Commonwealth and Comparative Politics*, Vol.39, No.1 (March), pp 42-65.

Hafsah Hashim (2014), *Business Ethics: The Real Deal.* Adjunct Lecture for Universiti Teknologi PETRONAS MBA Energy Management Students delivered on 22 June, 2014.

Integrity (2010). In Oxford Dictionaries Online. Retrieved August 11, 2010 from http://www.oxforddictionaries.com/definition/integrity?view=uk

Khalidah Khalid Ali (1983). M. Sc Research, *Malaysia- Looking East for Good Work Ethic* submitted as part requirement for the degree of M. Sc (Economics) Industrial Relations and Personnel Management of London School of Economics and Political Science, University of London, September 1983, pp. 5

Khalidah Khalid Ali, Satirenjit Kaur Johl, Lai Fong Woon, Rohani Salleh, Sofiah Molek Lope Aman Shah, Rahayu Abd. Rahman and Ilmiah Ibrahim (2012a). *Business Management: A Malaysian Perspective* (2nd Ed). Shah Alam, Selangor Darul Ehsan, Malaysia: Oxford Fajar.

Khalidah Khalid Ali, Rohani Salleh and Mashitah Sabdin (2012b). Ethical Values of Final Year Students at a Malaysian Private Higher Learning Institution, 2012 (1) ISSN: 978-1-61275-006-4, published by Information Engineering Research Institute (IERI), USA. *Proceedings of International Conference on Applied and Social Science (ICASS 2012),* Kuala Lumpur 1 – 2 February, 2012, pp. 119-131.

Khalidah Khalid Ali and Ahmad Murad Merican (2014a). Beyond the Malay Dilemma: Understanding Ethics and Integrity of the Malay Community in Malaysia (1970-2000), *ESTCON/ICHSS 2014,* Kuala Lumpur, 3-5 June, 2014.

Khalidah Khalid Ali (2014b), The Role of Malay Women in the Malaysian Workforce and its Impact on the Consciousness of Ethics and Integrity. *Proceedings of ESTCON/ICLM 2014, Kuala Lumpur,* 3-5 June, 2014.

Khalidah Khalid Ali, Zulkufly Ramly and Lau Teck Chai (2014c), *Business Ethics,* Oxford University Press.

Khalidah, K.A, Rohani, S. and Mashitah, S. (2014d) Ethical Values and Commitment Towards Achieving Excellence: A Study on Public Boarding School Students in Malaysia. *Pertanika Journal of Social Sciences and Humanities.* 22 (S), pp. 33–50.

Mahathir Mohamad (1970 and 1977), *The Malay Dilemma*, Kuala Lumpur: Pelanduk Publications.

Monthly Statistical Bulletin Malaysia, (On line), Available: http://www.statistics.gv.my/portal/index retrieved December, 2011.

Muhamad bin Ahmad (2009). *The Influence of Culture on Malay Muslim Business Practice: A Case Study of Malaysia with Special Reference to the State of Kelantan Dar al- Naim.* A Thesis submitted to University of Birmingham for the degree of Doctor of Philosophy 2009. pp 51-58.

Muhammad Rais B Abdul Karim (1995), Improving the Efficiency of The Public Sector: A Case Study of Malaysia, *United Nations.*

Nor' Azzah Kamri and Khairiah Salwa Haji Mokhtar (2004), The Role of Islamic Ethics in Organizations: An Experience in Malaysia. *Paper presented at 6th ASEAN Inter-University Seminar on Social Development* organized by School of Social Science, Universiti Sains Malaysia (USM) and University of Singapore at University Sains Malaysia, Penang, Malaysia, 14-16 May, 2004.

Noore Alam Siddiquee (2010). Combating Corruption and Managing Integrity in Malaysia: A Critical Overview of Recent Strategies and Initiatives, *Springer Science.*

Othman Mohd Yunus, Nur Hamizah Hamzah and Mohd Afandi Abu Bakar (2011), Religiousity and Work Values Orientation of Malaysian Malay Businesspersons. *Proceedings of Eighth International Conference on Islamic Economics and Finance - Sustainable Growth and Inclusive.* 19-21 December 2011. Doha Qatar.

Rahman, A.R.A (2003). Ethics in Accounting Education: Contribution of Islamic Principle of Maelaiah. IIUM *Journal of Economics and Management*, 11 (1): 1-18.

S. Awang, M. Maros, and N. Ibrahim (2012). Malay Values in Intercultural Communication, *International Journal of Social Science and Humanity*, Vol. 2, No. 3, May 2012.

Sharifah Hayaati Syed Ismail (2007), Values and Ethics Towards Quality Public Delivery System of Malaysia: An Islamic Perspective, *Jurnal Syariah,* Vol 15(2), pp. 25-43.

Sharifah Hayaati Syed Ismail (2011). Role and Transition of Public Service in Malaysia: An Effective Governance Perspective, *African Journal of Business Management,* Vol. 5(5), pp. 1963-1975, March 2011

Shaw, W.H. (2011). *Business Ethics* (7th Ed.). Belmont, CA: Thompson Wadsworth.

S.N.H. Naqvi (1981), Ethics and Economics: An Islamic Synthesis, *Islamic Foundation*, Leicester, UK.

S.H. Nasr (1984), Islamic Work Ethic, *Hamdard Islamicus,* Vol. 7, pp. 25-35.

Stanwick, P.A. and Stanwick S.D. (2009). Understanding Business Ethics, New Jersey: Pearson International Edition.

Syed Husin Ali (2008), *Ethnic Relations in Malaysia- Harmony and Conflict, Strategic Information and Research Development Centre*, Vinlin Press Sdn Bhd., Selangor, Malaysia.

Syed Husin Ali (2008), *Ethnic Relations in Malaysia- Harmony and Conflict, Strategic Information and Research Development Centre*, Vinlin Press Sdn Bhd., Selangor, Malaysia.

Syed Othman and H.G. Aidit, *Islamic Values and Management.* Kuala Lumpur, Malaysia. Institute of Islamic Understanding Malaysia (IKIM), 1994.

Weber, M. (1958). *The Protestant ethic and the spirit of capitalism.* (T. Parsons, Trans.). New York: Charles Scribner's Sons. (Original work published as two separate essays, 1904 -1905).

Zahazan Mohamed and Muhammad Zakaria, 2011) Hadith Riwayat Muslim no.8, Abu Daud no.4695, al –Tirmizi no. 2738 and al- Nasai no 8/97 with minor difference; sourced from *Hadith 40 Imam Nawawi*

Social media influencers - shaping consumption culture among Malaysian youth

Nur Leila Khalid[1,*], *Sheila Yvonne* Jayasainan[1], and *Nurzihan* Hassim[1]

[1] School of Communication, Faculty of Arts & Social Sciences, Taylor's University, Malaysia

Abstract. This study reveals the significance of social networking sites (SNS) that created vast opportunities for cultural awareness and consumption among youth SNS users in Malaysia. Simplicity and user-friendly features on Instagram had heightened interest and growth of its users where the ability to "hashtag" on Instagram identifies current trends based on user-generated images and videos that enable users to generate search results instantaneously. Researchers have argued that SNS users at present trust peer recommendations more than search engines and this phenomenon is evident from their resonance with "social media influencers" (SMIs) who are capable of influencing massive followers through reviews of products, services and social trends on their social media accounts; particularly on dining out and "cafehopping" as a social experience. This study examined the curation of Instagram posts by SMIs and found that their visibility of content created shared voices and reinforced cultural identities among their followers. The content analysis deployed in this study also investigated feedback on the selected SMI's curation of posts and established that peer engagement is a tangible socialization process that would contribute to the youths' future use of Instagram as a social, cultural and educational tool.

1 Introduction

Social networking is an essential tool for both businesses and media consumers in today's converged communication landscape. Aside from information search and knowledge dissemination, social networking sites (SNS) are known to reconnect users with former contacts and establish new connections on a personal level. According to Kemp (2017), more than 1.5 billion people across the Asia Pacific now use SNS on a monthly basis, with 95% accessing platforms via mobile devices; thus, becoming the largest segment of SNS consumers in the world. The increased mobility of these societies has transformed information-seeking habits of users when SNS platforms are frequently utilized compared to Internet search engines. In addition, information retrieval via SNS is specifically linked to the needs of youth consumers for inclusion in the emergent trends of cultural and consumption culture, where technology is agency to emotions, cognition and social patterns of Generation Y (Gen Y). This data savvy generation who are born after the year 1981 regularly interact, seek entertainment and moderate feelings with each other through their SNS of choice (Bolton et al., 2013; Brosdahl & Carpenter, 2011). Scholars argued that the desire to appear as independent thinkers or "post-subculturally progressive" had driven intentional use of SNS among them (Muggleton, 2000; Williams & Mohamed Nasir 2016).

Among the more preferred SNS of youths for this purpose is *Instagram*, a photo and video-sharing application (or known as app) that enables its users to upload insights on their daily routines and social activities to their "followers", to which public discourse is made possible from feedbacks of the shared post. Instagram was founded by Kevin Systrom and Mike Krieger in October 2010 and achieved 1 million users just two months after its launch, where Facebook later bought Instagram in April 2012 (Desreumaux, 2013). From the Malaysian perspective, Perception Media (2015) highlighted that internet and mobile data are most consumed by Gen Y at 43%, 40% by Gen X, while the Malaysian Communications and Multimedia Commission (2016) reported approximately 46.7% of Malaysians

* Corresponding author: NurLeila.Khalid@taylors.edu.my

own an Instagram account. Instagram's simple and interactive features that allow users to express "likes" with a red-coloured heart symbol for selected content by double-tapping the photo or video had contributed to higher ratios of engagement from the users (Gallagher, 2017). This can be seen as an opportunity for the users to interact with their followers on a more intimate level compared to other SNS; as the layout and gallery interface of the app has made it more personal for them. It can be said that the rise in the popularity of Instagram is also contributed from social and cultural engagement of the active youth segment through the app. Furthermore, the platform allows users to seek a specific keyword on its search function using "hashtags", where users will find themselves directed to a list of related posts and a pool of other like-minded users. With this advent of technology, social media users are no longer dependent on literal word-of-mouth recommendations but instantaneous display of information from social media through the use of hashtags. A hashtag is a type of label or metadata tag primarily used on social networking websites and microblogging services that allows multiple users to find content of the same topic. Hashtags are created by inserting the hash character '#' in front of a keyword or phrase and can be used to complement postings on social networking sites (SNS). On Instagram, hashtags can be used to direct users to related accounts based on the topical search and also leads to gaining followers should the account display the needful information that the user can follow on a regular basis. This also works as an archiving function where users can use the hashtag as a reference keyword in the future.

Subsequently, posting on Instagram have become a practice of educating and creating awareness about food products and services to followers, especially for users that have a high number of following. A study by the University of Washington (2017) described how users turn to posting photos on Instagram to track food intake or to be held accountable by followers in meeting healthy eating for weight loss goals. The study validated that social and emotional support from other Instagram followers by means of clicking the "like" button on a visual image help them stick to their goals. Currently, using these unique tags in social media works as the fastest and most convenient methods in seeking new eateries or dining experiences, thus mobilizing social trends. At the same time, attitudes about food and beverage are rapidly becoming points of deliberation on social media. Ariyasriwatana & Quiroga (2014) suggested that a consumer's dining behaviour captured through their online reviews or social media posts leave digital footprints that affects attitudes of observers in their network. According to Malaysian Digital Association (2016), food and beverages has increased number of revenue as many start-ups and online entrepreneurs have realized that social media platforms create opportunities to expand their businesses. Mejova et al. (2016) established that Instagram is a forum highly suitable for food logging and research where the obsession on foods such as desserts, savouries as well as healthy-eating were referred to as *#foodporn*. In addition, the distinctiveness of culinary experiences in Malaysia, particularly Kuala Lumpur have elicited a pleasant image in media for Malaysian eateries to have repeat visitations from consumers (Leong et al. 2010).

With that in mind, this preliminary study explores Instagram hashtags as a main source of SNS search that unites Malaysian youths in seeking novel food experiences. The study explores narratives of visual content on Malaysian Instagram users who are known as "social media influencers" (SMIs) to ascertain meanings of their posts on food experiences and its influence upon their large number of followers through the use of specific hashtags. Qualman (2014) argued that social media users at present trust peer recommendations more than search engines and this phenomenon is evident from the resonance of social media users with SMIs who review their experiences in product consumption, services and social trends on their social media accounts, particularly on dining out and "cafehopping" as a social experience. The study will analyse these posts with theoretical foundations derived from the Uses-and-Gratifications (U&G) model by Katz et al. (1974) that argued media users are motivated by their personal and social needs in their use of the media; and the concept of media affordance that proposed the relation between technology and its users as key to understanding and analysing social media interfaces (Bucher & Helmond, 2017). This study evaluated Instagram posts of five chosen influencers that embedded the *#cafehopkl* hashtag to explore the narratives of the user-generated media messages and SMI influence upon social media users. The analysis was also aimed to further understand cultural meanings and consumption patterns that emerge from the Instagram curation of the selected SMIs.

2 Literature Review

The uses-and-gratifications model (U&G) were among the earliest paradigms that considered the active role of the audience in choosing their communication media, where individuals identify with and employ specific platforms to fulfil personal gratification (Ku et al. , 2013). Previous applications of U&G theory on brand communities showed that consumption of informative content is an important factor that increase user participation (Raacke & Bonds-Raacke 2008), whereas a study by Park et al. (2009) found that content with entertainment value had a stronger effect. The U&G perspective on active recipients of media can now be applied in the social media context, wherein social networking sites (SNS) provide users with an interactive avenue to create value and engage with influencers and brands (Brodie et al.,2013). Users create social media content through their contributions, comments and likes, and subsequently, their input facilitates responses from other members. Influencers and companies encourage their followers to not just passively consume content, but actively exhibit social media engagement behaviours in such a way that they drive new content (Baird & Parasnis, 2011). Foundational to this is the argument on "agencies of socialization" posited by Dominick (2011, p.445), where media consumers are largely influenced by their communication with friends, personal experiences and mass media are relied upon as main sources of information. As identified by Cvijikj & Michahelles (2013), digital engagement on social network depends on several elements; namely the nature and content of the message, the post format which could be understood as the message's packaging (i.e. vividness and interactivity of the post) and posting time (i.e. during peak hours on workdays). Park, et al. (2009) further suggested that knowledge acquisition and sharing are social needs that are met by using social media. Similarly, Lee & Ma (2012) found that information-seeking is a significant factor that affects users' intention to share information.

West & Turner (2010) earlier elaborated that the use of media among audiences are driven by personal motivations; which in turn attains personal fulfilment. These dimensions comprise of firstly, *cognitive needs* in order for social media users to understand and identify a phenomenon. Media content are also able to appease users with their emotions that fall under the category of fulfilling *affective needs*. With social networking sites (SNS), users are acknowledged for their status, stability and most importantly their credibility, belonging to the U&G dimension of *personal integrative needs*. Similarly, the use of media content posits belonging for the social media user by fulfilling their *social integrative needs* in their engagement with others. Social media users are also granted a form of escapism in their media use the U&G dimension of *tension release needs*. Katz et al. (1974) in the pioneering studies of U&G prescribed that media users are active in choosing and using diverse platforms of media where they are able to exercise choice and control to accomplish personal goals regardless of their selection (Dominick, 2011, p.39).

Considering the discussion of U&G in relation to uses of SNS, it can be said that the sharing of information and curation of photos on Instagram is a new way of articulating brand credibility through the use of opinion leaders, who are now referred to as "social media influencers" or SMIs. "Hashtagging", in this sense would resolve the gaps of media exposure experienced by traditional advertising and marketing, while peer reviews are more trusted and is justified in the conversation among SNS users that discuss the brand, product or service compared to word-of-mouth. Consumers' desire for social interaction and a sense of belonging while seeking support for their personal growth are key motivators in their continuous access of SNS platforms. In relations to the U&G theory, the need for self-expression through content and activity disclosure using SNS often results in positive valence. Users create SNS content through their contributions, comments and "likes", which is known on Instagram as "love". SNS features carry symbols and connotations that mediate, communicate and relate to different affordances for users (Bucher & Helmond, 2017). Subsequently, these inputs facilitate the engagement and interaction of other members within the virtual yet social experience.

3 Methodology

This preliminary study examined how social media influencers (SMIs) create cultural identities and consumption patterns through the curation of their Instagram posts thus shaping social trends among Malaysian youths. To do so, the study analysed Instagram posts of five Malaysian SMIs in creating meanings and encouraging feedback from their followers. At present, there is limited literature on the study of social networking sites (SNS) in terms peer-generated posts with hashtags and its influence upon other media consumers' preference and perceptions on food and culinary experiences. In addition, the curation of visuals and narratives behind postings of social media influencers is relatively a growing phenomenon that incite further exploration to ascertain the dimensions of uses-and-gratifications experienced by the media consumer. Therefore, the population sampled for this study are actively involved in curating interesting and trending latest food culture in Malaysia through the use of their personal Instagram accounts, thus accessible to their followers.

To investigate the curation patterns of selected SMIs, the study deployed content analysis on images and captions. Previously scholars have challenged content analysis as an outdated yet systematic data organization, more recent studies have contended that content analysis can be used beyond conventional media and have progressed beyond the analysis of print media. In light of this, the perspective of "texts" have advanced into the social media realm, and therefore, online discussions are considered as public conversations that allow self-expression, similar to traditional content analysis (Adler & Clark, 2011; Donelle & Hoffman-Goetz, 2008).

3.1 Dataset and Data Sampling

Earlier on, Instagram was described in various sections to have been selected as the focus of this study due its popularity among users in Asia Pacific and Malaysia. Apart from its flexible platform that enables photo and video-sharing with captions, Instagram also allows the facilitation of "geotags" where users are able to identify the locations where the food featured on user posts can be found. The study compared datasets that are available from the hashtag or keyword search of *"cafehopping"* on Instagram, where the search results compiled posts by users that enjoyed either new or repeated food and beverage experiences in food outlets largely situated within Southeast Asia. In comparing the data retrieved, *#cafehopping* is observed to have 505,486 posts; *#cafehopkl* approximately 203,909 posts and #cafehopmy approximately 180,860 posts. The researchers later determined to focus on *#cafehopkl* to narrow the geographic scope within the capital city of Kuala Lumpur where the population of suburban Gen-Y youths are concentrated.

In order to examine the phenomenon of SNS hashtags as mediator to social engagement in the Malaysian context, this study examined food photos posted by five selected social media influencers (SMIs) with more than 5000 followers on Instagram. The SMIs were selected from the hashtag *#cafehopkl* to determine whom had the highest impact among urban audiences of Kuala Lumpur; in order to ascertain the influencer and followers' similar interest in gastronomic adventures.

Next, the gathered dataset shortlisted the top three posts based on the frequency of "likes" obtained from each the chosen influencers' postings. The criteria applied for selecting the set of posts on Instagram consisted of; (1) food postings with the use of hashtag *#cafehopkl*; (2) posts from the months of January to March 2017; and, (3) English language used for communication. The study sample included a total of 15 posts from *#cafehopkl,* collected on 1 May 2017. After *#cafehopkl* hashtag was filtered, the following units of analysis were established.

Table 1: Selected social influencers based on set criteria (as at 1 May 2017)

Influencers	Number of followers (approximate figure as of 1 May 2017)	Total number of posts (approximate figure as of 1 May 2017)
carol_eats	28600	2018
leochan_kh	9026	567
sxnflxwer_	64700	262
hafiqhasanbasri	5035	1506
dududedoodle	30500	1044

The Instagram posts were also analysed based on the style and presentation of photos posted by these five Malaysian influencers under the *#cafehopkl* hashtag. Besides the frequencies and percentages of "likes" which were analysed, the comments (words and emoticons) posted by followers were also examined. The comments were categorized under similar themes and the frequencies of these similar themes (words) and the emoticons were looked into to gauge the engagement of followers with the selected influencers.

4 Findings

4.1 Percentage of "likes" for Instagram postings

The samples comprised of 15 postings from January to March 2017. Table 2 details the top three posts from month of January to March 2017 for each of the five social media influencers (SMIs). In order to analyse the highest percentage of "likes"; the total number of "likes" for each of the influencer's top three postings were divided against the total number of followers. Influencer: *leochan_kh* is seen to have the highest percentage of "likes" for three consecutive months (January to March 2017) amongst all influencers. For the month of January 2017, influencer: *leochan_kh* obtained a total of 2126 "likes" for his post on 24 January, at 23.6%. Subsequently, in February 2017, 2256 "likes" were obtained on 8 February 2017, amounting to 25% whilst in March 2017, 2370 "likes" were received for his

Instagram post on 14 March 2017, at 26%. His post on 14 March 2017 gained the highest percentage of 26% "likes" compared to all other SMIs in the 3 months when these data were analysed.

Other SMIs, despite having substantially larger number of followers such as *sxnflxwer_* (64,700 followers) and *carol_eats* (28,600 followers) did not have a higher ratio of "likes" compared to influencer: *leochan_kh*.

Table 2: Top three posts from January to March 2017 for five selected influencers

Influencers	January 2017	February 2017	March 2017
1. carol_eats (28.6k Followers)	16 January 2017 Total likes: 1,366 (4.78%) Comments: 3	22 February 2017 Total likes: 1,647 (5.76%) Comments: 5	26 March 2017 Total likes: 1,367 (4.78%) Comments: 12
2. leochan_kh (9,026 Followers)	24 January 2017 Total likes: 2,126 (23.6%) Comments: 12	8 February 2017 Total likes: 2,256 (25%) Comments: 13	14 March 2017 Total likes: 2,370 (26%) Comments: 13
3. sxnflxwer_ (64.7k Followers)	28 January 2017 Total likes: 2,857 (4.41%) Comments: 17	22 February 2017 Total likes: 3,120 (4.82%) Comments: 29	11 March 2017 Total likes: 3,616 (5.59%) Comments: 24
4. hafiqhasanbasri (5,035 Followers)	6 January 2017 Total likes: 171 (3.4%) Comments: 0	5 February 2017 Total likes: 141 (2.8%) Comments: 4	2 March 2017 Total likes: 134 (2.66%) Comments: 17
5. dududedoodle (30.5k Followers)	2 January 2017 Total likes: 3,041 (9.97%) Comments: 42	26 February 2017 Total likes: 2,156 (7.07%) Comments: 30	22 March 2017 Total likes: 1,845 (6.05%) Comments: 26

Figure 1 exhibits the percentages of "likes" amongst the five influencers from the months of January to March 2017. As a result, this distinctly indicates influencer: *leochan_kh* as having the highest number of likes per ratio for all three months (January to March 2017) when samples of postings were selected and analysed.

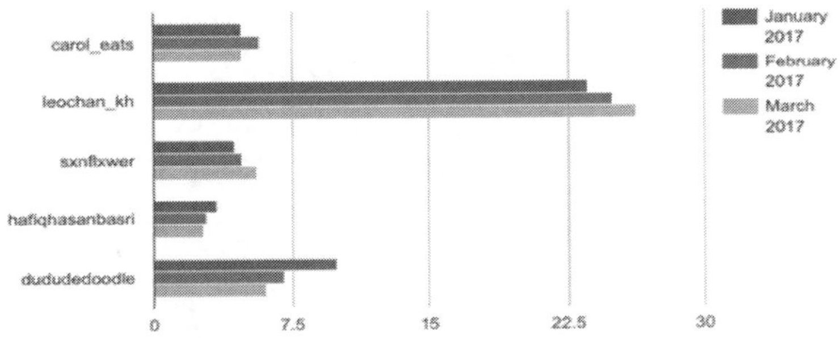

Fig.1. Summary of "likes" for the five influencers (January - March 2017)

4.2 Curation of Postings

To further understand how *leochan_kh* obtained not only the highest percentage of "likes" in March 2017, but also the highest percentage for all five influencers as indicated in the earlier section; the study further examined the curation or arrangement of the influencer's posts on his Instagram account. It was observed that *leochan_kh* had beautifully curated his Instagram feed to showcase food and cafe options around Kuala Lumpur. In doing so, followers of his account are able to discover many different cafes and food they can try and discuss about both online and offline. For a featured account like *leochan_kh*, it is all about curating the best of the best food and cafe photos in Kuala Lumpur through carefully planned photography but still appear spontaneous and candid. It is an excellent Instagram marketing strategy that brands and business proprietors can benefit from, that seems non-obtrusive and natural to the eyes of his followers.

Figure 2: Thumbnails of *leochan_kh*'s Instagram account

Fig. 3: Selected posts of *leochan_kh* that included *#cafehopkl*

One of the main criteria that was noticeable in *leochan_kh*'s Instagram posts is his consistent curation that has kept his posts standardized with similar styles of photography. Figure 3 shows some examples of his Instagram postings that features the hashtag *#cafehopkl* apart from other hashtags that were used. *leochan_kh* often use flat-lays for layout arrangements, which are shots from top view similar to the 'bird's eye view' that displays the overall table settings. The concept of flat-lay in photography was first known as 'knolling', a term coined in 1987 by Andrew Kormelow, a janitor at Frank Gehry's furniture shop (Fitzpatrick, 2016). It started off as arrangements of his displayed tools at the right angles on all surfaces. The idea of flat-lays is to declutter the Instagram feed, which viewers often pay attention on objects neatly laid out against neutral background.

All four samples of Instagram posts (in Fig.3) shows consistency; where aside from food and props, he adds in some human elements to his composition so that it has a natural feel to it. Most of his posts include partial shots of legs and hands so that it captures the moment of himself and others enjoying the feast. In addition to that, he would like his viewers to pay attention to the food details and arrangements so that it may attract viewer's interest to know more about the place that he has been. In this case, the usage of *#cafehopkl* will bring other Instagram users to his page and help to increase his number of followers. Based on the consistency in his posts, the followers will have a certain expectation for the posts to be in a similar style and theme. In one of his many posts, he did an experiment which he tried to move away his normal style which involve flat-lays. That posts triggered many comments that says that it is not his style and it was indeed something different.

leochan_kh also uses the same type of filter or image editing with grey, black, brown and white overtone. This is to ensure that the main focus of the composition are the food placements and props that surrounds that exudes vibrant colours. To capture this type of posts requires natural lighting to brings out the colours of the object in the composition. *leochan_kh* had carefully selected certain type of cafes that is able to fit his style of photo composition as he needs to maintain his style for that resembles his identity. He had also stated in his set of hashtags that he is only using an iPhone to capture his photos. This can be seen as a sense of belonging of a certain community of iPhone users which was labelled in his list of hashtags as *#iphoneonly*.

 https://doi.org/10.1051/shsconf/20185302008

Fig. 4. Example of tags and hashtags used by *leochan_kh*

Upon further analysis of his Instagram posts, *leochan_kh* often used many types of hashtags. *#cafehopkl* is one of the hashtags that he often uses for Instagram users to discover his account. This works as one of the marketing strategy to lead Instagram users of certain interest to find out if others have posted using the same hashtags. The more users use the same hashtags, the source will be regarded more trustworthy. According to Eventys Partners (2015), brands want their fans to share their personal thoughts, comments and pictures on social networking sites (SNS), it shows that consumers care about the brand and feel which has successfully build solid relationships between the consumers and the brand. Not only limited to hashtags, *leochan_kh* also use tagging of other Instagram users and pages on his posts as seen on Figure 4 above. He maximizes his exposure and make it noticeable. Most of the tags that he use is related to coffee enthusiasts which he would like to be linked to the curation of his posts.

5 Gratifications Achieved from Media Uses and Affordances

The key aspect discussed in this study is the significant growth of social networking sites (SNS) in Malaysia through Instagram, particularly from the lens of user engagement. It was observed from the posts uploaded by the media influencers selected in this study, that the engagement between influencer (source) and receiver (SNS user) is two-prong; hashtagging with *#cafehopkl* being more impactful than professional advertising and marketing by businesses themselves. The posts analysed that uploaded food experience photos with at least one hashtag had experienced 12.6% more engagement than those with none. Secondly, the study is significant in highlighting new trends on SNS. It can be summarized that the efforts of "hashtagging" on Instagram has vastly contributed to the appeals of retail establishments, particularly in the food and beverage industry.

The analyses have observed the importance of visuals as well as responses through amounts of "likes" that determine the relationship of "social media influencers" and the pertinence of their roles towards purchasing behaviours that lead to brand loyalty and credibility. From the perspective of the media user, the gratifications achieved from their information-seeking behaviours are fulfilled by peer-generated media content, where word-of-mouth is considered more credible than product marketing. The analyses also significantly highlighted the role of user-generated content in igniting constructs for cultural consumption among Malaysian users of Instagram.

5.1 Cognition

The research of U&G by Katz et al. (1973) had observed interchangeability of print and broadcast media that can be applied to social media in the digital context. This standard of consumer behaviour was agreed by several authors that the gratifications of the media users and their intentions of using social media apps are strongly dependent on peer recommendation and the access to interactivity (Logan, 2016; Oh et al.2015; Wang & Chou, 2016). This study further identified several usage habits by social media influencers (SMIs) that are perceived as elements of gratification for the media users or predicted intention to use the hashtags featured in the posts available. The data obtained from the Instagram account of *leochan_kh* purports that the social media influencer is using the SNS with relevant hashtags as a prevalent medium to deliver information, in which he expects the audience to obtain information from the curated posts. From the engagement point of view, the responses obtained was more visual than textual as well as a combination of both, through the use of varying emoticons.

5.2 Affiliation

Further to this, *leochan_kh*'s efficient use of hashtags saw affinity of both the SMI as well followers of his account from their information-seeking behaviour in pursuing the *#cafehopkl* hashtag. Feedback of the user with the postings are expressed minimally, but meaningfully through the use of limited words and "likes". This is assumed that the communal use of the hashtag is already understood as a representation of shared interest and relatability from the like-minded thoughts shared by the media consumers to the consistency of *leochan_kh*'s posts, he provides the sense of belonging or involvement within a social group. In this case all of his posts are related to food and drinks (mostly coffee) and is related to cafe settings. *leochan_kh* also used a lot of hashtags and tagged various Instagram accounts and common nomenclature such as #coffeeaddict, #foodie and #flatlays. Among the popularly used terms are short acknowledgments on food experience are such as ***"nice pic"***, ***"really impressive feeds"*** and ***"omg so yummy"*** and several others. When such recurring and positive-acknowledging emoticons are applied, there is a common ground where followers experience intimacy through their feedback with the SMI.

5.3 Expression

In this study, *leochan_kh* used Instagram as a medium to express his inner thoughts, feelings and opinions via images captured and captioned in his Instagram posts. He used his creativity in creating his own style of layout arrangements and image editing so that he stands out from the rest of other Instagram account users and social media influencers. This can be correlated to earlier research done by Whiting and Williams (2013) that found that social networking sites are definitive in expressing the thoughts and opinions of its users. This is illustrated in the capacity of *leochan_kh* as an SMI who is vulnerable to the keenness of his followers in making comments on his updates, reposting as well sharing comments to others. Both SMI and followers have to freedom to criticize others, and are able "to vent" their emotions (both textually or visually) on social media.

6 Discussion

In the local media sphere, impactful Instagram posts are carefully composed and curated through eye catching visuals, captions and hashtags to increase the appeal of the featured establishment, product or service. Food composition or food styling further enhances its likeability and visibility in Instagram, hence evoking curiosity that would lead followers to seek more information about the uploaded visual. This is where the clickable hashtags offered would assist the followers in examining crowdsourced posts by other Instagram users that would; first, indicate the popularity of a food establishment, product or service based on hashtag mentions; secondly to determine its location, contacts as well as accessibility; and lastly and most importantly; to evaluate what is said by other users about the chosen establishment, product and service in their peer reviews.

The result from this study shows that social media influencers are significant in sharing new knowledge through the display of food tasting experiences that is highly curated and often emphasize on aesthetics. These experiences are optimally displayed on Instagram and are easily accessible through mobile gadgets prevalently used by Gen Y who are given agency and affordance to this content through overflowing access to the social media platform. While the influencers are not always actually seen eating the food, the geotags and hashtags highlighted in the posts not only create a benchmark of "must-dos" for other users, but also crowdsources social approval from followers. The materials analysed in the study can be seen as a predictor of how other users should anticipate the meal from the same location, and highlighting food also demarcates special occasions, activities and even emotions as suggested by Holmberg et al. (2016).

7 Conclusion

The conversations in hashtagged posts are meant for a specified genre of users, namely Gen-Y youths; therefore relatability is only for those who are familiar with the use of SNS. Social media influencers include 'a brand voice' to all Instagram postings by including parts of their personality, be it physical or metaphorical. By combining humour and whimsy or playful expression to tell a brand's story and showing personality, the social approval (in terms of likes and comments) has constructed and re-constructed the identities of both social media influencers and users that subconsciously interchange roles in the process of interactions. Wilson's study of hermeneutics (2016) could be applied as an extension to this present study of hashtagging on social media to make sense of how the phenomena had changed and transformed the dissemination of information. In an earlier claim by Wilson (2012), using and transitioning to different media horizons or "second screens, cultures are "...anticipated, actualized and asserted..." (p.5). Future research can also explore on the effective posting characteristics in this consumption culture and the varied gratifications of followers that respond to them.

8 References

Adler, E., & Clark, R. (Eds.). (2011). *How it's done: An invitation to social research, 4th Edition*. Belmont, CA: Cengage Learning.

Alch, M. (2000), "The echo-boom generation: a growing force in American society", *The Futurist*, Vol. 5, September, pp. 42-48.

Ariyasriwatana, W., & Quiroga, L. M. (2014). A thousand ways to say "Delicious!"—Categorizing expressions of deliciousness from restaurant reviews on the social network site Yelp. *Appetite*, 104, pp. 18–32. http://doi.org/10.1016/j.appet.2016.01.002.

Baird, C., & Parasnis, G. (2011). From social media to social customer relationship management. *Strategy & Leadership*, 39, pp. 30–37.

Bolton, R. N., Parasuraman, A., Hoefnagels, A., Migchels, N., Kabadayi, S., Gruber, T., Solnet, D. (2013). Understanding Generation Y and their use of social media: a review and research and research agenda. *Journal of Service Management Marketing*, *24*(2), pp.328–344.

Brodie, R. J., Ilic, A., Juric, B., & Hollebeek, L. (2013). Consumer engagement in a virtual brand community: An exploratory analysis. *Journal of Business Research*, 66, pp.105–114.

Brosdahl, D.J. & Carpenter, J.M. (2011), Shopping orientations of US males: a generational cohort comparison, *Journal of Retailing and Consumer Services*, 18, pp. 548-554.

Bucher, T. & Helmond, A. (2017). 'The Affordances of social media platforms.' In Burgess, J., Poell, T. & Marwick, A. (Eds.), *The SAGE Handbook of Social Media*. London & New York: SAGE Publications, Ltd.

Cvijikj, I. P., & Michahelles, F. (2013). Online engagement factors on Facebook brand pages. *Social Network Analysis and Mining*, 3, pp.843–861. DOI 10.1007/s13278-013-0098-8.

Desreumaux, G. (2014, January 3). *The Complete History of Instagram*. Retrieved April 25, 2017 from http://wersm.com/the-complete-history-of-instagram/

Dominick, J.R. (2011). *The Dynamics of Mass Communication: Media in Transition (11th ed)*. New York, NY: McGraw-Hill.

Donelle, L., & Hoffman-Goetz, L. (2008). Health literacy and online health discussions of North American black women. *Women & Health*, *47*(4), 71-90.

Eventys Partners (2015, March 3). *How to Use Hashtags Correctly Across Social Networks for Marketing*. Retrieved August 28, 2017 from https://enventyspartners.com/blog/how-to-use-hashtags-correctly-across-social-networks-for-marketing

Fitzpatrick, E. (2016, January 29). *The Surprise History Behind Instagram Flat Lay Shots*. Retrieved August 28, 2017 from http://www.whowhatwear.com/instagram-flat-lay-knolling

Gallagher, K (2017, May 23). *Instagram Tops Facebook on Engagement*. Retrieved August 30, 2017 from http://www.businessinsider.com/instagram-tops-facebook-on-engagement-2017-5/?IR=T

Hiscott, R. (2013, October 8). *The Beginner's Guide to the Hashtag*. Retrieved April 20, 2017 from http://mashable.com/2013/10/08/what-is-hashtag/#JcRrVBiiWPqS

Holmberg, C., E. Chaplin, J., Hillman, T., & Berg, C. (2016). Adolescents' presentation of food in social media: An explorative study. *Appetite*, 99, pp.121–129, http://doi.org/10.1016/j.appet.2016.01.009

Katz, E., Haas, H., & Gurevitch, M. (1973). On the use of the mass media for important things. *American Sociological Review*, pp.164-181.

Katz, E., Blumler, J.G., & Gurevitch, M. (1974). 'Utilization of mass Communication by the Individual'. In J.G. Blumler & E. Katz (Eds.), *The uses of mass communication: Current perspectives on gratifications research* (pp. 19-32). Beverly Hills, CA: SAGE Publication Ltd.

Katz, E., & Lazarsfeld, P. F. (1995). Between media and mass/the part played by people/the two-step flow of communication. *Boyd-Barrett, Ch. Newbold (red.), Approaches to Media, London*, 124-134.

Kemp, S. (2017, January 24). *Digital in 2017: Global Overview*. Retrieved April 20, 2017 from https://wearesocial.com/special-reports/digital-in-2017-global-overview

Ku, Y.C., Chu, T.H. & Tseng, C.H. (2013). Gratifications for using CMC technologies: A comparison among SNS, IM and e-mail. *Computers in Human Behavior, 29*(1), pp. 226-234.

Lee, C.S. & Ma, L. (2012). News sharing in social media: the effect of gratifications and prior experience. *Computers in Human Behavior, 28*(2), pp.331-329, DOI: 10.1016/j.chb.2011.10.002.

Leong, L.Q., Ab Karim, M. S., Othman, M., Mohd Adzahan, N., & Ramachandran,S. (2010). Relationships Between Malaysian Food Image, Tourist Satisfaction and Behavioural Intention. *World Applied Sciences Journal*. (Special Issue of Tourism & Hospitality), 10, pp.164–171.

Logan, K. (2016). Attitudes towards in-app advertising: a uses and gratifications perspective. *International Journal of Mobile Communications, 15*(1), pp.26–48. http://doi.org/10.1504/IJMC.2016.10000002

Malaysian Digital Association (2016, August 3). *2016 Malaysia Digital Landscape*. Retrieved April 25, 2017 from http://www.malaysiandigitalassociation.org.my/wp-content/uploads/2016/08/Malaysia-Digital-Landscape-August-2016.pdf

Malaysian Communications and Multimedia Commision (2016). *Internet Users Survey 2016*. Retrieved April 25, 2017 from https://www.mcmc.gov.my/skmmgovmy/media/General/pdf/IUS2016.pdf

Perception Media. (2015). *Media Planning Guide Malaysia 2015*. Kuala Lumpur: Perception Media.

Mejova, Y., Abbar, S., & Haddadi, H. (2016). Fetishizing Food in Digital Age: #foodporn Around the World. *In Tenth International AAAI Conference on Web and Social Media*, pp.250–258.

Muggleton, D. (2000). *Inside subculture: The postmodern meaning of style*. Oxford:Berg Publishers.

Oh, S., Baek, H. and Ahn, J. (2015) 'The effect of electronic word-of-mouth (WOM) on mobile application downloads: an empirical investigation', *International Journal of Mobile Communications*, *13*(2), pp.136–156.

Park, J. and Gursoy, D. (2012), "Generation effects on work engagement among US hotel employees", *International journal of Hospitality Management*, *31*(4), pp. 1195-1202.

Park, N., Kee, K.F. & Valenzuela, S. (2009). Being immersed in social networking environment: Facebook groups, uses and gratifications, and social outcomes. *CyberPsychology Behavior*, *12*(6), pp. 729–733, DOI: 10.1089/cpb.2009.0003

Perception Media. (2015). *Media Planning Guide Malaysia 2015*. Kuala Lumpur: Perception Media.

Qualman, E. (2014). Socialnomics: How social media transforms the way we live and do business. Retrieved from http://www.socialnomics.net/2010/05/05/social-mediarevolution- 2-refresh/

Raacke, J. & Bonds-Raacke, J. (2008). MySpace and Facebook: Applying the uses and gratifications theory to exploring friend-networking site. *Cyberpsychology & Behavior*, *11*(2), pp.169-174, DOI: 10.1089/cpb.2007.0056.

University of Washington. (2017, April 26). Food photos help Instagram users with healthy eating. *ScienceDaily*. Retrieved August 17, 2017 from www.sciencedaily.com/releases/2017/04/170426130410.htm

Wang, E. and Chou, N. (2016) 'Examining social influence factors affecting consumer continuous usage intention for mobile social networking applications', *International Journal of Mobile Communications*, *14*(1), pp.43–55.

West, R. L., & Turner, L. H. (2010). Uses and gratifications theory. *Introducing Communication Theory: Analysis and Application*, 392-98.

Whiting, A., & Williams, D. (2013). Why people use social media: a uses and gratifications approach. *Qualitative Market Research: An International Journal*, *16*(4), 362-369.

Williams, J. P., & Mohamed Nasir, K. (2016). Muslim girl culture and social control in Southeast Asia : Exploring the hijabista and hijabster phenomena. *Crime Media Culture*, (January), pp.1–18. http://doi.org/10.1177/1741659016687346.

Wilson, T. (2012).What can phenomenology offer the consumer?:Marketing research as philo- sophical, method conceptual. *Qualitative Market Research: An International Journal*, *15*(3), pp.230–241. doi: 10.1108/13522751211231969.

Wilson, T. (2016). Hermeneutics. *The International Encyclopedia of Communication Theory and Philosophy*, pp.1–9. http://doi.org/10.1002/9781118766804.wbiect009

Stakeholders Perspective on Communicative Competence in Industry 4.0: Walk the Talk of Informative Technologists

Ena Bhattacharyya[1,*]

[1]21-03-26 Management and Humanities Department, Bandar Seri Iskandar, 32610 Tronoh, Perak Darul Ridzuan, Malaysia

Abstract. In facing the challenges of era 4.0, graduates are undoubtedly faced with language and communicative competency struggle to ensure that there is ease of transition from that of the academia to the workplace. This transition would entail innovative insight in language learning to equip graduates with apt technical and non-technical competencies for purposes of employability and work readiness. Graduates will need to familiarize themselves with the professional environment and be familiar with the language of the community of practice. To be considered a legitimate member of a community, the situated learning theory postulates that continued participation and engagement over a period of time with a community, provides one the ability to be considered as member of a community. For this purpose, a mixed method research was conducted in a private university to obtain the views of graduates and employers involved in student internship programs at an Information Technology (IT) company. Respondents were required to provide their feedback on communicative competence expected of interns for employability and work readiness. The study aims to share the communicative competencies and struggles as well as pedagogical implications expected of IT graduates in Industry 4.0. to be deemed employable and work ready.

Keywords: Communicative Competence; Technical Oral Presentation; Industry 4.0; Education; Employability; Work Readiness

1 Introduction

The fact that supply of human capital exceeds employers' demands casts a grave concern over the employability of our graduates. This concern is further aggravated by the

* Corresponding author: ena_bhattacharyya@utp.edu.my

mismatch between academia and industry driven skills requirement which calls for a re-look at aligning human capital skills with nation-building efforts. Scores of studies affirm such mismatch and the need to align graduate attributes with employment requirements (Bridgstock, 2009; Rosenberg, Heimler, & Morote, 2012). Cappelli (2008) makes mention that of the interdependence between employer-employees and the society at large where employers require affordable skills at specific moments of the industry, while employees seek career advancement and human capital development in the society. For this to occur, skills need to match industry requirement which must be translated to the community.

With globalization and the rapid technological borderless competition, it is without a doubt that employer's expectation of employability skills far differs from that of just mere technical skills (Fackler & Funabashi, 2018). Studies on the term employability skills is closely related and frequently associated to similar skills such as generic capabilities, transferable skills, basic skills, essential skills, work skills, soft skills, core skills, core competencies and enabling skills or even key skills (Zaharim, Yusoff, Omar, Mohamed & Muhamad,2009). Yet, there is limited literature to indicate the expected set of core skill requirement required for an industry. Different skills are deemed important by different employers in today's industries (Robles, 2012).

What was considered as important employability skills or work related skills in the 80's has transformed over the centuries to more high skill demands in the advanced world (O'Neil & O'Neil, 2014). Although it is reported that an estimated number of 100,000 graduates leave Higher Education Institutions (HEIs) every year (Singh, Thambusamy & Ramly, 2014), yet the alarming rate of unemployed graduates leaves many to question the employability and work readiness preparedness of the said graduates. The Department of Statistics, Malaysia (2017) indicates the rate of unemployment at 3.3% which means that graduates may not obtain immediate career placement upon graduation.

Universities and education training needs to be improve in order to lessen the gap between current employer needs and training that the graduates receive (Hernández-March, Martín del Peso & Leguey, 2009). Employers express the need for graduates to be more confident, critical and creative in competitive global work environment. Universities need to produce graduates to meet employer needs in the industries. The exposure and teaching of soft skills requirement need to be aligned to employer needs to ensure employability.

One particular communicative event that occurs in workplace is technical oral presentations or final year project. It is this communicative event that will be investigated to comprehend stakeholders' perceptions of communicative competence necessary to gain employability and work readiness befitting of graduates in era 4.0.

2 Literature Review

One of the essential soft skill expected of any employer is the need to communicate successfully in the workplace environment (Kay & Greenhill, 2011). Communication skills are acknowledged as a fundamental and imperative 21st century skill which propagates the success or failure of sustainable nation-building negotiations and efforts (Gokool-Ramdoo & Rumjaun, 2017; Kereluik, Mishra, Fahnoe & Terry, 2013). In the effort to ensure continued sustenance of human capital, mastery of the English language as the language of the global world, is essential to indicate ones' readiness to be a member of the professional community, i.e. professional engineering community (Arkoudis, Hawthorne, Baik, Hawthorne, O'Loughlin, Leach & Bexley, 2009).

One needs to be able to participate effectively as a member of the said community of practice in various types of oral and written communicative events, such as technical oral presentations, business meetings and sales presentations that take place in an organization

(Lave & Wenger, 1991). In the context of workplace communicative competence, graduates not only need to exhibit confidence in the mastery of the technical jargon but possess the ability to deliver, receive information and critique to and from an audience based on current scientific evidence (Celce-Murcia, 2007; Dannels, Gaffney & Martin, 2008; Bhattacharyya, 2014).

Graduates need to exhibit confidence in technical knowledge as well as other soft skill traits which attributes to effective workplace participation. An understanding of communicative competence requirements in technical oral presentations stems from the lack of linguistic and rhetorical competency studies required among English as Second Learners (ESL) in a Malaysian setting; employer discontent over graduates' apparent lack of communication skills and limited literature on linguistic and rhetorical features necessary in scientific and engineering oral presentations (Dannels, 2009; Hafizoah Kassim & Kassim Ali 2010; Morton, 2009). This becomes even more apparent in Industry 4.0 where the oral presentation requirements of today far differ from previous requirements. It is important that an investigation of such gap is conducted to resolve academia-industry practitioner gap over communicative competence requirements necessary for employability and work readiness of graduates in tomorrow's industry.

Workplace employability centers on the graduates' agility and ability to utilize his/her achievements skills, understandings and personal attributes that appeal to employers, to gain employment and be successful in their chosen occupations, which in turn benefits themselves, the workforce, the community and the economy (Yorke, 2008; Omar, Manaf, Mohd, Kassim & Aziz, 2012). What this means is the employee's flexibility and ability to utilize his own skills to meet the workplace demands as required by the employers of the company (Meister, Willyerd & Foss, 2010; Eisner, 2005; Crebert, Bates, Bell, Patrick & Cragnolini, 2004).

Clearly, the debate over what constitutes employability skills from the perspective of the academia and industry practitioners remains a struggle yet to be resolved. Workplace readiness, on the other hand, is a relatively new concept in predicting graduate potential. To Caballero, Walker, & Fuller-Tyszkiewicz (2011) work readiness is exemplifies the desired degree that graduates actually possess the required attitudes and attributes that enable them to be ready to face the real workplace environment. For graduates to be deemed as being work ready, one exhibits the potential, ability and rigor to overcome challenges in the workplace and are able to show their ability, success and career progression in their jobs.

2.1 Communicative Competence Framework and Situated Learning Theory

In relation to gaining employability and acquiring work readiness of graduates, the theory of communicative competence (Hymes, 1964) and theory of situated learning (Lave & Wenger, 1991) aptly looks at communication practiced within a group of practitioners who share common knowledge on discourse practices in a community of practice.

Communicative competence essentially views the act of communication "as a relative and dynamic, interpersonal construct among individuals with focus on functionality and adequacy of communication, sufficiency of knowledge, judgment, and skills in linguistic competence, operational competence, social competence, and strategic competence (Light, 1989 as cited in Bhattacharyya, 2014). Grammatical competence looks at the syntactic and semantic competencies in the use of the tools of communication; while social and strategic competencies reflect functional knowledge and judgement in interaction" (Light, 1989 cited in Bhattacharyya, 2014).To Hymes, appropriacy of every grammatically correct utterance is taken into consideration as human interaction occurs in a heterogeneous speech community.

The theory stipulates the need for a combination of competency requirement such as grammatical competence, socio-linguistic competence, strategic competence as well as discourse competence, as essential linguistic competency requirement for graduates to utilize such knowledge and skills in its appropriate context, to be considered communicatively competent. This is clearly illustrated in the framework in Figure 1 on the Theory of Communicative Competence (Canale & Swain, 1980).

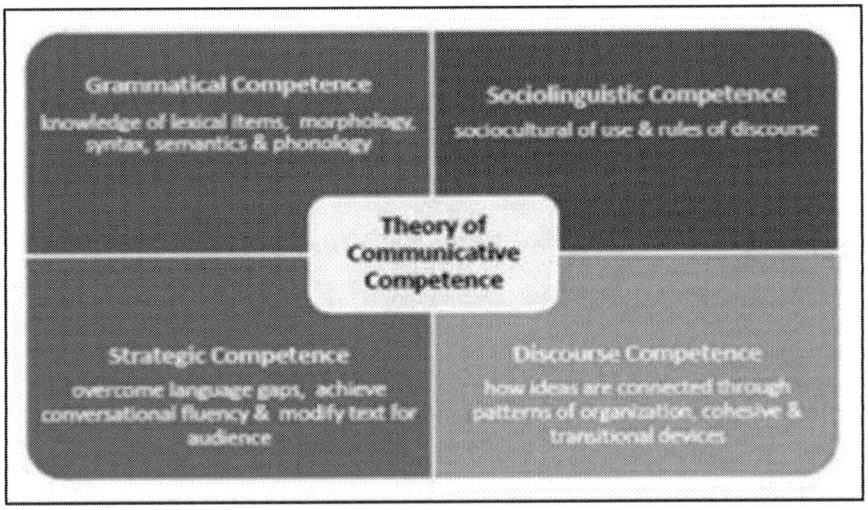

Fig. 1. The theory of Communicative Competence (Canale & Swain, 1980)

The origins of communicative competence as stipulated in the above Figure 1 clearly illustrate the emphasis on four different linguistic competences. However, we ask ourselves, if these competencies are sufficient in today's working world and Industry 4.0 era? What other communicative competencies are necessary in today's world of work to be considered communicatively competent? Are the above competencies sufficient to ensure employability and work readiness of graduates? What other competencies are then deemed necessary in era 4.0?

To understand the notion of communicative competence for graduates' employability and work readiness in a community of practice, it is necessary to investigate if there exist other competencies which challenge graduates to rise above the expected norm of communicative competence as propagated by Hymes (1964).

3 Methodology

The methodology utilized in this study is that of mixed method research design. In this study, a quantitative and qualitative analysis is carried out to extrapolate a general consensus and followed by qualitative findings of selected respondents. For this study, a questionnaire was distributed among 207 respondents who comprised of Information Technology and Engineering Interns in various local, regional and multinational companies affiliated as industrial intern partners of the research university. Qualitative interview was further conducted at the collaborating Information Technology (IT) Industrial Internship Company which was mentoring and supervising 6 IT students who were attached as interns.

The selected respondents in the qualitative study included employers who were also intern supervisors to the respective interns in the company.

These interns were attached for a duration of 8 months as part of the graduation requirement of the degree program. Ethical consent was obtained from the respective Heads of Department within the research university as well as the Manager of the Industrial Internship Company which was training the interns. The objective and the purpose of the study was clarified to the said personnel prior any data collection. It was also clearly stipulated that data collected was to be utilized only for academic publication purposes. The mixed method was employed as both quantitative and qualitative methods seek to complement each other and make the research more credible when they are effectively used (Creswell, 2008; Johnson & Onwuegbuzie, 2004). Thus, it is for this reason that a mixed-method design was employed to obtain views with selected participants on communicative competence requirement necessary in technical oral presentations or final year project to meet employability and work readiness demands for Industry 4.0.
lace the figure as close as possible after the point where it is first referenced in the text. If there is a large number of figures and tables it might be necessary to place some before their text citation.

4 Quantitative and Qualitative Findings and Discussion

The summary for the mean rating on communicative competence criteria exhibited in final year project is illustrated in Table 1. The average mean score indicates that the graduates rated the communicative competence criteria exhibited in final year project as being relatively high. For example, the results can be interpreted to mean that the graduates are aware of the importance to "understand the purpose of the presentation" (M=1.73) or "apply suitable formal salutation or greetings to the audience" (M=1.76).

Table 1. Communicative Competence Criteria Exhibited in Final Year Project

COMMUNICATIVE COMPETENCE CRITERIA EXHIBITED IN FINAL YEAR PROJECT	MEAN	MEDIAN	STD. DEVIATION
Understand the purpose of the presentation	1.73	2	0.625
Understand the scope of the presentation	1.79	2	0.633
Apply the correct organization pattern. i.e.: problem-solution order, cause-effect	2.04	2	0.652
Apply suitable formal salutation or greetings to the audience	1.76	2	0.675
Analyse the presentation material prior the real presentation	1.88	2	0.715
Memorize the points of the presentation	2.15	2	0.854
Anticipate the possible type of questions asked	2.15	2	0.856
Clarify any technical terms used in the presentation	2.04	2	0.775

Table 2 show the mean score of communicative competence response in final year project for workplace readiness which shows a high score. Some of respondents stated that during final year projects, respondents need to "defend ideas when questioned" (M=1.89). They also realise the importance to "listen attentively to questions posed by the audience"

(M=1.68). In other words, to a competent communicator of Industry 4.0, one not only defends but also listens to the audience. All of the stated items within the said construct need to be practiced by the graduates to be considered competent in era 4.0.

Table 2. Communicative Competence Response in Final Year Project for Workplace Readiness

COMMUNICATIVE COMPETENCE RESPONSE IN FINAL YEAR PROJECT FOR WORKPLACE READINESS	MEAN	MEDIAN	STD. DEVIATION
Be analytical when answering the audience queries	2.13	2	0.772
Be flexible in answering the audience's questions	1.96	2	0.726
Be courteous to the audience	1.95	2	0.698
Give time to the audience to reflect on the topic	1.98	2	0.797
Invite the audience to ask questions	2.02	2	1
Listen attentively to questions posed by the audience	1.68	2	0.636
Incorporate audience feedback for further improvement	2.1	2	0.878
Respond to sudden unexpected queries	2.12	2	0.776
Defend ideas when questioned	1.89	2	0.705
Total Mean	1.9791	2	0.50192

In terms of language competence, Table 3 showed the mean score of Language Competence in Final Year Project for Workplace Readiness. All of elements indicated the high score and illustrated that they are highly presenting workplace readiness elements in final year project presentations. However, element of "use correct grammar at all times" acquired M=2.43 which means nearest to "undecided" scale value. Therefore, the responsible party need to more attention on this element in order to improve workplace readiness.

Table 3. Language Competence in Final Year Project for Workplace Readiness

LANGUAGE COMPETENCE IN FINAL YEAR PROJECT FOR WORKPLACE READINESS	MEAN	MEDIAN	STD. DEVIATION
Use correct grammar at all times	2.43	2	0.982
Use appropriate language throughout the delivery	1.92	2	0.823
Ensure language used is easily understood	1.78	2	0.687
Avoid using complex language	1.86	2	0.736
Articulate the words well	2.15	2	0.827
Enunciate the words clearly	2.15	2	0.833
Pronounce words and names clearly	2	2	0.815
Choose the appropriate words	1.89	2	0.787
Total Mean	2.0223	2	0.62769

The mean score of Presenting Workplace Readiness Elements in Final Year is indicated in Table 4. The findings denote that there is a need to improve in the said items related to work readiness component. To be competently ready for an industrialized era, there needs to be emphasis critical and creative analysis, possess presentation skills necessary for global diverse type environment, team spirit as well as being computer literate. Graduates of era 4.0 need to be digitally literate to be considered employable and work ready.

Table 4. Presenting Workplace Readiness Elements in Final Year Project

PRESENTING WORKPLACE READINESS ELEMENTS IN FINAL YEAR PROJECT	MEAN	MEDIAN	STD. DEVIATION
Present goal-relevant activities, rank them, allocate time, prepare and follow schedules	2.02	2	0.788
Present on budgets, makes forecasts, and deliberate on adjustments to meet objectives	2.32	2	0.839
Present information on assessment of skills and provide feedback	2.07	2	0.741
Present to group effort	2.02	2	0.753
Present and teach others new skills	2.22	2	0.917
Present and communicate ideas to justify position	2	2	0.766
Present in dynamic groups with men and women from diverse backgrounds	2.05	2	0.805
Present, interpret and communicate information	1.94	2	0.644
Present Information by use of computers	1.82	2	0.745
Involved in presenting trends, prediction of impacts on system operations, diagnoses deviations in systems' performance and corrects malfunctions	2.34	2	0.855
Presents suggestions on modifications to existing systems	2.07	2	0.757

and develops new or alternative systems to improve performance			
Presents information on identification, or provide solutions on problems with equipment, including computers and other technologies	2.04	2	0.793
Total Mean	2.0765	2.0833	0.5653

The above findings as indicated in the mean score generally denote that there is a need for improvement to be considered viable graduates of industry era 4.0. In general, these findings indicated that in the current age of Industrial revolution 4.0 demand, the language and communicative competencies for workplace readiness of graduates' today's is relatively high. The respondents perceived the importance of language competency to be relatively high while the work readiness at this stage is average which needs to be improved in order to prepare graduates for better workplace readiness.

5 Thematic Findings

The managers provided various insights from the employers' perspective to ensure employability and work readiness. The themes derived from the findings include the importance of global competitiveness, self-accountability and ownership, time management, sense of alertness and self-reliance, accountable, self-motivated, self-driven and compassionate when assigned to tasks.

Theme 1: Competitive Global Market Player

From the perspective of an employer, trainings are provided to staff in order for staff to be a global competitive player, and be ahead of competitors. With a borderless world in the IT industry, staff must be value added with necessary technical and non-technical expertise (Rasul, Rauf, Mansor & Puvanasvaran, 2012). Manager A mentioned,

"…We have now, you know competitors, so we need to make sure that our talent is valued and the investors are willing to come because we have competent talents…"

The employer shares the knowledge where neighboring competitors in China provide training in English language programs as part of the company's need to be communicatively competent. For the said company, the seriousness of being a global market player is indicated when a walk around the company premise indicated the presence of a "corporate academic university" and training rooms available at one of the levels of the company. Staffs are provided with daily training sessions in various skills competency programs. The employer's sentiments are mirrored in the government's nation building initiative of developing human capital (Ministry of Education Malaysia, 2012; Gibson, 2015). The willingness to equip staffs with up-to-date training and skills competencies is necessary for companies in era 4.0.

Theme 2: Ecosystem for Talent Sourcing and Mobility

The sustenance of employability and work readiness in era 4.0 is not only instrumental among the manpower but that of the organizational set up. Employers are vigilant of the staff's talents and address such competencies via mutual discussion. This means there is a need for both employers and graduates to have a mutual and open discussion on matching the capabilities with rightful job specifications. Manager B mentions,

"…Talents are recognized. Mutual discussions are held so that staffs are rightfully matched to the area of interest. The right ecosystem is set. There is certification for in-house training courses…"

This clearly mirrors the company's commitment to equipping its staff with the required competencies so that staffs enjoy higher remuneration in relation to the skill competency. To be employable, both employers and graduates or staffs need to match skills with specific interest areas to be able to create staff mobility within departments and outside of the company (Qomariyah, Savitri, Hadianto & Claramita, 2016).

Theme 3: Personality Trait: Self-Motivation and Self-Driven

Employers gauge the interns' employability and work readiness for era 4.0 when staff (inclusive of interns and fulltime staff) take ownership of the task delivered. This is reflected in Manager C's response who stated the criteria of work commitment and ownership of the individual to the said task. This is clearly indicated if the individual invested in the task productively or otherwise. Manager C mentions the importance of employability and work readiness is judged by the individual's

"…work knowledge, ability, work habit, performance, you can see on the knowledge ability, job knowledge, whether its minimum, or over work productivity, the work quality,…whether they are really productive or not productive…"

Thus, the employability and work readiness of an individual is gauged by his or her well preparedness of the individual in taking ownership of acquiring more knowledge and performing productively on the task (Billett, 2001; Helyer & Lee, 2014; Adnan, Daud, Alias & Razali, 2017).

Theme 4: Accountability and Time-Management Skills

For Manager D, part of employability and work readiness ethic requirement is the importance for individuals

"to note the urgency and the importance to deliver within the timeline…if a task is delivered to the individuals on a certain time, we expect the graduates or staff to complete it urgently as clients' demand solutions for their problems. This is critical for graduates to understand…"

To Manager D, time management is crucial as it companies revolve around datelines. The essence of time management is concurred in other employability studies (Moore & Morton, 2017). Pedagogically, there is a need to be timely and accountable for responsibilities as individuals and in teams (Burrus, Jackson, Xi & Steinberg,2013).

Theme 5: Goal Driven and Result Oriented

To be employable and work ready, an individual must be pro-active, talented and passionate about what he or she does. For Manager E,

"...one must be passionate as we are a business entity, you must indicate the passion...to get things done using your skill related to the software for sure..."

This finding concurs with employability studies which states the importance of passion and motivation of graduates to attain gainful employment and ensure career success (Hurst, Fowler & Scapens, 2017). Pedagogically, one needs to have a sense of business acumen. Thus, speed and result driven analysis are important competencies expected of era 4.0. The outcome of the results from the study indicate the need to beef up on added technical and non-technical skills necessary for gainful employment and work readiness in era 4.0. As technological innovation occurs at a rapid pace, there is a need to improve and equip graduates with the apt skills. For this to materialize, educators need to be aware and equip graduates with skills of the digital era.

6 Conclusion

To be considered viable and competent as indicated by the findings, it is clear that graduates of today cannot solely be assured by linguistic competence. More importantly for one to be considered employable and work ready, various other attributes and skills have been perceived as determinants in era 4.0. It is skills such as adaptability, critical and innovative entrepreneurial mind-set, accountability, goal driven and passionate are some of the many skills deemed pertinent to be employable and work ready. Both educators and industry practitioners need to be engaged in more mutual learning platforms that share mutual input on employability skills and work readiness. The synergy between academia-industry partnership platforms need to be enhanced to meet the demands of era 4.0 and beyond. Through such methods, pedagogical innovation in classroom input on employability and work readiness will gradually be enhanced and thus meet industry demands to allow graduates to eventually walk the talk of industry practitioners.

The researcher wishes to acknowledge the funding provided by the University Internal Research Funding and sincere thanks to all respondents involved in the study.

7 References

Adnan, Y. M., Daud, M. N., Alias, A., & Razali, M. N. (2017). Importance of soft skills for graduates in the real estate programmes in Malaysia. *Journal of Surveying, Construction and Property, 3*(2).

Arkoudis, S., Hawthorne, L., Baik, C., Hawthorne, G., O'Loughlin, K., Leach, D., & Bexley, E. (2009). The impact of English language proficiency and workplace readiness on employment outcomes and performance of tertiary international students. *Melbourne: Centre for the Study of Higher Education, University of Melbourne.*

Baird, A. M. (2016). *21st Century Workforce: Employers' Ratings of Importance and Satisfaction of Skills and Competencies College Graduates Need to Get Hired in the Northeast* (Doctoral dissertation, Johnson & Wales University).

Bhattacharyya, E. (2014). A case study of stakeholder perceptions on communicative competence in engineering technical oral presentation (Doctoral dissertation, University Malaya).

Bridgstock, R. (2009). The graduate attributes we've overlooked: Enhancing graduate employability through career management skills. *Higher Education Research & Development*, *28*(1), 31-44.

Burrus, J., Jackson, T., Xi, N., & Steinberg, J. (2013). *Identifying the most important 21st century workforce competencies: An analysis of the Occupational Information Network (O*NET* (Research Report No. RR-13-21). Princeton, NJ: Educational Testing Service.

Caballero, C. L., Walker, A., & Fuller-Tyszkiewicz, M. (2011). The work readiness scale (WRS): Developing a measure to assess work readiness in college graduates. *Journal of teaching and learning for graduate employability*, *2*(2), 41-54.

Canale, Michael, & Swain, Merrill. (1980). Approaches to communicative competence (pp. 63). Singapore: SEAMEO Regional Language Centre.

Cappelli, P. (2008), "Talent management for the twenty-first century", Harvard Business Review, Vol. 86 No. 3, pp. 74-81.

Celce-Murcia, M. (2007). Rethinking the role of communicative competence in language teaching. In Eva Alcón Soler & Maria Pilar Safont Jordà (Eds.), *Intercultural Language Use and Language Learning* (pp. 41-57). Dordrecht, The Netherlands: Springer.

Crebert, G., Bates, M., Bell, B., Patrick, C. J., & Cragnolini, V. (2004). Developing generic skills at university, during work placement and in employment: graduates' perceptions. *Higher Education Research & Development*, *23*(2), 147-165.

Creswell, John W. (2008). *Educational Research: Planning,conducting and evaluating quantitative and qualitative research* (3rd ed.). Upper Saddle River, New Jersey: Pearson Prentice Hall.

Dannels, Deanna P. (2009). Features of success in engineering design presentations: A call for relational genre knowledge. *Journal of Business and Technical Communication, 23*, 399-427.

Dannels, Deanna P., Gaffney, Amy , & Martin, Kelly. (2008). Beyond content, deeper than delivery: What critique feedback reveals about communication expectations in Design Education. *International Journal for the Scholarship of Teaching and Learning, 2*(2), 1-16.

Department of Statistics, Malaysia (2017), Key Statistics of Labour Force in Malaysia, November 2017, Ministry of Human Resource.

Eisner, S. P. (2005). Managing generation Y. *SAM Advanced Management Journal*, *70*(4), 4.

Fackler, M., & Funabashi, Y. (Eds.). (2018). *Reinventing Japan: New Directions in Global Leadership*. ABC-CLIO.

Gibson, J. A. (2015). Reimagining the Rhetorical Canons for Professional Communication Pedagogy.

Gokool-Ramdoo, S., & Rumjaun, A. B. (2017). Education for sustainable development: Connecting the dots for sustainability. *Journal of Learning for Development-* JL4D, 4(1).

Hafizoah Kassim, & Fatimah Ali. (2010). English communicative events and skills needed at the workplace: Feedback from the industry *English for Specific Purposes, 29*(3), 168-182.

Hernández-March, J., Martín del Peso, M., & Leguey, S. (2009). Graduates' skills and higher education: The employers' perspective. *Tertiary education and management*, *15*(1), 1-16.

Helyer, R., & Lee, D. (2014). The role of work experience in the future employability of higher education graduates. *Higher Education Quarterly*, *68*(3), 348-372.

Hurst, C., Fowler, J., & Scapens, G. (2017). Sustainable employability in higher education: Career development outside of the curriculum. In *Success in Higher Education* (pp. 217-228). Springer, Singapore.

Hymes, Dell. (1964). Introduction: Toward ethnographies of communication. *American Anthropologist, 66, Number 6* (Part 2), 1-34.

Johnson, R.B., & Onwuegbuzie, A.J. (2004). Mixed methods research: A research paradigm whose time has come. *Educational Researcher, 33*(7), 14 - 26

Kay, K., & Greenhill, V. (2011). Twenty-first century students need 21st century skills. In *Bringing schools into the 21st century* (pp. 41-65). Springer Netherlands.

Kereluik, K., Mishra, P., Fahnoe, C., & Terry, L. (2013). What knowledge is of most worth: Teacher knowledge for 21st century learning. *Journal of Digital Learning in Teacher Education, 29*(4), 127-140.

Lave, J., & Wenger, E. (1991). *Situated learning: Legitimate peripheral participation*. Cambridge, UK: Cambridge University Press.

Meister, J. C., Willyerd, K., & Foss, E. (2010). *The 2020 workplace: How innovative companies attract, develop, and keep tomorrow's employees today* (p. 5). New York, NY: Harper Business.

Ministry of Education Malaysia. (2012). Malaysia education blueprint 2013–2025.

Moore, T., & Morton, J. (2017). The myth of job readiness? Written communication, employability, and the 'skills gap' in higher education. *Studies in Higher Education, 42*(3), 591-609.

Morton, Janne. (2009). Genre and disciplinary competence: A case study of contextualisation in an academic speech genre. *English for Specific Purposes, 28*, 217-229.

Omar, N. H., Manaf, A. A., Mohd, R. H., Kassim, A. C., & Aziz, K. A. (2012). Graduates' employability skills based on current job demand through electronic advertisement. *Asian Social Science, 8*(9), 103.

O'Neil Jr, H. F., & O'Neil, H. F. (2014). *Workforce readiness: Competencies and assessment*. Psychology Press.

Qomariyah, N., Savitri, T., Hadianto, T., & Claramita, M. (2016). Formulating Employability Skills for Graduates of Public Health Study Program. *International Journal of Evaluation and Research in Education, 5*(1), 22-31.

Rasul, M. S., Rauf, R. A. A., Mansor, A. N. & Puvanasvaran, A. P. (2012). Employability skills assessment tool development. *International Education Studies, 5*(5), 43.

Robles, M. M. (2012). Executive perceptions of the top 10 soft skills needed in today's workplace. *Business Communication Quarterly, 75*(4), 453-465.

Rosenberg, S., Heimler, R., & Morote, E. S. (2012). Basic employability skills: a triangular design approach. *Education+ Training, 54*(1), 7-20.

Singh, P., Thambusamy, R. X., & Ramly, M. A. (2014). Fit or unfit? Perspectives of employers and university instructors of graduates' generic skills. *Procedia-Social and Behavioral Sciences, 123*, 315-324.

Yorke, M. (2008). Employability in higher education: What it is and what it is not. *Higher Education Academy*: ESECT.

Zaharim, A., Yusoff, Y., Omar, M. Z., Mohamed, A., & Muhamad, N. (2009, July). Engineering employability skills required by employers in Asia. In *Proceedings of the 6th WSEAS international conference on Engineering education* (pp. 195-201).

Non-linguistics Problems in Teaching of Indonesian Language

Rustam Effendi[1], and *Fatchul* Mu'in[2*]

[1]Lambung Mangkurat University, Department of Language and Arts Education, Jalan Brigjend H. Hasan Basry-Banjarmasin 70123 Indonesia
[2]Lambung Mangkurat University, Department of Language and Arts Education, Jalan Brigjend H. Hasan Basry-Banjarmasin 70123 Indonesia

Abstract. There are two kinds of problems in the teaching of a language, namely: linguistic and non-linguistic problems. These two problems must be handled in balance so that the teaching of language can reach the objective optimally. It is a fact that the non-linguistic problem is not yet seen as a problem in the teaching of a language. Every strategy of a language teaching is done entirely in order to handle the linguistic problems. The consequence of strategies which neglect non-linguistic problems is a less satisfactory result of the language teaching. There are some non-linguistic problems in the teaching of Bahasa Indonesia, some of them are elaborated in this paper, namely: (i) the interlanguage contact in Indonesia, (ii) the attitude of Indonesian people to Bahasa Indonesia, (iii) feeling that Bahasa Indonesia has been spoken well, (iv) legal sanction to those who look down upon Bahasa Indonesia is unavailable yet. (v) teaching materials, especially teaching materials of other than Bahasa Indonesia pay less attention to good and right language, and (vi) not many people can be followed as example in speaking Bahasa Indonesia.

Key Words: linguistic problems, non-linguistic problems, language teaching

1 Introduction

There are two problems on language teaching, namely: problems related to linguistics (linguistic problems) and non-linguistics (non-linguistic problems). The linguistic problems are problems which exist or come from inside the language itself. Non-linguistic problems are the problems that derive from the outside of the language, but come from outside the language or language external factors.

In simple way we say that linguistics is defined as the scientific study of language. Linguistics or language science is always related to the rules or principles of a language.. From different viewpoints, as a science, linguistics can be divided into several branches, among others, descriptive linguistics and historical/comparative linguistics (if it is based its methodology), synchronic and diachronic linguistics (if is based on its aspect of time), and phonetics, phonology, morphology, syntax and semantics (if it is based on a language as a system).

* Corresponding author: muin_sihyar@yahoo.com

A language itself refers to a systematic sound symbols (see Soeparno, 2013; Chaer, 1994). The more complicated definition of language can be seen in the following. Language is a system of arbitrary vocal symbols, which permit all people in a given culture, or other people who have learned the system of that culture to communicate or to interact (Finocchiaro, 1964).

Related to that, a language is not just spoken or written as the speaker or writer like. The speaker or writer cannot assume that the language that he or she uses can always be understood by other people as the listeners and/or the readers. The expected language is a good and right language. Good because the language is suitable with the situation when it is being used and right because the language is suitable to the grammatical principles of the language (see Alwi, Dardjowidjojo, Lapoliwa, & Moeliono, 2000).

A language is not only studied based on the internal language itself but also based on the external perspective. From the internal side, it is discussed based on its internal structures of the language; while from the external perspective, the study is based on the linguistic perspectives related to the non-linguistic phenomena influencing the use of language. A study of internal language structures (or, it is based on the sub-systems of a language) will result sub-discipline of linguistics such as phonetics, phonology, morphology, syntax and semantics. It is conducted through theories and procedures belonging to the discipline of linguistics; it is not related to the problems beyond the language (Mariani & Mu'in, 2006).

Making people speak well and right happens in schools through the teaching of *Bahasa Indonesia*. Children in schools are taught the standard Bahasa Indonesia, the language which language systems have been standardized. The standardization of language in Indonesia comprises the standardization of the vocabulary by which *Kamus Besar Bahasa Indonesia* (The Dictionary of Bahasa Indonesia) is the reference of the standard vocabulary, the spelling of Bahasa Indonesia by which *Pedoman Umum Ejaan Bahasa Indonesia, Edisi Keempat* (the General Guidelines of Bahasa Indonesia Spelling, Fourth Edition) as the spelling reference, and standardized grammar by which *Tata Bahasa Baku Bahasa Indonesia,* (the Standardized Indonesian Grammar) by Hasan Alwi and friends as the reference for the formation of standard words and sentences.

All the instruments for the language standardization above are used as teaching materials for the teaching of *Bahasa Indonesia* at schools. The objective of *Bahasa Indonesia* teaching is the use of good and right (standard) language use. The teaching of Standardized Indonesian Language (*Bahasa Indonesia*) cannot be reached easily without extraordinary efforts by teachers and students. The good and standardized language with several systems intended becomes linguistic problems in the teaching of Indonesian Language (*Bahasa Indonesia*). Students must know and master the standardized vocabulary, diction, correct pronunciation, correct word formation, the use of phrases, clauses and correct sentences, and all other things related to the language systems.

The main objective of the learning of Indonesian Language (*Bahasa Indonesia*) from the elementary to senior high schools is the language skills. Related to this objective, the learning of Indonesian Language *(Bahasa Indonesia)* is concentrated on the learning of listening, speaking, reading, and writing skills (see Tarigan, 1994). The expected result from the learning of all the language skills is that students have skills in using oral as well as written language well and correclt. It means, although the learning objective and the implementation looks like neglecting the grammatical aspect, the final objective of all the language skills is that students are skillful in using Indonesian Language *(Bahasa Indonesia)* with correct system of the language. For this objective, the learning material for listening, speaking, reading and writing must be in the forms of linguistically guaranteed sentences and/or texts.

This paper does not talk about the linguistic problems as what has been stated above. This paper wants to explain non-linguistic problems, problems of language teaching which come from outside the language itself. Non-linguistic problems mean talking about language teaching problems without relating it with the lingulistic principles. This paper has been written based on the result of research in which the writers used the qualitative research approach. Denzin & Lincoln (1994:2) said, "Qualitative research is multimethod in focus, involving an interpretive, naturalistic approach to its subject matter. This means that qualitative research study in their natural setting, attempting to make sense of interpret phenomena in terms of the meanings people bring to them. Qualitative research involves the studied use and collection of a variety of empirical materials—case study, personal experience, introspective, life story, interview, observational, historical, interactional, and visual text—that describe routine and problematic moment and meaning in individuals'lives.

Data were collected through listening, observation, introspection, and documentation. The researchers conducted the research : (1) through listening and observing the speakers' talks in the various language activities, such as at the office, school, and drinking shops, (2) through documentation based on which we collected data using documents, such as newspapers, papers, advertisements posted on the roadsides, and so on (see Mahsun, 2011), and (3) through introspection by which we utilized the language intuition to obtain data (Chelliah, 2011: 400).

There are some problems in the learning of non-linguistics of learning Bahasa Indonesia. Among the non-linguistic problems which will be explained in this paper are: (i) contact among languages in Indonesia, (ii) the attitude of Indonesian people to Bahasa Indonesia, (iii) feeling that Bahasa Indonesia has been mastered and spoken well , (iv) there isn't legal sanction yet to those who look down upon Bahasa Indonesia is of, (v) teaching material, especially for the teaching subjects other than the Bahasa Indonesia subject have less attention to good and right language, and less sample speakers.

2 Non-linguistic problems in teaching of Bahasa Indonesia

This part discusses about: (i) interlanguage contact among languages in Indonesia, (ii) the attitude of Indonesian people to *Bahasa Indonesia*, (iii) feeling that *Bahasa Indonesia* has been mastered and spoken well , (iv) there is no legal sanction yet to those who look down upon *Bahasa Indonesia*, (v) teaching material, especially for the teaching subjects beyond the subject of *Bahasa Indonesia* with less attention towards good and right language, and less samples of speakers.

2.1 Interlanguage Contact in Indonesia

At least there are 3 categories of languages living and developing in Indonesia, namely: (1) Indonesian Language *(Bahasa Indonesia)* as the national language and formal/official language, (2) local languages (such as *Bahasa Banjar* or Banjarese language and (Dayak language) as those that are living and developing in local areas or regions, being used for local communication and (3) foreign languages (such as English, Arabic, and Mandarin) as those that have been taught in schools since elementary school until senior high schools, even higher education (see Suherdi, 2012).

As a result of language contact between Indonesian and local language (e.g. Banjarese language), from day to day the latter language increasingly leads to death. Many vocabularies of Banjarese language are no longer known by the younger generation of Banjarese at the present time. Crystal, (2003: 79) points out, "... in which the younger generation is increasingly proficient in the new language. Identifying more with it, and finding their first language less relevant to their needs. "The fear of the local language death is also expressed by some linguists who attended the Forum of Language Diversities (the Asia-Europe Meeting) on 4-5 September 2012. One of them said "Strengthening the function of the Indonesia language in its capacity as the national language and official language, this condition indirectly placing local languages on the marginal position. In addition, most of the local language has a relatively small number of speakers and have no written tradition. Therefore, it is not surprising that the problem of extinction of local languages becomes serious problems faced by Indonesia (Mahsun, 2012:2).

In the age of 6 years old, the Indonesian children have known only one language, the language of the local region in which they live with their family. Banjarese children who live in Banjarmasin master Banjarese, Javanese who live in part of Javanese island master Javanese language, Madurese children in Madura island, Sundanese children in West Jawa master Sundanese language , etc. *Pusat Pembinaan dan Pengembangan Bahasa* (now *Badan Pengembangan dan Pembinaan Bahasa*) (the Center of the Developing and Founding of Language) inform that results of researches up to 2012, that there are 546 local languages in Indonesia. According to those researches, there are more local languages which are unidentified and it is estimated that the local languages in Indonesia can reach 600 local languages (see Sariono, 2016:4).

After the age of 6, when children enter the elementary schools, they are introduced to one language, Indonesia Language (*Bahasa Indonesia).* They try to master Indonesia Language because it has been used as a medium of instruction and mostly the textbooks as the sources of learning materials are written in Indonesian Language. When the mastery of local languages was obtained by the Indonesian children naturally as the result of language acquisition through their experience in using the language in their social life in their surrounding speech community, the mastery of Indonesia Language is obtained by the Indonesian children through education in schools (see Dardjowidjojo, 2016). That is to say that the local language is acquired, whereas Indonesian Language is learned. Indonesia Language is taught in schools since Education of Early Age (*Pendidikan Anak Usia Dini)* until Senior High Schools (SLTA). The total years of learning Indonesian Language experienced by Indonesian children are fourteen years (two years in PAUD, six years in the elementary school, three years in junior high school, and three years in senior high school).

English begins to be taught to the Indonesian children since they go to the Junior High School (for three years) until the Senior High School (for three years), so is Arabic, The Indonesian children who go to religious schools begin to learn Arabic since they go to Sanawiyah until Aliyah.

Based on the explanation above, all the Indonesian people who have education up to senior high schools are multilinguals. They do not only know their local language (mother tongue) but also know Indonesian Language and a foreign language (English, Arabic and/or other foreign language).

Indonesian Language (*Bahasa Indonesia*) has two functions, namely : the national language and the state language. The function of Indonesian Language as the National Language was obtained since *Bahasa Indonesia* was mentioned in the *Sumpah Pemuda* (Youth Pledge). *Sumpah Pemuda* was declared by the Indonesian Youth, who built Indonesian Nation in October 28 and the third pledge is about *Bahasa Indonesia. Bahasa Indonesia* is stated by the pledge as a language which was prepared to be the national language when Indonesia gets its independence (see Alwi, 2011; Alwi & Sugono, 2011). The content of the third pledge becomes the beginning of the proclamation of *Bahasa Indonesia* as the language of unity for the Indonesian nation. *Sumpah Pemuda* (Youth Pledge) written in *Ch. A. van Ophuijsen* spelling is as follows:

Kami poetra dan poetri Indonesia mengakoe bertoempah darah jang satoe,
tanah air Indonesia
Kami poetra dan poetri Indonesia mengakoe berbangsa jang satoe, bangsa

Indonesia

Kami poetra dan poetri Indonesia menjoenjoeng bahasa persatoean, Bahasa

Indonesia

The second status of *Bahasa Indonesia* is as the official language of the state of Republic of Indonesia. The status as the official language of the state was obtained by Bahasa Indonesia since it was written or recorded by the language builder of the Indonesian nation in the 1945 Constitution, Chapter XV, article 36, and by the statement of this article Bahasa Indonesia was stated and inaugurated as the state language.

Three languages that are in Indonesia (the local language, *Bahasa Indonesia*, the foreign language) are expected to keep their own status and function, do not influence each other and do not compete one another among languages. Seminar Politik Bahasa Nasional/ Indonesia (Seminar of National Language Politics) conducted in Jakarta from 25 to 28 of February, 1975 had formulated the status and function of Bahasa Indonesia, local language and foreign language in Indonesia so that languages in Indonesia do not show contact and can live together (see Hasil Perumusan Seminar Politik Bahasa Nasional, 1975). The formulation of the status and function of the languages are:

As the National Language, *Bahasa Indonesia* has its function as (1) the symbol of national pride, (2) the symbol of national identity, (3) as unifying force several different communities with different social cultural backgrounds and languages, and (4) connecting tool among cultures among regions.

As the state language, *Bahasa Indonesia* has the functions of (1) an official state language, (2) an official language for communication in education institutes, (3) an official language in national communication nationally for the planning and executing development and government administration, and (4) as an official language in cultural development and use of science and modern technology.

Local languages have the status of local languages. A local language is one of the elements of national culture protected by the government (Explanation of article 36, Chapter XV, 1945 Constitution. In the status of the local languages in Indonesia function as (1) symbols of local prides, (2) symbols of local identity, and (3) communication tools in the family and the local community. Related to the function of *Bahasa Indonesia*, the local languages also function as (1) supports to the national language (2) language of communication in the elementary schools in certain regions in the beginning, to accelerate the teaching of Bahasa Indonesia and other subjects, and (3) tool for the development and support local culture.

All languages other than *Bahasa Indonesia* and local languages (like English, Arabic and the others) are called foreign language. Foreign languages functions as (1) media communication among nations, (2) development tool to make possible for *Bahasa Indonesia* develops to a modern language, and (3) as tool to use in science and modern technology for national development.

With the status and function of languages above, it is expected that three languages living in Indonesia consistently make use of the formulation. With the status and function of the language, it is expected that the three languages living in Indonesia consistently use the formulation as a guide. Bahasa Indonesia does not take the function of the local language and/or the foreign language, and vice versa. All languages live with their nature directed by the national language politics.

National language political wills that all languages living in Indonesia can live side by side peacefully, cannot be fulfilled entirely. By the Indonesian community, being realized or not, Bahasa Indonesia is considered to be a more prestige than the local languages, This makes local languages, day by day, being chaos, buried and abandoned by their speakers because of the intervention of Bahasa Indonesia. Then, English, by some of the Indonesian community, English is considered to be more prestige compared to Bahasa Indonesia. English shows more and more power and goes to the social life of the community where Bahasa Indonesia should have its status and function. As the result of this contact, local languages become mixed language (seen from the mixed vocabulary) and there are worries that someday local languages will extinct because the vocabulary has been replaced by the vocabulary of the more prestige language, Bahasa Indonesia. So is bahasa Indonesia, although not as serious as local languages (because *Bahasa Indonesia* is always being guarded by *Badan Bahasa* (Language Authority), the consequence/result of the intervention of English to the status and function of *Bahasa Indonesia*, some people's loyalty of the Indonesian people will be crushed more and more. The aspiration of *Sumpah Pemuda* that all Indonesian people 'menjoenjoeng' the language for Indonesia will be more and more difficult to reach.

Languages supported by less people in the community, by experts, are called minority language and languages with more loyal users are called majority language. Majority languages tend to influence even kill the minority languages. The majority and minority language and cultural issues have become global issues. Gibbons and Ramirez stated, "A major challenge for linguistik minorities, whether indigenous, refugee or migrant, is the maintenance of their language and culture. In almost all societies around the world, the nationalist myths that societies are (or can be) homogenous culturally, linguistically and ethnically, have led to the overt or covert suppression and linguistic difference, and sometimes the 'cleansing' of ethnic difference -including genocide- (see Gibbons & Ramirez, 2004)."

Among the proofs that there is contact between Bahasa Indonesia and English is the great number of articles written by Indonesian scientists using English, even if the writings are put in journals in Indonesia. It also the case in several journals with articles in several languages (like Bahasa Indonesia, English, and Arabic), at least the writer of the article must write the abstract of the article in English. Actually the Indonesian Law No. 24, 2009 Chapter III, article 35 has prohibited the use of foreign languages in writing and publishing articles.

Language contact seen from outside education field has very big impact in the teaching and learning Bahasa Indonesia. Some teachers, moreover those who do not teach Bahasa Indonesia, do not care to give help in reaching the objective of teaching and learning Bahasa Indonesia. At schools, actually teachers and students must use Bahasa Indonesia; but some of them prefer to use the local language. In the classroom, consciously or not, often use the local language to explain. Some schools also open special class using English as the medium of communication to explain the subject. This situation makes the objective of the teaching and learning Bahasa Indonesia is difficult to reach.

2.2 Speakers' Attitude to Bahasa Indonesia

The non-linguistic problem of the teaching of Bahasa Indonesia other than contact among languages is the attitude of the speakers of the language, like considering certain language lower than the other prestige languages. It is fact that cannot be denied some Indonesian people consider Bahasa Indonesia lower than foreign languages especially English. They are less loyal to their own language. This unloyalty or negative attitude can be seen in the commercials which are put along the street. Almost 90% of the commercials use English or local language, although the words can be changed to Bahasa Indonesia.

The unloyal attitude to Bahasa Indonesia is not only seen in the big business with big capitals, but also in small businesses such as beauty salons in small places (kampung). Among the words written are *ladies and gent* which can actually be written in Bahasa Indonesia as *pria dan wanita*. The unloyal attitude is seen more in businesses with big

capital like cigarettes, big trade business, housing complexes etc. Actually the Indonesian Law Number 24, 2009, Chapter III, article 36, point 3, written:
"Bahasa Indonesia wajib digunakan untuk nama bangunan atau gedung, jalan, apartemen, atau permukiman, perkantoran, kompleks perdagangan, merek dagang, lembaga usaha, lembaga pendidikan, organisasi yang didirikan atau dimiliki oleh warga negara Indonesia atau badan hukum Indonesia." (*Bahasa Indonesia* must be used to name building, streets, apartments, housings, trade complexes, business marks, business institutes, and education institutes, organizations built or owned by Indonesian citizen or legal organization)

The negative attitude of speakers to Bahasa Indonesia also enters the education field. Some teachers and also students feel attracted more to those who can speak well in a foreign language especially in English. Naming rooms at schools is also done in English, for examples: *Headmaster* for the headmaster's room, *Teachers' Rooms, Laboratory, Library, etc,* written in English. Because of that situation, feeling love for Bahasa Indonesia is becoming less and less, and reaching the objective of learning Bahasa Indonesia becomes more difficult.

2.3 Feeling Satisfied to be Able to Use *Bahasa Indonesia*

Some of the other Indonesian people consider that Bahasa Indonesia does not need to be seriously learned. *Bahasa Indonesia* is an easy language and it will be mastered by itself by every Indonesian . This consideration may be true if what is meant by *Bahasa Indonesia* is ungrammatical, and "as it is like". It is true that almost all Indonesian people are able to understand what is spoken in *Bahasa Indonesia* and are able to communicate in Bahasa Indonesia (speak). But, again, their ability to speak *Bahasa Indonesia* is very low, moreover if they must use it in writing.

As the result of considering or feeling that they are able to speak Bahasa Indonesia, the Bahasa Indonesia school subject at school is considered to be very easy and often be neglected. Often time the subject of *Bahasa Indonesia* is placed at the last periods because it is considered to be an entertainment and relaxation after dealing with other subjects considered to be the main subjects before. Students are competing to chase other subjects considered to be more important like: mathematics, physics, or English although at the end some of them fail to reach what have been left behind.

The consideration that *Bahasa Indonesia* learning is easy because for some languages originated from Melayu languages, like Banjarese, there are similarities from several sides of the languages. In fact the similarities in most of Melayu languages often cause carelessness in students using *Bahasa Indonesia*. In Banjar language (Banjarese) there are many Indonesian words that are exactly the same as Banjarese language. However, the similarities are only in words. When the words are formed by putting affixes, then we can find the differences (see Kawi, 2011).

There are words like gula, *cari, beri, hadang,* and *lampu* in *Bahasa Indonesia* and in Banjarese. The similarities are only in the level of words, and after they are being affixed, the meanings are quite different. All the same words above, In Banjarese can be added by suffix *–i*, so they become *gulai, carii, berii, hadangi, dan lampui*. So the similarities of words often " *cheat"* the students who are learning *Bahasa Indonesia*. They think that the words with the suffix -i above are *Bahasa Indonesia*. A teacher, from Banjarese etnics, for example, may sometimes make sentences:

Teacher: "Nak, tolong *carii* Ibu yang mengajar di kelas lima."
Student: "Ya, Bu! Sudah saya *carii*, Ibunya tidak ada."
Teacher: "Gambar yang kamu buat warna kuningnya kurang tebal. Coba *kuningi* bagian yang di ujung itu."
Student: Baik Bu. Sudah saya *kuningi*."

From the language data above us can see the interference of Banjarese to Bahasa Indonesia, in the words *carii* and *kuningi*. The teachers as well as the students are not aware that they have made Banjarese interference to *Bahasa Indonesia* that they are using. Interference refers to "the use of features belonging to one language while speaking or writing another" (Mackey in Fishman, ed. 1972 :569).

It cannot be denied that the responsibility of using *Bahasa Indonesia* for the Indonesian people in certain places, for examples in the work place of government or private places of works are not entirely be fulfilled. This is not apart from the attitude of some Indonesian people who tend to look down upon *Bahasa Indonesia*; they don't have attitude yet to fulfil governments' demand. The heroic event of October heroik and no reason whatsoever to neglect the *Sumpah Pemuda*. Of course every person who loves and respect the heroes will go along and guard the oaths they stated until it comes true. Not too much if it is said that people who do not guard Bahasa Indonesia are those who cannot respect their heroes.

The responsibility of using bahasa Indonesia in the work place of government and private sectors written in 1945 Constitution of the Republic of Indonesia Number 24 , 2009, Chapter III, Article 33 point 1, "Bahasa Indonesia is compulsory used in official communication in the government and private work places."

2.4 The Need for Legal Sanction to the Law Breaker of Language Function Stipulation

Historical background and regulation which tie people to be loyal to *Bahasa Indonesia* has been very suitable. *Sumpah Pemuda* of October 28, 1928 was a historical heroic event and no reason whatsoever to neglect to their oath, Of course, everybody who loves and respects their heroes will keep guarding the oath they utter until all the oaths come into reality. It is not too much to say that those who do not guard *Bahasa Indonesia* are those who do not respect their heroes.

The heroic event of October 28. 1928 was heroically welcomed by the Indonesian fighters. *Bahasa Indonesia* which was fought and became an oath in the event of *Sumpah Pemuda* was welcomed by the independence fighters by putting *Bahasa Indonesia* into the 1945 Indonesian Constitution, Chapter XV, article 36. At last, the Government of Indonesia The government of the Republic Indonesia publish again the Legislation of the Republic of Indonesia No 24, 2009, which contain the rule of the use of bahasa Indonesia, the development, , restoration, and protection of bahasa Indonesia, and the improvement of the function of bahasa Indonesia to be an iinternational language.

All the legal products above have been suitable to be the basic to give sanction to those who underestimate Bahasa Indonesia. Facts show that there are still quite many Indonesian people who do not have loyalty to use *Bahasa Indonesia*. To those who haven't mastered bahasa Indonesia and use it ambivalently, it is the government's duty to make them speak *Bahasa Indonesia* well, as the message in the Legislation Number 24, 2009, Chapter III, Section 33 Article 2, which says, *"Pegawai di lingkungan kerja lembaga pemerintah dan swasta yang yang belum mampu berbahasa Indonesia wajib mengikuti atau diikutsertakan dalam pembelajaran untuk meraih kemampuan berbahasa Indonesia."*/ "The government and private employees in governments and private institutions who are not able to speak ***Bahasa** Indonesia* must follow or be included in the *Bahasa Indonesia* learning to get the ability to speak *Bahasa Indonesia*." For those who are already speaking *Bahasa Indonesia* well, but not being loyal to *Bahasa Indonesia*, as is seen on the information board for the commercials of an industrial product are entitled to the legal sanction".

The punishment is based on the 1945 Constitution and Legislation Number 24, 2009 as the operational guidelines. In Laws Number 24, 2009 in:

a. Article 32 (1),"*Bahasa Indonesia* is compulsory to be used in national forum and international in Indonesia."
b. Article 32 (2),"*Bahasa Indonesia* is compulsory to be used in formal communication in government and private work place."
c. Article 35 (1), "*Bahasa Indonesia* is compulsory to be used in writing scientific articles and publication in Indonesia."
d. Article 36 (3)," *Bahasa Indonesia* is compulsory to be used to name buildings, street, apartment or housing, offices, trading complex, trade mark, business institutions, educational institutions, organizations owned by the Indonesian citizen or law institutions in Indonesia.

The unavailability of legal sanction for those who underestimate Bahasa Indonesia also has impact to the teaching –and learning of Bahasa Indonesia. All the students see examples of wrong sentences and maybe think of them as correct sentences, students who have been loyal to Bahasa Indonesia see how those who do not obey the language rules are able to express Bahasa Indonesia as they want.

3 Conclusion

The non-linguistic problems have influenced the results of teaching and learning Bahasa Indonesia. Non-linguistic problems influence the learning result of *Bahasa Indonesia*. Consequently, these problems need to be thought and solved by Bahasa Indonesia experts working together with the government of the Republic of Indonesia.

Until now, a lot of efforts have been made by the government through the Language Institution which is related to the non-linguistic problems solution. The latest was the issue of The Republic Indonesia Law Number 24, 2009. One of the content is about the instruction of how to use *Bahasa Indonesia* in the country: Indonesia as well as the use of *Bahasa Indonesia* when an Indonesian is outside the country.

Insya Allah, in the future bahasa Indonesia hand in hand with Bahasa Melayu in Malaysia and Brunei Darusslam will become a big language and an international language or formal language in the United Nations (see Collins, 2011).

4 References

Alwi, Hasan. 2011. *Bahasa Indonesia Pemakai dan Pemakaiannya*. Jakarta: Badan Pengembangan dan Pembinaan Bahasa.

Alwi, Hasan & Dendy Sugono. 2011. *Politik Bahasa*. Jakarta: Badan Pengembangan dan Pembinaan Bahasa.

Alwi, Hasan. dkk. 2000. *Tata Bahasa Baku Bahasa Indonesia*. Jakarta: Balai Pustaka.

Chaer, Abdul. 1994. *Linguistik Umum*. Jakarta: Rineka Cipta.

Collins, James T. 2011. *Bahasa Melayu Bahasa Dunia*. Jakarta: Yayasan Pustaka Obor Indonesia.

Dardjowidjojo, Soenjono. 2016. *Psikolinguistik. Pengantar Pemahaman Bahasa Manusia*. Jakarta: Yayasan Pustaka Obor Indonesia.

Denzin. 2005, N.K. & Lincoln, Y.S. 1994. *Handbok of Qualitative Research*, Thousand Oaks, CA: Sage

Chelliah, L., Shobhana & J. De Reuse, Willem. 2011. Handbook of Descriptive Linguistic Fieldwork. New York: Springer.

Crystal, David. 2003. Language Death. Cambridge: Cambridge University Press.

Ellis, Rod. 1995. *The Study of Second Language Acquisition*. Oxford: Oxford University Press.

Finocchiaro, Mary. 1989. *English as a Second/Foreign Language From Theory to Practice.* Fourth Edition. Englewood Cliffs, new Jersey: Prentice Hall Regents

Fishman, J.A. 1972. *Language in Sociocultural Change.* California : Stanford Universut Press.

Fromkin, Victoria and Robert Rodman, 1978. *Introduction to Language.* New York : Holt, Rinehart and Winston.

Gibbons, J. & Ramirez, E. 2004. *Maintaining a Minority Language: A Case Study of Hispanic Teenagers.* Clevedon, GBR: Multilingual Matters Limited.

Hasil Perumusan Seminar Politik Bahasa Nasional, Jakarta: 25—28 Februari 1975. Jakarta: Pusat Pembinaan dan Pengembangan Bahasa.

Kawi, Djantera. 2011. *Telaah Bahasa Banjar.* Yogyakarta: Scripta Cedekia.

Mu'in, Fatchul & Sirajuddin Kamal, 2006. *Sociolinguistics: An Introduction.* Banjarmasin : Jurusan Bahasa dan Seni FKIP Universitas Lambung Mangkurat.

Mahsun. 2011. Metode Penelitian Bahasa: Tahapan Strategi, Metode, dan Tekniknya. Jakarta: Rajawali Pers.

Mahsun. 2012. The management of Language Diversity in Indonesia: Issues and Policies. Asia Europe Meeting (ASEM). Jakarta, 4-5 September 2012.

Mariani, Nanik & Mu'in, Fatchul. 2007. *Introduction to Linguistics.* Banjarmasin : Jurusan PBS FKIP Universitas Lambung Mangkurat.

Sariono, Agus. 2016. *Pengantar Dialektologi Panduan Penelitian dengan Metode Dialektometri.* Yogyakarta: CAPS (Center for Akademic Publishing Service).

Suherdi, Didi. 2012. *Rekonstruksi Pendidikan Bahasa Sebuah Keniscayaan bagi Keuggulan Bangsa.* Bandung: CELTICS Press.

Suparno. 2013. *Dasar-Dasar Linguistik Umum.* Jakarta: Tiara Wacana.

Tarigan, Henry Guntur. 1994. *Menyimak Sebagai Suatu Ketrrampilan Berbahasa.* Bandung: Angkasa.

Undang-Undang Republik Indonesia Nomor 24 Tahun 2009 Tentang Bendera, Bahasa, dan Lambang Negara, serta Lagu Kebangsaan. Jakarta: Lembaran Negara Republik Indonesia.

Cultural Violence represented in Indonesia and American Literatures

Fatchul Mu'in[1*], and *Rustam* Effendi[2]

[1]Lambung Mangkurat University, Department of Language and Arts Education, Jalan Brigjend H. Hasan Basry-Banjarmasin 70123 Indonesia
[2]Lambung Mangkurat University, Department of Language and Arts Education, Jalan Brigjend H. Hasan Basry-Banjarmasin 70123 Indonesia

Abstract. This article is aimed at describing the lives of dominated people of both Indonesia and America. Among the dominated people in Indonesian community are Indonesian Chinese people, and those in American community are African American people. The discussion on some Indonesia novels of post tragedy of 1998 shows that personally Chinese faced a hard life; and socially they were dominated. Therefore, it can be concluded that: (1) the personal behaviour of Indonesian Chinese is represented through the hard life, (2) social behaviour of Indonesian Chinese is represented through the dominated social life, and (3) cultural behavior is represented through the religious life with many problems.This is say that the cultural behaviour of Indonesian Chinese in Indonesia novels is represented through cultural violence. The similar result of discussion on some American novels of post slavery shows that (1) the black man as the representation of Black people (African-Americans) was always in a dilemmatic condition leaving him without any options. Whatever he chose, will have negative consequences, (2) the struggle for 'equality' through 'violence' will result in a 'tragic fate', and (3) the novel reflected the black people who yearned for freedom from white domination and expected to have good education, good employment, and equality in political opportunity, law enforcement/law protection, and in other sociocultural life.

Keywords: dominated people, Indonesian Chinese, African American, and cultural violence

1 Introduction

Indonesian Chinese literary authorship activity stalled. The cessation of activities of authorship was caused by a number of factors, among others: (1) the emergence of Balai Pustaka to control the reading material for society (Rukardi, 2010), (2) the strengthening of the Indonesian language as the national language in 1928, (3) an end of education separation for Indonesian Chinese (Sumardjo, 2004: 2-3), and (4) prohibition against Indonesian Chinese to express their culture in the era of New Order (Soyomukti, 2012: 203). The researches on literature with Indonesian Chinese"sl ife background were conducted by, among others: (1) Sainul Hermawan, (2005), which discussed the bad stigma of the Chinese in Indonesian Literature, (2) Alfiah Chusjairi (2005), which constructed Identity of Indonesian Chinesein Post New Order through the media, (3) Woodrich Christopher Allen (2011), which connected the tragedy of 1998 in the writing of novel *Putri Cina (*Princess of China), and (4) Kurniati Christina (2013), which examined the dynamics of the violence experienced by Chinese women.

* Corresponding author: muin_sihyar@yahoo.com

Meanwhile, the history of the United States has recorded African-American experiences, especially grievances. The spirit of independence has been marred by the Americans enslaving the black people. Even after the Civil War, when the slavery was abolished, the black people were regarded as inferior citizens; they have still been treated unfairly by the white people (Martin, et al., 1989; Franklin & Moss,1988).

Although the slavery was abolished after the Civil War, and slaves were proclaimed free, they were still regarded as inferiors who were discriminated and segregated by the white people. Facing this fact the black people struggled for equality, African-American writers raised those subjects in hopes of contributing to a social revolution. These writers' commit-ments to the political implication softheir creative work are all the m are apparent when we note that several of them maintaineds imultaneously as activists and as literary figures (Smith, et al.,1993: xi). They are, among others, Langston Hughes, Toni Morrison, Zora Neale Hurston, and Richard Wright (Smith, et al.,1993). In *Native Son* Richard Wright portrays the white domination and promotes human rights through the Max's defense of Bigger Thomas who was accused of raping and killing Mary, a white woman (Wright, 1966).

Based on the above description, both Indonesian Chinese and African Americans were under domination of the powerful parties or groups. As consequences, they physically and culturally experienced the bad treatments. Both faced what we call violences: physical and cultural violences.

2 Method of Study

This study uses a comparative method, trying to find out the similarities between Indonesian literature and African American literature having the same themes. There are some Indonesian novels with the theme of violence; one of those chosen as the source of data is *Putri Cina*. There are also some African American novels with the same theme as that of *Putri Cina;* one of those chosen as the source of data is *Native Son*. Both are studied on the basis of interdisciplinary approach.

3 Findings and discussion

The research findings show Chinese group : (1) are well-known to have a high working spirit, (2) have a view to be superior, resulting socio-cultural problems, and (3) Indonesian Chinese live under the domination of mainstream group. Their inferiority causes them to face some prejudice, discrimination, segregation, expatriation, and to become the object of violence, pillage, robbery, murder, and rape

3.1 Work Ethic

Indonesian Chinese have a high working ethics. The high working ethics are indicated by : (1) seriousness, (2) perseverance (3) hard work, and (4) never giving up as presented as "*Bekerja dengan memeras keringat* "Working with sweat" . "*Bekerja tanpa kenal lelah* "Working tirelessly" (Putri Cina, P. 74). To be successfull, they must work hard.

3.2 Having problems for being superior

Indonesian Chinese had been positined at a higher social class than native Indonesian people in the colonial era "*Kompeni menjunjung orang-orang Cina secara istimewa* (Dutch's company upholds the Chinese in a special way) (Putri Cina, p.105). "*Kelihatan berbeda dan lebih daripada kaum pribumi* (They seem to be different and to have higher social position that the native Indonesian people) (Putri Cina, p.105). After independence of Indonesia, the feeling of superiority resulted in sociocultural problems.

3.3 Stereotype, Prejudice, and Discrimination

Indonesian Chinese people have a label of the economic animal "[Mereka] ***Tergila-gila akan dagang*** *juga tak bisa disalahkan pada orang Cina*. "Loving business cannot only directed to Indonesia Chinese" (Putri Cina, p. 80). This label is rejected "*[Dibantah karena di antara]**Mereka menjadi tukang kayu, tukang batu, pandai besi, bahkan juga banyak yang masih menjadi kuli, yang semiskin kaum pribumi*** (They become carpenters, masons, blacksmiths, even many are still coolies, which are as poor as the native people" (Putri Cina, p. 105). They are regarded as the resource of chaos "*[Dianggap] Jadi biang keladi kekacauan di Pedang Kemulan ini adalah orang-orang Cina* (They were regarded as the main causes of chaos in Pedang Kemulan) (Putri Cina, p. 150). They are discriminatively treated "[Mereka] ***Dilarang berdagang di kota kecamatan dan kota kabupaten.*** Orang Cina hanya boleh *berdagang di kota-kota besar* (The Indonesian Chinese businessmen were banned to trade in the centre of subdistrict and country town) (Putri Cina, p. 144).

In relation to the aspect of stereotypes, the data presentation give the meanings that Indonesian Chinese accept the stereotypes as economic animals and treasure lovers. The way they have taken to get the treasure is the trade. Through the trade, they accumulate wealth that bring them into the wealthy; and for avoiding their domination in trade and in some aspects, they are banned to trade in the strategic areas.

3.4 Segregation, Violence, Robbery, Rape, Expulsion and Murder

Furthermore, Indonesian Chinese have the problems of residential segregation, violence, robbery, rape, expulsion and murder, as summarized in the following. They are segregated *"Tempat tinggal mereka pun **dipisahkan dari penduduk pribumi**, supaya mereka kelihatan berbeda dan lebih tinggi daripada kaum pribumi* (Their houses were also segregated from the native indonesian community, in order that they looked different from and higher than the native ones) (Putri Cina,105)*"*. They are violently treated "*Mereka (pribumi)* ***menyerbu*** *permukiman orang-orang Cina, menjarah hartanya, dan membantainya* (They [native Indonesian people] raided the settlement of the Indonesian Chinese, looted their property, and slaughtered them) (Putri Cina, 107).

Chinese women are the victim of violence "Bila pertikaian itu pecah, maka dia dan ***kaumnyalah yang menjadi korbannya*** (When the riot broke out, she [Putri Cina] and her people were the victim) (Putri Cina, 124). They are the victim of robbery "*Harta mereka dijarah. Rumah-rumah mereka dibakar. Tempat-tempat berdagang mereka dibumihanguskan* (Their property was looted. Their houses were burned. Their trade center was burned down) (Putri Cina, 150). The women are the target of rapping "Ketika akhirnya terpegang, para lelaki itu dengan beringas *menelanjangi wanita-wanita Cina itu, merebahkan mereka, dan melampiaskan nafsu mereka, sepuas-puasnya* (When she was finally held in check, the man violently stripped the Indonesia Chinese women, laid them, and lusted heart's content up to the point of satisfaction) (Putri Cina, 119). *They are the victim of explulsin "Malah mereka **mengusir** sebagian orang orang Cina dari Batavia. Banyak orang Cina waktu itu **terpaksa pulang** ke Negeri Cina* (Instead, they drove some Indonesian Chinese from Batavia. At that time, many Indonesian Chinese were forced to return to the Land of China) (Putri Cina, 106). They are the victim of murder "***mayat-mayat orang Cina*** *bergelimpangan di mana-mana.* (Indonesian Chinese dead bodies were everywhere) (Putri Cina, 107). "*Cina, kamu! Kamu **harus mati** di tangan kami* (You are Chinese! You must die in our hands) (Putri Cina, 150)

This segregation was not just physical separation (in the case of separating residences) but it was also social one (in the case of positioning Indonesian Chinese community on the higher level position than the indigenous peoples). In the present context, the Indonesian Chinese community are getting a negative stigma as an exclusive community.

In connection with the violence, the above data serve the meanings that the tragedy of May 1998. This tragedy was triggered by national economic crisis. Toward the crisis, mass demonstration that was initially held peacefully, turning into a huge riot and as a consequence many people died. When a mass demonstration turns into a riot, Indonesian Chinese women: (1) were in the condition of great fear, and (2) were subjected to rape, and (3) many of them were brutally murdered. In the riot, demonstration mass, that had been provoked, destructed whatever belonging to the Indonesian Chinese, looted, robbed, and burned their properties. This makes the Indonesian Chinese be in a state of extraordinary fear. The state of the fear is influenced by the past experience, namely: mass riot turning into the tragedy.

In connection with the act of expulsion, the above data show that the expulsion of Indonesian Chinese community took place since the Dutch colonial era. In the Dutch period, although under the pressure, the Chinese had the strong mentality, high work ethic, and unyielding attitude. The condition was regarded as harming the interests of the Netherlands, thus Dutch colonial expelled them from Batavia and returned them to the land of China. In the reign of the Old Order, Indonesian Chinese traders were not allowed to trade in the rural areas and subdistricts. They should move to trade in big cities. In the period of leading up to the collapse of the New Order, a number of Indonesian Chinese preferred to leave Indonesia and live abroad because of unsafe condition.

The above data also show the mass killings againt Indonesian Chinese and the followers of the Indonesian Communist Party in Batavia occurring in the Dutch colonial period, in the Tragedy of 30th September, 1965, and the Tragedy of May 1998. Mass killings against them in the Dutch colonial period because they were considered to launch rebellion ; the mass killings in the Tragedy of 30th September, because they were considered to be involved in the rebellion conducted by the Communist Party of Indonesia; and massacres in the Tragedy of May 1998 because they were considered to be the prime cause of the monetary crisis.

Meanwhile, the history of the United States of America shows that black Americans or African-Americans and the other minority groups are positioned as the secondclass citizens.In the past, most African Americans were brought, sold, and then enslaved to work on plantations. *As* slaves, they were badly treated and severely punished whenever they did wrong. When they did something wrong, they were severely punished.

When slavery was abolished, the freed blacks did not automatically obtain equal rights as the whites. In almost all the aspects of life, the Black American or African American people were under the White domination. Such condition continued until the appearance of Richard Wright's *Native Son* in 1940. In *Native Son*, the whites' assessment of the blacks depicts their racial prejudice. In this novel, Richard Wright portrays the blacks' inferiority through Bigger, and the Daltons. Aware of being inferior, Bigger felt uncomfortable in the white family's family's house. He addressed Dalton and his wife using 'Mr.' And 'Mrs.', `Sir' and 'Mare, Mary Dalton as 'Miss.' And Jan as 'Mr.' Or 'Sir'. The Daltons except Mary addressed Bigger as `boy' or calling his first name. Other terms symbolizing the blacks' inferiority are 'sonofabitch' [son of a bitch) (Wright, 1966: 253 and 289), 'black ape' (Wright, 1966: 312), 'apelike animal' (Wright, 1966 : 256), and 'dog' (Wright, 1966: 235). Bigger had to behave in such a way that he could be accepted by the whites.

Richard Wright portrays racial discrimination in housing through Bigger Thomas' statements: "We live here and they live there. We black and they white. They got things and we ain't. They do things and we can't. It's just like living in jail. Half the time I feel like I'm on the outside of the world peeping in through a knot-hole in the fence" (Wright, 1966: 23).

Residential segregation implied by Richard Wright in his Native Son can be pictured that black people "live here [in ghetto area} and they [white ones] live there [in the other area specified to them] (Wright, 1966: 23). Between the areas of black people and of white people there was a "line"; a black man, Bigger, could cross the "line' (Wright, 1966 : 234); the "line" was sometimes called a shadowy region, a No Man's Land, the ground that separated the white world from the black one (Wright, 1966 : 67).

Residential segregation seems to be planned by white people, "for they have agreed among themselves to keep Negroes with the ghetto- areas of cities ((Wright, 1966: 363), and it is that "residential segregation is imperative. Such measure tends to keep black people as much as possible out of direct contact with white women and lessen their attack against them" (Wright, 1966: 261).

Through Native *Son*, Richard Wright pictures white domination upon black people manifested in the ways of racial prejudice, discrimination and segregation. These manifestations result in the ill treatment of the blacks by the whites.

White domination can destroy all aspects of life such as cultural, social, educational, occupational, and political aspects, and in law protection or law enforcement between black and white people. As a consequence, the domination of the white people has negative impacts for the black people in almost all the aspects of life. Thus, the injustice toward the people in the black skin showed that democracy in America has not been completely developed. Democracy is not based on equal rights in all aspects of life.

White domination also causes unfair treatments in education which Richard Wright (1966) portrays in his *Native Son*. He writes:"If you wasn't black and you had some money and if they'd let you go to that aviation school, you could fly a plane." "I could fly a plane if I had a chance" (p. 20). "I wanted to be an aviator once. But they wouldn't let me go to the school where I was suppose' to learn it. They built a big school and then drew a line around it and said that nobody could to it but those who lived within the line. That kept all the colored boys out." (p. 327). The statements imply that Richard Wright wants to indicate that there is discrimination in education. A black man, like Bigger Thomas, is exempted from a certain school because he is black. The whites consider blacks as inferiors who do not have the right for equality in education. As stated above, the obstructions to maintain black people's inferiority through education can be in the educational fund, facilities, programs, opportunities.

Black workers can easily lose their jobs when they do wrong. Bigger was afraid of losing his job when he was befriended Mary and Jan. Whereas, Jack, a black worker, stated that he lost his job because Bigger killed Mary. This implies that if black man does wrong, another black man also endure the effects. The whites consider that he would commit the same crime against other white woman (Wright, 1966, p. 244).

The black people's social protest toward social injustice caused by white domination is related to their difficulties in achievingtheir rights in education, employment, and political participation, and in other aspects of social life. The history of the United States of America also shows that the whites commit violence against the blacks such as beatings, whipping, and lynching. Violence committed by the blacks is a response to the violence done by the whites upon them in the past. The death of Mary Dalton in the hands of Bigger Thomas is a reaction of violence committed by the whites to whom he hates and fears of. Mary Dalton symbolizes 'white power' which Bigger regards as an oppressor upon black people.

4 Conclusion

Based on the discussion above, it can be concluded that: (1) the cultural behaviour of Indonesian Chinese in Indonesian novel is represented through cultural violence, and (2) the black man as the representation of Black people (African-Americans) was always in a dilemmatic condition leaving him without any options.

Whatever both Indonesian Chinese and African American chose (1) will have negative consequences, (2) the struggle for 'equality' through will result in a 'tragic fate', and (3) the novels reflected that both faced the different treatments in education, employment, and in political opportunity, law enforcement/law protection, and in other sociocultural life.

5 References

Chusjairi, Juni Alfiah. 2005. *Konstruksi Identitas Etnik Cina Pasca Orde Baru Melalui Media : Studi Pemaknaan Film Cina : Wo Si Ni Indonesia, Jangan Panggil Aku Cina dan Ca-Bau-Kan*. Jakarta : Universitas Indonesia.

Franklin, John Hope & Moss, Alfred A.(1988). *From Slavery to Freedom: A History of Negro Americans*. Sixth Edition. New York : McGraw-Hill, Inc.

Hermawan, Sainul. 2005. *Tionghoa dalam Sastra Indonesia*. Yogyakarta : IRCiSoD

Kurniati, Christina. 2013. *Putri Cina Dan Gundik Orang Lewi Suatu Upaya Cross-textual Interpretation Terhadap Dinamika Kekerasan Yang Dialami Oleh Perempuan Asing Sebagai Warga Kelas Dua* . Salatiga UKSW.

Rukardi. 2010. "Tionghoa dan Subversi Sastra Melayu-Rendah" http://media-sastra-nusantara.blogspot.com/2010/03/ accessed at 2 June 2013

Sindhunata, 2007. *Putri Cina*. Jakarta : PT Gramedia Pustaka Utama.

Smith, Valerie, et al. (1993). *African American Writers*. New York : Collier Books Macmillan Publishing Company

Soyomukti, Nurani. 2012. *Soekarno & Cina*. Jogyakarta : Garasi.

Sumardjo, Jakob. 2004. *Kesusastraan Melayu-Rendah Masa Awal*. Yogyakarta : Galang Press.

Woodrich, Chistoper Allen. 2011. Pengaruh Tragedi 1998 Dalam Novel Putri Cina. Yogyakarta : Universitas Sanatha Darma.

Wright, Richard. (1966). *Native Son*. New York : Harper & Row Publishers, Inc.

The role of interdiscursitity in learners' discussion when writing a formal letter

Daljeet Singh Sedhu[1,*] and *Raja Nor Safinas Raja Harun*[2]

[1] Department of Social Science and Hospitality, Tunku Abdul Rahman University College.
[2]Faculty of Languages and Communication, Universiti Pendidikan Sultan Idris.

Abstract. Writing is the foundation of real formal task which allows writers to engage in real communication outside the classroom. Thus, formal letter writing is a range whereby learners get engaged in a wide range of writing tasks not limited to corrective facts but also the language used outside of the classroom. This study emphasises on the use of interdiscursivity through learners' group discussion when writing a formal letter. A qualitative method was employed in this study whereby the participants' discussions on the process of formal letter writing were recorded. The recordings were then transcribed and analysed. The transcriptions revealed how interdiscursivity could be used as an efficient strategy for formal letter writing. The findings of this study can aid academicians in devising instructional strategies and materials for the teaching of writing formal letter writing. This study implicates that interdiscusivity has a role in learner's discussion when writing a formal letter.

* Corresponding author: daljeetsingh@tarc.edu.my

1 Introduction

Major conformist models of language teaching have been found to be inadequate to address the current interdisciplinary anxieties and practices in the field (Bhatia & Anthony, 2017). Several learners who enrol in institutions of tertiary learning have lack of prior knowledge in writing, native and non-native speakers of English alike (Macbeth, 2006). Learners face difficulties in many aspects of their writing skill such as the use of proper grammar, punctuation, capitalisation, vocabulary, and even the fundamental aspects of initiating writing, which is also a challenging task of output for both educators and learners in ESL classes (Ghabool & Kashef, 2012). The authors confirm that the lack of English language proficiency is the leading source of difficulty that Malaysian ESL learners face in their writing tasks. Prospective employers often commented that despite being academically excellent, Malaysian graduates lack soft skills, particularly communication and logical-thinking skills (Ngang & Chan, 2015).

In extension to the above argument, tertiary learners were found to have lack of vocabulary to explain their writing tasks in the form of sentences (Halim, Ahsan & Munir, 2011). Formal letter writing is a field whereby learners get engaged in a wide range of writing tasks that require corrective facts of the language. Shafie and Nayan (2010) state six problems faced by novice writers: (i) inability to work together with associates, (ii) low language proficiency, (iii) inadequate time to discuss, (iv) limited research skills, (v) lack of ideas and (vi) pressure during two-way writing.

Since the researchers have been lecturing in higher learning institutions for more than ten years, we have observed that learners are not familiar with the context and content of formal letter writing when introduced to them. Learners are unable to understand written features of some of the expert genres, and they are still inexperienced in the discursive realities of the professional world (Bhatia, 2016).

Interdiscursivity is the characteristic of discourse that relates to other discourses. Fairclough (2003) prefers the concept 'orders of discourse' which means it is often an analytic concept. Interdiscursivity has close affinity to recontextualisation because interdiscourse often implies that elements are imported from another discourse. The meaning of interdiscursivity varies and it can be divided into three levels.

Firstly, Courtine (1981) refers interdiscursivity as a discourse which has a relation to another discourse, a meaning which is close to the meaning of intertextuality. Secondly, Linell (1998) refers interdiscursivity as a relation between types of discourse such as genres. Thirdly, Fairclough (2003) refers interdiscursivity as relations between discursive formation that is between large heterogeneous discursive entities, which can be related to natural arguments and oral communication. In this study, Faircloughs' (2003) definition of interdiscursivity will be used to look into the diverse oral communication between students.

In this study, the researchers focused on learners' knowledge in formal letter writing of the academic discourse. Knowledge in the ESL writing of formal letter discourse is fundamental to the learners' academic survival, as it is important in assessing the learners' academic progress. As learners of ESL, writing a formal letter has often been central to the discussions on this subject. The researchers, thus, considered that it is vital to focus on the interaction practices in which learners participate in the writing process. These practices can provide insights into developing a potentially effective academic writing model for learners. As such, the researchers attempt to find out how the use of interdiscursivity through learner-learner interaction in a group discussion can assist learners in writing a formal letter.

2 Objective and Research Question

2.1 The objective of this study is:
To study the role of interdiscursivity in learners' interaction when writing a formal letter.

2.2 The research question that underpins this study is:
How does the role of interdiscursivity through learner-learner interaction in a group discussion assist learners when writing a formal letter?

3 Methodology

The research method used in this study was document analysis. Transcriptions of learners' discussions during the formal letter writing tasks were analysed. A qualitative analysis method was used to find out how interdiscursivity could support learners when writing a formal letter through learner-learner interaction. The purpose of using this method is to enable the researchers to answer the research question as stated in this study.

The details of how the research instruments were used to collect data are explained further in this section. First, the students were divided into six groups and each group consisted of four members. Each group was given the same task that was to write a formal letter based on the outcomes of their group discussion. Secondly, each group was told to do a voice recording of their discussion from the moment they started attempting the task until it was completed. The recordings of the discussions were between 10 to 15 minutes. The recordings were then collected by the researcher to be transcribed for an in-depth data analysis purpose. The transcriptions were analysed to observe the process of learner-learner interaction in each group. This was done to determine how interdiscursivity could be used by learners to assist them in writing a formal letter. The main areas that were looked into were enquiries, negotiations, agreements and disagreements, and clarifications done by each of the group members.

Each group was required to engage in a discussion pertaining to the task given before they started writing the formal letter. The task given was for them to a write a formal letter to conduct a charity fund raising event on behalf of a society. The outcomes of the discussion will be the learners' interaction with each other and their discursive arguments towards learning. Interaction and discursive arguments are linked because when learners interact with each other, they will voice out their opinions related to the given task. These learners presented their agreements and disagreements of each other's discursive arguments. In other words, interdiscursivity was evident in the learner-learner interaction.

4 Findings and Discussions

Research Question: How does the role of interdiscursivity through learner-learner interaction in a group discussion assist learners when writing a formal letter?

Based on the document analysis, there were emerging themes on the role of interdiscursivity through learner-learner interaction, which can assist learners when writing a formal letter. These are detailed as follows:

4.1 Discursive Understanding of the Task

The learners developed a discursive understanding of the task given to them during the course of the discussions. At the beginning of the discussions, it was apparent that they were confused about the task prerequisite. In an attempt to address this confusion, they engaged in debates and resorted to linguistic adaption during their discussions. They learnt to modify ideas to fit the requirement of the given task. They became conscious of what they learned, more focused on completing the task and not on language use, used the Internet to search for ideas, and developed diverse ideas in the discussions.

Each of the learners had their discursive understanding after reading the task given. They gave their views on the task given to them. This can be seen in the following excerpts:

Learner F: "From my point of view, we should choose which society to be represented before we think of the fundraising. I think by doing this, we can decide what kind of activity we can plan for the fundraising."

Learner N: "After I have read the question, I think let's look into what activity we come up for the fundraising then think of a society. It does not matter of taking any society to conduct this activity."

Each learner came up with their individual idea of completing the task given to them. They shared with each other their point of view and discussed among themselves how to incorporate everyone's ideas to suit the requirement of the task. It is essential to analyse each learner's ideas critically as this will help learners to stay on topic. This will also help teachers to easily identify and correct the discursive views given by learners by providing immediate feedback. This supports the findings of Bakhtin (2010) in which he argues that utterances in a language are always dialogised, and this changes based on how a reader interprets what they read and after that describe what they have understood through verbal discussion.

4.2 Confusion of Task Requirement Through The Discussion Process

Learners became confused with the spoken genre when they communicated with other group members. One member might influence others to stray away from the task requirement, as demonstrated in the following excerpts:

Learner C: "I think we can plan fundraising as a secondary event, let us focus on another event that will be benefit for the Society to organise".

Learner A: "We should have two different dates for the event so that we can gain more money. We write the formal letter to mention we want do this event and tell the participants to register then later we will inform them about the particulars if it is confirmed."

It is crucial for learners to synthesise what they read in order to fulfil the requirement of the task. The act of misleading by one group member may cause the whole group to produce an out of topic formal letter, which did not fulfil the task requirement. Each member should play their roles to ensure that every group member gets a proper understanding of the task given to them. This supports the finding of Bakhtin (2010) in which he states that the relationship of genre involved in the mixing, in a way, can cloud a clear distinction and may perhaps mislead the action of the doer of a task. Corresponds as this is a 'primary genre' as mentioned by Bakhtin (2010), which means that oral dialogue distinguisher to the listener of how the speaker presents the definition of genre discussed. He elaborates by saying that the speaker might also produce a different definition when they speak of what they have read from any text genre and present an idea based on their understanding and interpretation. This is known as 'secondary genre'. This also supports the findings of Wu (2011) whereby in some specific cases it is the mixing of discourses or styles that makes sense in the formation of interdiscursive relations.

4.3 Negotiation and Linguistic Alteration

Learners negotiated with each other on the points to be included in the formal letter. They gave their ideas and negotiated with the group members for their ideas to be included in the formal letter. This was revealed in the following excerpts:

Learner M: " I think we should have an event like Canteen Day and have other special events like lucky draw, clown performance and much more for this fundraising, what do you think of my idea?"

Learner F: "I think the event should be organised on a weekend or public holiday which will be easier for everyone to attend as I think this event should open to all public."

Learners were able to use proper vocabulary through negotiation among group members. This allowed members to think of the rational way of writing the content of a formal letter and lead them to write more creatively. This supports the findings of Wu (2011), whereby interdiscursivity can also act as a product of dynamic negotiation, indicating the process of making a linguistic choice. Here, the doer of a task will use negotiation to voice out his or her opinion so as to allow listeners to think rationally before accepting the views and ideas directed at them. According to Verschueren (2015), language is adjustable from our daily use of language in the process of learning. Interdiscursivity, thus, enables the speakers to convey the message to the listeners by strengthening the ideas and opinions delivered, eventually leading to a successful communication.

4.4 Revision of Ideas

Learners constantly revised the ideas they had initially conveyed as the discussion went on. They tended to refer to what they had written down earlier in the formal letter and changed them to relate to the new points written in the content of the formal letter. The following excerpts illustrate this revision of ideas.

Learner A: "I think we should look back at the number of programmes involved in this fundraising event. From my point of view, it should not have too many programmes."

Learner E: "We should see whether these programmes can be conducted in this venue. Otherwise, we need to reconsider of a different idea for this formal letter."

Learners often looked back at the content of a formal letter that they had written to check whether the points were coherent, and to match the points in the content with one another. This allows learners to understand the changes needed in ensuring that a formal letter can be used as a successful communication tool to avoid confusions among readers. This process is known as "hybridisation" (Bakhtin, 2010), in which changes take place in the task when there is an on-going discussion. This all depends on the speaker's utterance by noticing that there is a need for changes to be done and notifying the listeners to identify the changes. Marchenkova (2005) mentioned that people normally reinterpret what they have uttered or heard earlier into giving it a meaningful definition so as to achieve successful communication.

4.5 Sensible of Learning

Learners were able to recall and connect the discussion to their social context. The discussion on the content of the formal letter by these learners was often based on their personal experience. This is apparent in the following excerpts:

Learner C: "Let's have this fundraising by having recycling campaign. When I was in my secondary school, we did a fundraising activity by doing recycling campaign, and we gained much money by selling old newspapers, aluminium cans and other recycling products."

Learner G: "I think we should have something like a Family Day and invite all the parents of the learners to come and participate in this event. I have seen many this kind of activities work well."

Learners often participated in a discussion by placing the topic in their social context; a simple example is by relating the topic to their personal experience. Kubli (2005) mentioned that consciousness of learning is based on the exchange of ideas among learners. In a learner-learner interaction, every word used is based on the learners' exposure limit within their personal social context. Such exchange of knowledge is considered as interdiscursivity in which Bakhtin theory could be matched with the mentioned process (Kubli, 2005). According to Kuo and Lin (2015), this way also encourages learners to participate independently, thus enabling them to manage a learning process on their own without the aid of a teacher.

4.6 Attentive on Task and not on Language

Learners were more focused on completing the task at the expense of language precision, as revealed in the following excerpts:

Learner L: "I think we have misspelled a few words in this letter, but let's look at it later, we think on how to complete it first."
Learner M: "I do not think it is necessary to see what words important to be used in this formal letter task, we focus on finishing otherwise we will not have enough of time."

Learners concentrated less on producing a formal letter that follows grammatical rules accordingly. They were more focused on trying to complete the task. This supports the findings of Mathews and Jacquet (2013) that learners can convey their ideas anonymously and focus on the content of a writing task. They can, in fact, achieve a successful communication despite the lack of grammatical precision.

4.7 The Idea of Using the Internet

Students came up with the idea of using a computer to complete the formal letter writing task. They preferred using a computer with an Internet connection to complete the task. This is evident in the following excerpt:

Student F: "I think it will be better if we could use the computer. We can find more information and discuss among ourselves with different ideas."

Students preferred using a computer to complete a discussion-based task. It is convenient for them to gather different ideas from each member by using the Internet. Smith (2005) mentioned that computer could enhance students' competence in reading and allow them to convey ideas effectively. He also suggested that students should be encouraged to use a computer for them to write more and produce more creative output. Mathews and Jacquet (2013), however, highlighted the need for computer-literate instructors too, without whom the interactive and collaborative tasks made possible with the use of computers, will be rendered useless.

4.8 Improved the Ability to Speak Based on Task Given

Students were able to improve their ability to speak to each other in a group discussion. They overcame factors such as shyness and unwillingness to speak as elucidated in the following excerpts :

Student D: "I feel more confident to speak in this kind of group discussion. I can listen to everyone's idea and try to think how to relate to other points."

Student O: "I prefer to do this way of discussion because it helps me to speak out what I feel like sharing with everyone."

Students were able to freely voice their opinions when interacting with other group members, thus building their confidence to communicate. They were also able to improve their communication skills, hence allowing them to think critically of other group members' ideas and to give comments. This finding echoes that of a research done by Javid (2014), in which he found that by using learner-learner interaction as a strategy in a learning process, the student will be able to develop their own mode of learning. Furthermore, this can be used as an aid for students to learn a second language such as the English language. Javid (2014) too proposed that students often find it much easier to express themselves verbally in front of a small group when learning to work in a group. By using this strategy in a learning process, it will also improve students' interpersonal communication skills (Lee, 2017).

4.9 Diverse Communication

Students conducted brainstorm session to develop ideas for the content of the formal letter. The group leaders in each group decided to choose this strategy to get every member to participate in the discussion so as to complete the task given, as evident in the following excerpt:

Student P: "Let's sit around and discuss the content of the formal letter. I suggest we do brainstorming on the points that can be written for the formal letter."

The students used brainstorming as a part of their discussion process in completing the formal letter writing task. This can be considered as an aspect of group dynamics role in a classroom interaction. The group leaders were able to provide guidelines and helped the members to maintain their energy and interest to participate in the discussion process. These findings also support the statement by Curry, Jiobu and Schwirian (2005) that there are few characteristics must be shown by a group leader that is authoritarian, style of communication shown and being democratic to each member in a group. A circular arrangement of a group is adaptable because it assists the best interaction among members during the discussion as mentioned by Winston and Patterson (2006).

5 Conclusion

The purpose of this study is to determine the use of interdiscursivity through learner-learner interaction among tertiary-level learners when writing a formal letter.

Most of the time, learners do not understand what they are writing in a formal letter because they do not know the structures and conventions involved in formal letter writing. Improving formal letter writing is a big challenge for both teachers and learners. Many teachers may feel that there are many challenges in implementing effective formal letter writing based on the past studies discussed in Section 1.0. As for the learners, there are various strategies available for them to learn yet it takes time for them to be able to apply the appropriate strategy to a specific writing task. Thus, their motivation to write and their discursive ideas are limited. It is the researcher's opinion that efforts should be directed towards supporting interdiscursivity through learner-learner interaction among tertiary level learners when writing a formal letter.

Future research on using interdiscursivity as a strategy for writing skills should not only focus on the result of this group discussion. The emphasis should be on the process of learning as learners would be able to use the social and learning skills, acquired through interdiscursivity, in their everyday life.

As such, it can be surmised that it is vital for future researchers to continue their efforts in promoting interdiscursivity as an alternative learning strategy and maximising its potential as an instructional intervention.

6 References

Bakhtin, M. M. (2010). *Speech genres and other late essays*. University of Texas Press.

Bakhtin, M. M. (2010). *The dialogic imagination: Four essays* (Vol. 1). University of Texas Press.

Bhatia, V. K. (2016). *Critical genre analysis: Investigating interdiscursive performance in professional practice*. Taylor & Francis.

Courtine, J.-J. (1981). *Analyse du discours politique.* Paris: Langages.

Curry, T. J., Jiobu, R. M., & Schwirian, K. P. (2005). *Sociology for the twenty-first century*. Pearson College Division.

Fairclough, N. (2003). *Analysing discourse: Textual analysis for social research*. Psychology Press.

Ghabool, N., & Kashef, S. H. (2012). Investigating Malaysian ESL students' writing problems on conventions, punctuation, and language use at secondary school level. *Journal of Studies in Education*, *2*(3), 130-143.

Javid, C. Z. (2014). Perceptive determination of Saudi EFL learners about the characteristics of an ideal English language teacher. *Research on Humanities and Social sciences*, *4*(8), 42-53.

Kubli, F. (2005). Science teaching as a dialogue–Bakhtin, Vygotsky and some applications in the classroom. *Science & Education*, *14*(6), 501-534.

Kuo, K. Y. C., & Lin, C. S. (2015). Studies on Patterns and Variations of Learning Community of Practices in Global Context. In *Learning and Teaching in Computing and Engineering (LaTiCE), 2015 International Conference on* (pp. 176-179). IEEE.

Lee, M. K. (2017). To Be Autonomous or Not to Be: Issues of Subsuming Self- Determination Theory Into Research on Language Learner Autonomy. *TESOL Quarterly*, *51*(1), 220-228.

Linell, P. (1998). *Approaching dialogue: Talk, interaction and contexts in dialogical perspectives* (Vol. 3). John Benjamins Publishing.

Marchenkova, L. (2005). Language, culture, and self: The Bakhtin-Vygotsky encounter. *Dialogue with Bakhtin on second and foreign language learning: New perspectives*, 171-188.

Mathews, T. J., & Jacquet, R. C. (2013). *Using Strategic Interaction in L2 Writing*. American Association of Teachers of Spanish and Portuguese.

Ngang, T. K., & Chan, T. C. (2015). The importance of ethics, moral and professional skills of novice teachers. *Procedia-Social and Behavioral Sciences*, *205*, 8-12.

Halim, N. B. A., Ahsan, N. I. A., & Munir, F. A. (2011). Writing Final Year Project Report in English: Problems, Motivating and Hindering Factors among UTeM Final Year Engineering Learners. Asian Social Science, 8(1), p27.

Shafie, L. A., & Nayan, S. (2010). Employability awareness among Malaysian undergraduates. *International Journal of Business and Management*, *5*(8), 119.

Smith, R. O. (2005). Working with difference in online collaborative groups. *Adult Education Quarterly*, *55*(3), 182-199.

Verschueren, J. (2015). Ideology in discourse. *The International Encyclopedia of Language and Social Interaction*.

Winston, B. E., & Patterson, K. (2006). An integrative definition of leadership. *International Journal of Leadership Studies*, *1*(2), 6-66.

Wu, J. (2011). Understanding Interdiscursivity: A Pragmatic Model. Guangzhou: *University of Cambridge*.

A FAHP approach to select students' performance assessment criteria in task-based English language teaching

Waseem Sattar*[1] , *Muhammad Ridhuan* Tony Lim Bin Abdullah[2] , and Fatemeh Mirzaei [3]

[1]Management and Humanities Department, Universiti Teknologi PETRONAS, Perak, Malaysia

[2]Management and Humanities Department, Universiti Teknologi PETRONAS, Perak, Malaysia

[3]Management and Humanities Department, Universiti Teknologi PETRONAS, Perak, Malaysia

Abstract. Assessing students' performance is a main means to improve students' learning and can play an important role in strengthening the evaluation process in task-based language teaching. This paper adopts a novel approach to select students' performance assessment criteria based on fuzzy AHP method. Specifically, after determining dimensions and sub-dimensions of learning, the students' performance assessment instrument can be established. In the instrument, the dimensions and sub-dimensions can be selected by experts' opinion. Employing the fuzzy AHP method in group decision-making can facilitate a consensus of decision-makers to reduce uncertainty. The application of this method can make the assessment results more scientific, accurate and objective. It is expected that this work may serve as an assistance tool for members of higher education institutions in improving the educational quality level.

1 Introduction

Assessing students' performance is an important aspect of learning and teaching, which includes several components employed by instructors to draw conclusions about their achievements (Ma & Zhou, 2000). In task-based approach, this component is more complex than standard language assessment as it requires observation in real-world context of students' language and communicative ability (Bachman 2002; Weigle 2002). The aims of assessing performance is to improve students' learning processes, motivation, self-esteem and thinking (EL-Koumy, 2004) that, preferably lead to productive citizens. In student-centred learning environment, language instructors prefer using criterion referenced techniques to assess the performance of students. In addition, these criteria and assigning weightage to these criteria is determined by language instructors themselves (Ma & Zhou, 2000).

Many frameworks have been suggested over the years on assessing performance of students. A framework provided by Ellis (2003) focuses on task evaluation based on the following three criteria: 1) student-based, which only measures the extent to which the task was useful and enjoyable; 2) response based, which compares actual task outcomes with predicted outcomes; 3) and learning-based evaluation, which measures the level of actual learning when the task is completed. However, these evaluations are limited to measuring teaching effectiveness which only determines whether a task was meaningful, useful and enjoyable. American Council for Teaching Foreign Languages scale (ACTFL) early proficiency scales were limited to describing learner behaviour (Bachman, 2002). Therefore, the current assessment models which are based on vague and ambiguous criteria are likely to limit language instructors understanding how the characteristics of task affect students' performance (Bachman, 2002). Apart from aforementioned studies, other related researchers focused on measuring abstract and psychological qualities related to the performance such as aptitude, motivation and language proficiency (Nunan, 2004). In other words, there is a need to address those aspects of students' learning performance which are important but missing and may play an important role to make judgments about their performance. Hence, to determine the quality of students' performance, using a scientific method plays an important role to draw true picture of their performance. Therefore, the main purpose of this paper is to explain the process and steps involved to select and prioritize the assessment criteria according to their relative importance.

Assigning weight to the factor is a multi-criteria decision-making problem and giving judgments in linguistic variables is more convenient for decision-makers rather than giving in the form of numerical values (as cited in Lee, Chen & Chang 2008).

* Corresponding author: swsbrothers11@gmail.com

In recent years, several researchers have focused on assessing students' performance. To produce 'fair' mark distribution, (Echauz & Vachtsevanos, 1995) presented a fuzzy grading method that utilizes students' and instructors performance measure. Kahraman (as cited in Lee et al., 2008) proposed a methodology based on FAHP to improve the quality of prioritization of human capital measurement indicators under fuzziness. The studies discussed above revealed the high applicability of fuzzy set theory. Therefore, to deal with imprecise and uncertain data, Zadeh introduced fuzzy set theory in 1965 (as cited in Lee et al., 2008) and is used in this paper to determine the dimension and sub-dimensions weights. It is an effective method to represent the uncertainty and fuzzy terms in the assessment environments (Ma & Zhou, 2000).

The reminder of the paper is organised as follows: Section 2 presents Zadeh fuzzy set theory and its application in the development of the instrument. Section 3 discusses the essential steps and sub-steps includes in the establishment of the performance assessment instrument. Section 4 presents an example of the instrument; and finally, the conclusions are given in section 5.

2 Fuzzy Set Theory

Fuzzy set theory was introduced to solve problems when the criteria is not clearly defined (Zadeh, 1965). To avoid misleading results, it is very important to consider fuzziness in human decision making. And this theory has been used in different fields to solve such kind of problems for last four decades. The two distinct directions of this theory are: treating fuzzy sets as precisely defined mathematical objects subject to the rules of classical logic, and the linguistic approach. The logic behind linguistic approach is that the truth values are fuzzy sets and rules inferences are approximate rather than exact.

A triangular fuzzy number is a special case of trapezoidal fuzzy number, is very popular in fuzzy applications. As shown is Figure 1, the triangular fuzzy number M is represented by (a,b,c) and memberships function is defined as

$$\mu_{\tilde{M}}(x) = \begin{cases} \frac{x-a}{b-a}, & a \le x \le b \\ \frac{c-x}{c-b}, & b \le x \le c \\ 0, & \text{Otherwise} \end{cases} \tag{1}$$

With $-\propto < a \le b \le c < \propto$

The strongest grade of membership is parameter b, that is

$$f_M(b) = 1$$

, while a and c are the lower and upper bounds.

An important step of fuzzy sets is the α-cut. For a fuzzy number M and any number $\propto \in [0,1]$, is the crisp set (Khr & Yan, 1995):

$$C_{\propto} = \{x \mid C(x) \ge \propto\} \tag{2}$$

The α- cut of a fuzzy number m is the crisp set M that contains all the elements of the universal set U whose membership grades in m are greater than or equal to the specified value of a, as shown in fig. 2.

By defining the interval confidence at level, a, the triangular the triangular fuzzy number can be characterized as (Cheng,1999; Cheng 1996; Cheng and Mon, 1996);

$$\tilde{M}^2 = [a^{\propto}, c^{\propto}] = [(b-a)\propto + a, -(c-b)\propto + c], \quad \forall \propto \in [0,1] \tag{3}$$

The distance between the 2-triangular fuzzy number can be defined by the vertex method (Chen, 2000). Let $\tilde{M}_1 = (a_1, b_1, c_1)$ and $\tilde{M}_2 = (a_2, b_2, c_2)$ be two triangular fuzzy numbers, the distance between them is

$$d(\tilde{M}_1, \tilde{M}_2) = \sqrt{\frac{1}{3}\,[(a_1 - a_2)^2 + (b_1 - b_2)^2 + (c_1 - c_2)^2]} \tag{4}$$

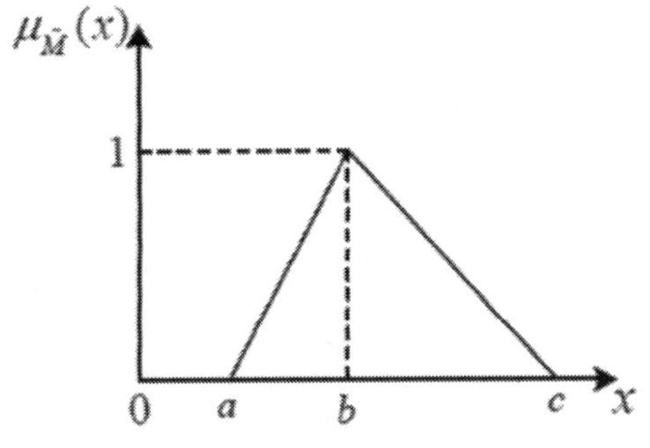

Fig. 1. Membership function of a triangular fuzzy number $\tilde{M} = (a, b, c)$

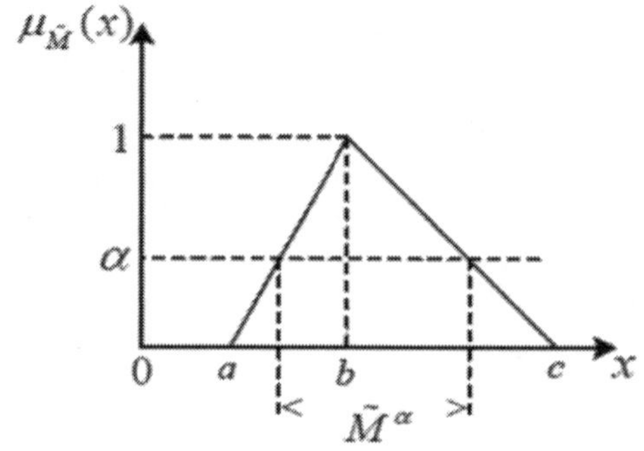

Fig. 2. α- cut of a fuzzy number $\tilde{M}$.

Different methods have been proposed to rank fuzzy numbers and each has its own advantages and disadvantages (Klir & Yan, 1995). Since the conventional method only used numerical equivalence method, that is why linguistic approach is used to consider any ambiguity when decision makers rate using linguistic values whose membership functions are usually characterized by triangular fuzzy numbers (as cited in Lee et al., 2008).

3 Steps involved in instrument development

In this paper, six perspectives of performance assessment dimensions and sub-dimensions are extracted from extensive review of literature. The next step is to get experts opinion to select final list of dimension and sub-dimensions. They can add or remove any perspective in the final list. According to Adler and Ziglio (1996) the selection of experts should be based on four 'expertise' requirements: I) knowledge and experience with the issues under investigation; ii) capacity and willingness to participate; iii) sufficient time to participate in the study; iv) effective communication skills. In the construction of framework, Expert Choice, 2006 software will be applied. Different software packages have been applied to solve AHP problems however, expert choice will be used in this research.

3.1 Data Collection

A conventional AHP method will be used to design a questionnaire based on nine-point Likert scale to select and then to finalize the selection by making pairwise comparisons. The questionnaire will then be distributed among forty experts to get their opinions. There are two phases of the data collection: In the first phase, experts will select the dimensions and sub-dimensions and in second phase, they will make pairwise comparisons. After the selection of dimensions and sub-dimensions and pairwise comparisons, the last step is to obtain final ranking of each dimension and sub-dimension based on their importance. The fig.3. below shows the dimensions and sub-dimensions extracted from the review of literature to develop performance assessment tool. These dimensions and sub-dimensions are selected from the studies by (Bachman, 2002), (Nunan, 2004), Ellis (2003) and Norris (2001) to develop performance assessment tool. However, the final selection for developing the assessment tool will be based on expert's opinion.

Goal	Performance Dimensions	Performance Sub-dimensions
Performance Assessment Instrument	Task components	Accuracy Complexity Fluency
	Task Difficulty	Code complexity Cognitive complexity Communication stress
	Task Types	Jigsaw task Information gap task Real-world task
	Task Selection	General interest Discipline specific Real-world focus
	Task Planning	Pre-task planning Within task planning
	Task phases	Pre-task Main-task Post-task

Fig. 3. The items for assessing students' performance

3.2 Fuzzy analytic hierarchy process (FAHP)

FAHP theory is used to generate the weights of performance dimensions. This includes six important steps:

1. After constructing the hierarchal structure which includes dimensions and sub-dimensions, then experts will be asked to assign the relative importance of two decision elements in the same level. Once pairwise comparison scores from the experts have been collected, the next step is to form comparison matrix for each decision maker.
2. Checking the consistency of decision maker is significant in FAHP and to get that decision makers priority of elements will be compared by computing eigenvectors and eigenvalues.
   $R \,.\, W = \lambda max$
   Where w represents the eigen vector, normally named as the weight vector matrix of the matrix. To ensure that judgments are consistent, consistency is then checked. Saaty (1980) has defined consistency ratio and consistency index. It can be written as (CI) and (CR).

$$CI = \frac{\lambda max - n}{n - 1} \tag{5}$$

$$R = \frac{CI}{RI} \tag{6}$$

Where n represents the total number of elements being compared and RI is random index. To generate random index, matrices of similar size can be compared randomly. If the consistency ratio is greater than 0.10 then the decision maker (expert) must have to revise the original values.

3. Construct positive matrices. Linguistic variables will be formed by transforming pairwise comparison scores and Triangular fuzzy numbers (TFNs) will represents these variables (Buckley, 1985).
4. Fuzzy weights will be calculated on the basis of (Csutora & Buckley, 2001) Lambda-Max method.
5. In this step, opinions of the decision makers will be integrated. And to combine fuzzy weights of the decision makers geometric mean is then applied.
6. The last step is to obtain final ranking of the decision makers and this will be done by following Chen (2000) proposed equation.

4 Example of the performance assessment instrument

Fig. 4. shows the hierarchal structure of performance assessment tool. However, this is not the final version of the instrument. This is to highlight how the comparison and pairwise comparisons will be made between dimensions and sub-dimensions to come up with a assessment tool.

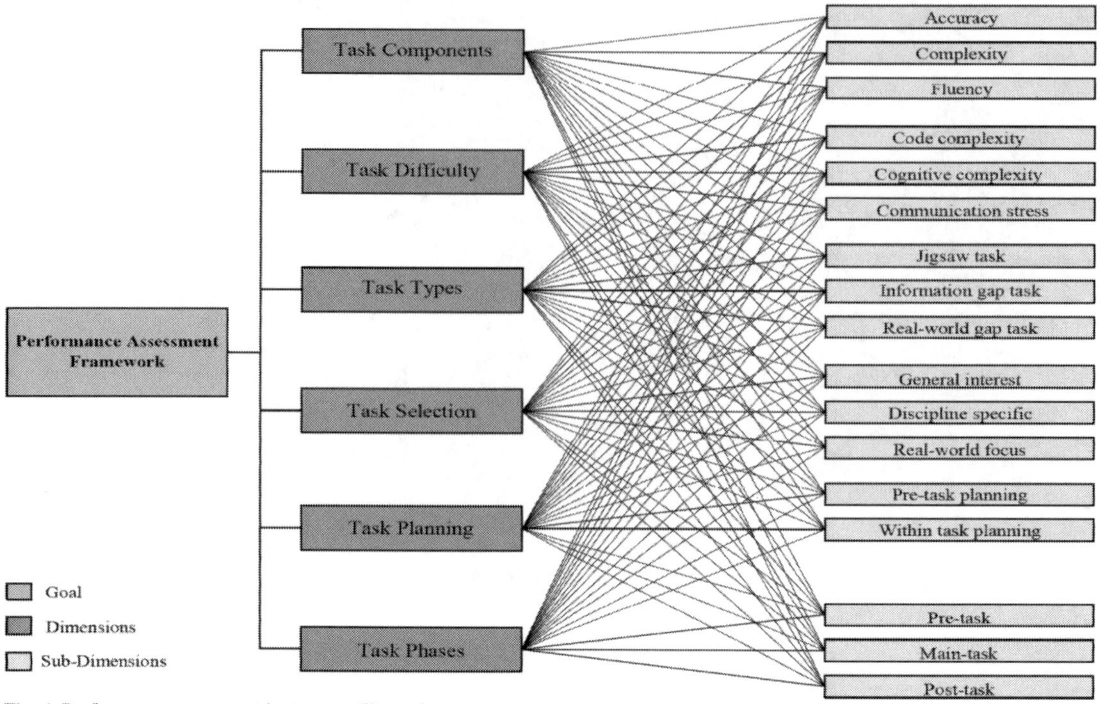

Fig. 4. Performance assessment instrument illustration

5 Conclusion

Students' performance assessment instrument can be an effective tool to promote student learning. This paper presents a systematic method to develop a comprehensive assessment instrument based on fuzzy analytical hierarchal process. The application of this method in task-based language teaching cannot only reflect the overall student performance, but also their achievement in each assessment criteria. In this way, the instructors can know what areas of performance really need improvement which ultimately will help students to enhance their learning quality. Introducing fuzzy analytical hierarchal process in TBLT to determine weights of dimensions and sub-dimensions based on experts' opinion is main contribution of this study. As fuzzy AHP has the capability to capture the vagueness of human judgments and makes the derived weights in the instrument more objective and reasonable. The approach can reduce subjectivity in the assessment process. Additionally, the application of this may provide an effective, scientific and objective measures to assess students' performance at higher education institutions. Hence, it may also be used as a reference for management practitioners when solving decision- making problems.

6 References

Adler, M., & Ziglio, E. (1996). *Gazing into the oracle: The Delphi method and its application to social policy and public health*. Jessica Kingsley Publishers.

Bachman, L. F. (2002). Some reflections on task-based language performance assessment. *Language testing*, *19*(4), 453-476.

Buckley, J. J. (1985). Fuzzy hierarchical analysis. *Fuzzy sets and systems*, *17*(3), 233-247.

Chen, J. F., Hsieh, H. N., & Do, Q. H. (2015). Evaluating teaching performance based on fuzzy AHP and comprehensive evaluation approach. *Applied Soft Computing*, *28*, 100-108.

Chou, Y. C., Sun, C. C., & Yen, H. Y. (2012). Evaluating the criteria for human resource for science and technology (HRST) based on an integrated fuzzy AHP and fuzzy DEMATEL approach. *Applied Soft Computing*, *12*(1), 64-71.

Csutora, R., & Buckley, J. J. (2001). Fuzzy hierarchical analysis: the Lambda-Max method. *Fuzzy sets and Systems*, *120*(2), 181-195.

El-Koumy, A. S. (2004). Language performance assessment: Current trends in theory and research.

Echauz, J. R., & Vachtsevanos, G. J. (1995). Fuzzy grading system. *IEEE Transactions on Education*, *38*(2), 158-165.

Ellis, R. (2003). *Task-based language learning and teaching*. Oxford University Press.

Klir, G., & Yuan, B. (1995). *Fuzzy sets and fuzzy logic* (Vol. 4). New Jersey: Prentice hall.

Lee, A. H., Chen, W. C., & Chang, C. J. (2008). A fuzzy AHP and BSC approach for evaluating performance of IT department in the manufacturing industry in Taiwan. *Expert systems with applications*, *34*(1), 96-107.

Ma, J., & Zhou, D. (2000). Fuzzy set approach to the assessment of student-centered learning. *IEEE Transactions on Education*, *43*(2), 237-241.

Nunan, D. (2004). *Task-based language teaching*. Cambridge University Press.

Norris, J. M. and Ortega, L. (2001), Does Type of Instruction Make a Difference? Substantive Findings From a Meta-analytic Review. Language Learning, 51: 157-213.

Sets, F. (1965). L. Zadeh. *Information and Control.–NY*, (8/3), 338-353.

Weigle, S. C. (2007). Teaching writing teachers about assessment. *Journal of Second Language Writing*, *16*(3), 194-209.

Zadeh, L. A. (1996). Fuzzy sets. In *Fuzzy Sets, Fuzzy Logic, And Fuzzy Systems: Selected Papers by Lotfi A Zadeh* (pp. 394-432).

SHS Web of Conferences **53**, 03006 (2018) https://doi.org/10.1051/shsconf/20185303006
ICHSS 2018

A case on Tamil education in Malaysia: Is Tamil education being sidelined?

Normardhiah Ibrahim[1]

MPhil Candidate, Department of Management and Humanities, Universiti Teknologi Petronas, 32610, Tronoh, Perak

Abstract. This paper is a part of an empirical research that investigates the status of Mandarin and Tamil language in Malaysia. The empirical research is cleaved into three research questions and hence, three fieldworks. This paper singles out the findings from one of the fieldworks of this study and seeks to lay out and discuss the specific findings with the aim to unveil issues that surround Tamil education in Malaysia. The findings highlighted in this paper involve an interview with a principal of a Tamil vernacular primary school in northern Malaysian state. With the centre of the topic polarized around the notion of Tamil education in Malaysia, the findings obtained from the interview will be presented and discussed as to attain some ideas not only regarding the status of Tamil education in Malaysia, but also the underlying factors that instigate the issues surrounding it.

1 Introduction

The objective of this paper is to explore the issues that revolve around the notion of Tamil education in Malaysia. The issues mentioned will be evidenced by findings obtained from one part of the fieldwork carried out for this study. This paper will touch on issues related to certain aspects such as monetary flow, students' enrolment, shortage of trained and graduate teachers and the development of the so-called minority language, as they revolved around the Tamil education in Malaysia. The analysis and the discussion of this findings are presumed to be able to illuminate on the problem statement of this study, which is to find out the status of Tamil education in Malaysia.

Holding the identity as a heterogeneous country with more than half of its people capable of speaking more than one language, it is safe to assume that the concern related to minority languages and vernacular education in Malaysia would not simply be overlooked. However, issues concerning Tamil language and Tamil vernacular school continue to be a gray area for researchers and scholars who seek to investigate the status of Tamil language and the development of Tamil education in Malaysia. Issues such as the constant low rate of enrolment of both Indian and non-Indian students to Tamil vernacular schools, the drastic reduction of the number of Tamil vernacular schools (http://www.malaysia-today.net/2011/11/05/tamil-schools-in-malaysia-denied-fully-financial-aided-status-is-unconstitutional-a-illegal/) and problems in obtaining fully-aided status are examples of persisting concerns. Findings obtained from an interview with the principal of a Tamil vernacular primary school in northern Malaysian state seem to prove that perhaps, Tamil education in Malaysia is in fact running on a satisfying pace.

[1] Corresponding author: normardhiahibrahim@yahoo.com

In the next section of literature review, issues surrounding Tamil education in Malaysia are briefly laid out and discussed.

2 Literature review

2.1 Issues surrounding Tamil education in Malaysia

The growing number of issues that revolve around Tamil education in Malaysia is the testimony to the criticality of the development and future of Tamil language. A surfeit of issues pertaining to Tamil education in Malaysia was listed by Paraman in Malaysia Today, where he emphasized how Tamil education in Malaysia continue to face dire issues amid rapid development in the Malaysian education system (Paraman, 2011).

2.1.1 Issue 1: Low student enrolment

One of the biggest issues that continues to haunt Tamil education is the issue of low enrolment of students into Tamil vernacular primary school (SJKT).

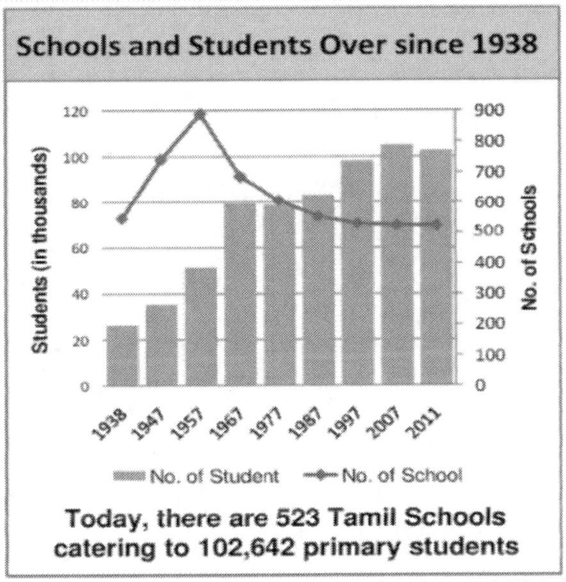

Source: EMiS, MOE; Tamil Foundation Research Unit

Fig. 1 Number of Schools and Students 1938-2011

Figure 1 depicts a continuous improvement in terms of the number of student enrolling into Tamil vernacular primary school from year 1938 until 2007, with slight decline in year 1977 and year 2011. Judging from the trends shown in Figure 1, the future of Tamil education seems to be promising in terms of the students' enrolment into the Tamil vernacular primary schools. However, year 2013 saw the consistent decrease of the number of students enrolling into Tamil vernacular primary schools. Figure 2 below shows the number of schools and students of SJKT from year 2013 until 2017.

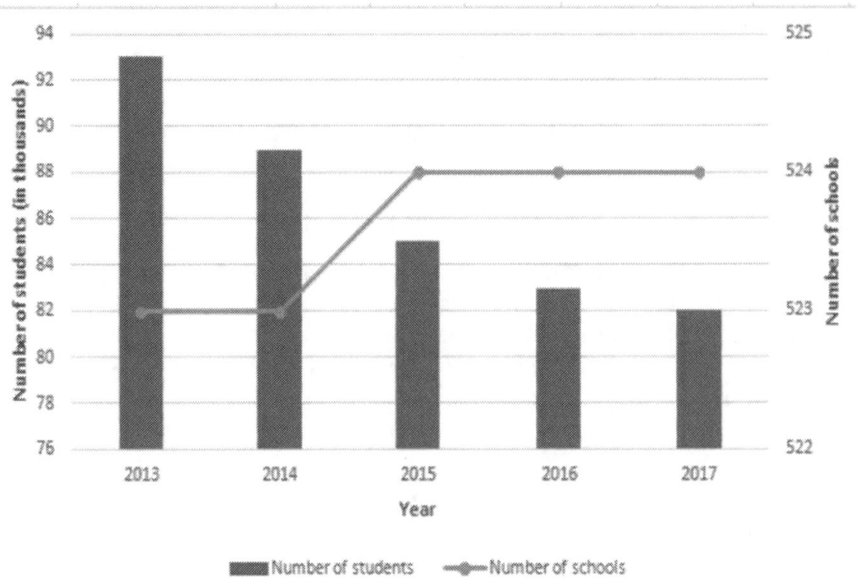

Source: Malaysia Educational Statistics 2013-2017

Fig. 2 Number of Schools and Students 2013-2017

In Figure 2, it can be clearly seen that the number of students enrolling into Tamil vernacular primary schools has been continuously decreasing from year 2013 until 2017 (Ministry of Education Malaysia, 2013, 2014, 2015, 2016, 2017). This, however, is not the only issue. For instance, it has been reported by the Tamil Foundation Research Unit that 63% of Tamil schools are under-enrolled (<150 students). In fact, there is a total of 68 schools (13%) claimed to have fewer than 25 students, and on the verge of closure (Elanjelian, 2011). In 2016, there was only one Malay student enrolled in Tamil school out of the 2,000-odd students entering Year One in 70 Tamil schools state-wide ("Malay kid enters Tamil school," 2016). Similarly, in year 2017, only one Malay student enrolled in Tamil school ("Malay boy only Year One pupil in Tamil school," 2017). Recent news reported that in year 2018, more than 20 Tamil vernacular primary schools in Kedah are in the midst of dire straits owing to the declining number of students' enrolment. In fact, it has also been reported that one of the oldest schools in the state, SJK (T) Ladang Dublin Bahagian 7, does not have any Year One pupil this year. The school also has no student in Years Two to Four. There are only five students in Years Five and Six.' ("20 Tamil schools in Kedah hit by low enrolment," 2018)

2.1.2 Issue 2: Number of schools

Looking from the angle of the number of Tamil vernacular primary schools in Malaysia, the situation is almost as ram shackled as the issue of low enrolment previously discussed. This phenomenon is starkly juxtaposed with that of the phenomenon of the number of schools shown within the same figure (Figure 1). The blue graph representing the number of schools from the year 1938 until 2011.

In 2015, there were only 523 SJK (T), which makes it the lowest amongst vernacular schools whereas, SJK (C) with 1296 schools and national (SK) with 5943 schools. This shows that Tamil schools, when compared to other vernacular schools, is the smallest and perhaps, least prioritized group. It is not surprising that in fact, Tamil education system is seen as a sham and a mockery by some, even suggesting Tamil education system the Cinderella of the whole Malaysian education system (Arasaratnam, 1970). While the number of SJK (C) schools and national schools has increased throughout the years, with SJK (T) it does not seem to be the case. The number of SJK (T) schools, unlike SJK (C) and national schools, remains unchanged since the last nine years (refer Table 1).

Table 1. Comparison between the three types of National-based primary schools in Malaysia, 2006 and 2015 *(number of schools)*

Type of Schools	2006	2015	-/+
National	5774	5943	169
National-type (C)	1288	1296	8
National-type (T)	523	523	0
Total	7585	7762	177

Table 1 shows the comparison between the three types of National-based primary schools in Malaysia (in terms of the number of schools), with the year of focus being the year of 2006 and 2015. The main purpose of presenting Table 1 is to highlight how there has been no changes in the number of Tamil vernacular primary schools from the year 2006 and 2015 (highlighted in red). Looking at this situation, it can be presumed that one of the factors might be lack of funding. This is because, if we go through the history of Tamil schools in Malaysia, we can see that most are run and funded by the community itself or by Christian missionary bodies such as the Ladies' Bible and Tracts Society, the Society for the propagation of Christianity, and the Big Church Missions. Figure 3 below illustrates the growth of National type schools with National Type-T way behind National and National Type-C:

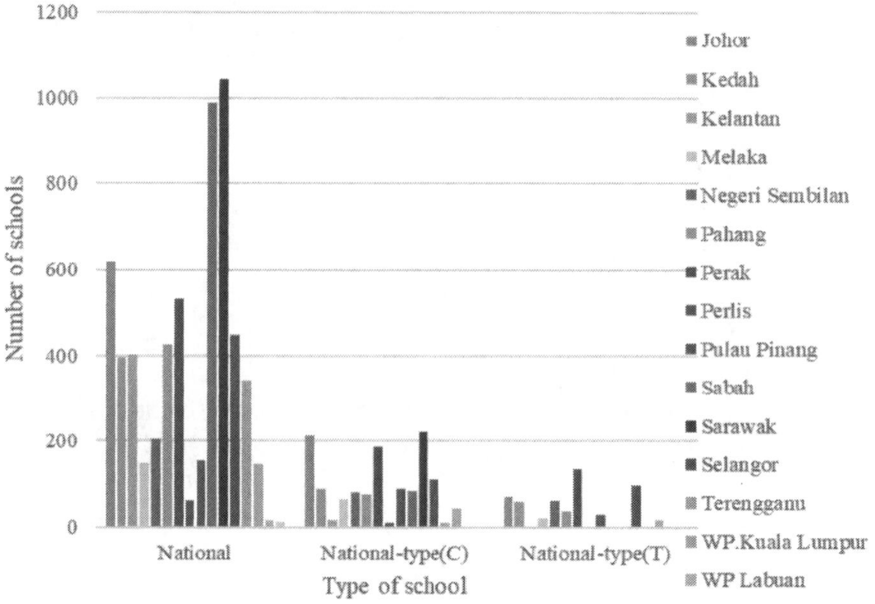

Fig. 3. Number of Schools According to States and Types of Schools

Figure 3 shows the trend of the number of schools (three types of National-based schools) according to the states in Malaysia. In Figure 3 above, the number of Tamil vernacular primary schools in Malaysia is dangerously small as compared to the National schools and Chinese vernacular primary schools. However, this is not the main emphasis in Figure 3 above. The main emphasis is to highlight the number of Tamil vernacular primary schools which dismally, out of the 14 Malaysian states, five states has zero number of Tamil vernacular primary schools. These states are Sabah, Sarawak, Terengganu, Wilayah Persekutuan (WP) Labuan and WP Putrajaya.

3 Methodology and method

3.1 Methodology

The methodology for this study is phenomenology. However, this study does not utilize the whole concept of phenomenology but only apply the perspective of phenomenology. In other words, this study is phenomenologically-informed. Phenomenology, according to (van Manen, 2007) can be defined as "a project of sober reflection on the lived experience of human existence–sober, in the sense that reflecting on experience must be thoughtful, and as much as possible, free from theoretical, prejudicial and suppositional intoxications" (pg. 12). This means to say, phenomenology is a methodology that utilizes human-lived experience as flexible foundation and source to investigate issues. Ultimately, the selection of the phenomenological perspective as the methodology for this study is rooted in the realization that this study revolves largely around the notion of 'human lived experience'. This core perspective of phenomenology is seen as a crucial perspective for this study. Through the process of analyzing the experience of the research participant, this study aims to obtain an understanding and answers for the presumption constructed on the research objectives and research problems. Another way of saying this is that, as regard to the objective of this study, human lived experience is a suitable way in making our steps towards seeking out the answers for the research questions of the study.

3.2 Method

Selection of methods is another step of carrying a study. Methods are one of the components in a paradigm (Scotland, 2012) and are described by Crotty (1998) as specific techniques and procedures used to collect and analyze data. As mentioned in the previous section, the selection of methods is established from the thread of the methodology selected. Since the method of semi-structured interview allows a researcher to listen to the voices of the people who has lived in the experience, this method is seen as a suitable method to investigate a complex notion such as language, status of language, minority language and language education.

The selection of a teacher as a participant for the fieldwork is grounded by the researcher's *a priori* idea of teachers being at the forefront of the act of implementing the language policies in schools. There are three primary schools involved in the semi-structured interview conducted for policy implementers, which in this study, is the specific label tagged on the teachers. The primary outline of the fieldwork is to conduct semi-structured interview with teachers from the selected vernacular schools. The schools involved are national primary school (SK), Chinese vernacular primary school (SJKC), and Tamil vernacular primary school. In this paper, however, the findings that will later be discussed is extracted from an interview with a principal of a Tamil vernacular primary school in northern Malaysian state. In other words, the findings from the remaining two fieldwork places; national primary school and Chinese vernacular primary school are not discussed in this paper.

During the initial attempt of entering a Tamil vernacular school, one Tamil school in a rural area was selected and the procedure was piloted. Nevertheless, upon meeting the principal of the school to arrange a scheduled interview with one of the teachers, the principal clarified that the school did not have any Malay or Chinese students studying in the school. This unexpected turnabout has resulted in an unplanned interview with the principal herself as the researcher believes there must be some underlying reasons as to why there has been no Malay or Chinese students enrolling in the school. This is important because this specific inquiry may offer answers to one of the fundamental questions behind this study; is Tamil language being side-lined?

Considering that a principal still falls under the same roof as a teacher since a principal is also an educator and in the context of this study, a policy implementer, data from this unplanned interview with the principal was analysed to provide answers that can illuminate some fundamental questions of this study.

4 Findings

In this section, the findings from one of the fieldworks in this study is laid out and discussed. These findings are revolved around the data obtained from an interview with the principal of a Tamil vernacular school as mentioned earlier.

During the interview with the principal, she assents there is indeed a decrease in the number of Chinese and Malay students enrolling in Tamil schools. When asked to provide a reason for the persistent decrease, the principal stated that it was majorly because parents refused to send their children to Tamil schools. She, however, declined to explain further. Following this, the researcher asked her opinion on the development of Tamil education in Malaysia. She confidently stated there have indeed been many improvements on Tamil education. She accentuated her answer by explaining these improvements include the syllabus for Tamil education and specifically, the appearance of Tamil schools. To prove her point, the principal referred to a book published by the Ministry of Education titled '200 Tahun Kegemilangan Pendidikan Tamil di Malaysia' (200 Years of the Glory of Tamil Education in Malaysia) and pointed to the improved physical appearance of Tamil schools. With the book playing the role of a proof, she continued showing pictures of the improvements, that is, differences in the old appearance of Tamil school in the 90's and compared them with pictures of obviously much improved appearance of the Tamil schools today.

For the next question in which the principal was asked why there are very small number of Malay and Chinese (non-Indians) enrolling in the school, she shrugged her shoulders and said it was probably because they did not want to attend the school. On her opinion regarding students in her school, she gave a simple answer saying that everything was okay. Again, she declined any further explanation. For the last question of the interview, the researcher decided to ask about the reported decreasing number of students enrolling into Tamil schools. There was a sense of unwillingness to agree in the answer given by the principal as she stated that, "yes, there is a decrease". She explained further by saying that the approximate decreasing rate of Tamil schools are from 1000 plus to 542 schools.

5 Discussion

This section presents the discussion on the paradox between the findings obtained from the interview with the principal of a Tamil vernacular primary school in northern Malaysian state and the issues surrounding Tamil education in Malaysia. The crux of the discussion will be regarding the issue of low rate of enrolment of the students to Tamil vernacular primary schools, the issue of monetary flow and the issue of physical appearance and infrastructures of Tamil vernacular primary schools in Malaysia.

5.1 The paradox: Is Tamil education being sidelined in Malaysia?

From the initial analysis of the data concerning the interview with the principal, it is pertinent to presume that the principal of the Tamil vernacular school is playing the role of an advocate of Tamil education. The findings obtained from the interview suggest that a lot has to do with the earlier mentioned issues that surround Tamil education. The principal indeed concurs some improvements in Tamil education in Malaysia are needed. This implies there is in fact more to the matter, essentially when we compare the nature of the issues with the role of the principal in Tamil education. While we continue to be very much concerned, it is also important to listen to the voices of the people who are the insiders of the vernacular schools themselves. In other words, the voice of the principal interviewed for this study can unearth some real condition and phenomenon of Tamil education in Malaysia.

Coming back to the statement given by the principal, her statement interestingly contradicts prevalent issues surrounding Tamil education system. This is because she feels that notion of improvements should equally take consideration the syllabus of Tamil education books and the appearance of Tamil vernacular primary schools. She was clearly determined to prove her point by showing me pictures in the book '200 Tahun Kegemilangan Pendidikan Tamil di Malaysia' (200 Years of the Glory of Tamil Education in Malaysia) how Tamil schools have better their syllabus and appearance. This assertion by the Principal that Tamil schools have improved over the years has inevitably stirred up lingering question as to whether Tamils vernacular primary schools are truly being sidelined. However, it can be argued, in the sense that perhaps the principal, being an Indian by ethnicity, culture, religion and language, subconsciously *chose* to believe that Tamil vernacular primary schools are indeed undergoing improvements. This argument is grounded on the fact that in the interview itself, she had admitted there was indeed a decrease in the number of students enrolling in Tamil schools. Juxtaposed against her statement that Tamil schools have undergone improvements and that everything was okay in her school, in any case, raises the question as why there continue to be a decline in enrolment of students in Tamil schools.

First, in relation to the point made by the principal regarding the physical appearance of the school, there is a point that stepped into the light. In this study, the context of the students mentioned is focused on Chinese and Malay students attending Tamil schools. Although there is a chance that an improved appearance of a school may indicate that the school is stable, there is slight possibility that enrolment of Chinese and Malay students into the school is not as 'improved' as the appearance. This is considered based on data gathered from interview sources and fieldwork conducted during the study that provided clear evidences of incremental decline in the number of Chinese and Malay students enrolling into Tamil schools. A statement by the State Tamil schools' coordinator, in which he hopes the school's new building will attract more parents to send their children to the school (Lian, Aravinthan, & Chiam, 2016) may be a cause for optimism. But his statement, when laid side by side with the statement by the principal regarding the appearance of the school seems to elicit distinctively different meaning concerning improved appearance of Tamil schools and improved enrolment. The principal and the State Tamil schools' coordinator appear to have different perception as to whether an improved appearance of the schools can have an impact on students enrolling into Tamil schools.

Second, still concerning physical appearance and the number of students' enrolment in Tamil schools, in 2016, there was also a similar disparity with a statement given by the Deputy Education Minister, Datuk P. Kamalanathan in which he mentioned the number of students in Tamil schools in Malaysia have undergone numerous improvements in terms of facilities and infrastructure. He claimed that over the years, Tamil education has grown in leaps and bounds and that apart from the physical changes to the schools, there were also many students, both Indians and non-Indians, enrolling into Tamil schools ("Tamil education goes online," 2016). Yet, his statement contradicted in the number of students enrolled in Tamil schools which showed a decline from about 97,000 in 2013 to 92,000 in 2014 ("Only 13600 Students Enrolled in Tamil School," 2015). This contradiction between the data and the statement by the Deputy Education Minister creates several presumptions. One of them being the possibility that language in Malaysia, although arguably treated solely as a language, is also essentially intertwined with other aspects such as religion, culture, political assertion and ethnicity.

Third, in term of monetary flow, undoubtedly there has been a reliable flow of monetary assistance for Tamil schools. Information obtained during fieldwork for example, indicates a sufficient flow of donations to develop Tamil schools in this northern Malaysian state. This includes the school of the Principal I interviewed earlier which is one of the recipients of the financial assistance. This information triggered a few thoughts on whether this specific school is facing financial instability; which if it is true, it would be a contradiction to the discussion about the appearance of the school. Although, the fact that the appearance of the school has improved purports that other important aspects of the schools such as education materials and utilities are also in good condition for them to spend the money on improving the physical appearance of the school. In other words, it can be concluded at this stage of the study that Tamil schools, at least the Tamil school that I investigated so far, does not suffer from financial instability. Donations to the school are possibly a gesture of gratitude from parties who are concerned with Tamil education in Malaysia.

Fourthly, in terms of the number of Tamil vernacular primary schools in Malaysia, there seems to be a new illumination from Asmah Haji Omar in her book in 2016, in which she mentioned that there was an increase in the number of Tamil vernacular primary schools in Malaysia in 2015. According to Omar, one Tamil vernacular primary school was opened on 1 March 2015 as a new initiative of the Prime Minister which added to the number of SJK (T) in Malaysia, with a total number of 524 schools in 2015. In addition, Omar also mentioned that there would be six more new schools on the way (Omar, 2016). This statement by Omar, when placed in parallel with that of the discussion for Table 1 in Section 2, creates another confusion. The confusion arises from the fundamental question of this study itself; is Tamil language and Tamil education in Malaysia being sidelined? While the statistical evidences shown in Table 1 and 2 imply that the development of Tamil education seems to be in the backwater of school education in Malaysia, the statement from Omar seems to prove otherwise. Omar highlighted that the addition in the number of Tamil vernacular primary schools is done in the interest of providing the Indian community in this country basic education through their mother tongue (Omar, 2016).

From the discussion, there is a possibility that perhaps, Tamil language is not being sidelined or marginalized in Malaysia, but rather the language is being deprioritized by the responsible parties. Also, it may perhaps be best to hold on to a statement given some time ago by the MIC president Datuk Seri G. Palanivel. He stresses that the government has promised the Indian community that not a single Tamil school will be closed. He emphasized that Tamil schools, where necessary, would instead be relocated to areas with higher Indian population densities ("Only 13600 Students Enrolled in Tamil School," 2015). Although there may be some amongst the Indian community who are pessimistic, it is a fact that so far, no Tamil school has been closed and that the future of Tamil language and Tamil education in Malaysia to date, is guaranteed.

6 Conclusion

The issues surrounding Tamil education are not a bullet-proof indication that Tamil education in Malaysia is at the brink of deterioration and disappearance. The findings extracted from the interview with the principal of a Tamil vernacular school revealed that there is more to the issues that surround Tamil language and Tamil education in Malaysia. The intricacies of the problem are presumed to be rooted in the fact that we are dealing with the notion of minority language in a heterogeneous country. Regardless, solving the issues that surround Tamil education in Malaysia can be attained. It is crucial that Malaysians from all walks of life adopt the issue of Tamil language and Tamil education as a collective concern that should continue be investigated and solved.

7 References

20 Tamil schools in Kedah hit by low enrolment. (2018). *The Star Online*. Retrieved from https://www.thestar.com.my/news/nation/2018/01/17/20-tamil-schools-in-kedah-hit-by-low-enrolment/

Arasaratnam, S. (1970). *Indians in Malaysia and Singapore*. London: Oxford University Press.

Crotty, M. (1998). *The Foundations of Social Research*. Thousand Oaks: Sage.

Elanjelian. (2011). Future of Tamil Schools in Malaysia. Retrieved September 2, 2018, from https://www.scribd.com/document/58799035/Future-of-Tamil-Schools-in-Malaysia

Lian, V., Aravinthan, R., & Chiam, C. (2016). 10 Tamil schools do not have Year One pupils. Retrieved from http://www.thestar.com.my/news/nation/2016/01/01/10-tamil-schools-do-not-have-year-one-pupils/

Malay boy only Year One pupil in Tamil school. (2017). *The Star Online*.

Malay kid enters Tamil school. (2016, January). *The Star Online*.

Ministry of Education Malaysia. (2013). *Malaysia Educational Statistics 2013*.

Ministry of Education Malaysia. (2014). *Malaysia Educational Statistics 2014*.

Ministry of Education Malaysia. (2015). *Malaysia Educational Statistics 2015*.

Ministry of Education Malaysia. (2016). *Malaysia Educational Statistics 2016*.

Ministry of Education Malaysia. (2017). *Malaysia Educational Statistics 2017*.

Omar, A. H. (2016). *Language in the Malaysian Education System: Monolingual Strands in Multilingual Settings*. New York: Routledge.

Only 13600 Students Enrolled in Tamil School. (2015). Retrieved from http://itimes.my/headline-only-13600-students-enrolled-in-tamil-school/

Paraman, V. (2011). Tamil Schools In Malaysia Denied Fully Financial Aided Status. Retrieved August 3, 2018, from www.malaysia-today.net/2011/11/05/tamil-schools-in-malaysia-denied-fully-financial-aided-status-is-unconstitutional-a-illegal/

Scotland, J. (2012). Exploring the philosophical underpinnings of research: Relating ontology and epistemology to the methodology and methods of the scientific, interpretive, and critical research paradigms. *English Language Teaching*, *5*(9), 9–16. https://doi.org/10.5539/elt.v5n9p9

Tamil education goes online. (2016). Retrieved from https://www.pressreader.com/malaysia/new-straits-times/20161030/283154313210940

van Manen, M. (2007). Phenomenology of Practice. *Phenomenology & Practice*, *1*(1), 11–30.

https://doi.org/10.1051/shsconf/20185303006

Exploring Undergraduates' Judgements on the Qualities of Corrective Feedback for English Pronunciation Errors

Zulqarnain Abu Bakar[1], *Abdul Mutalib* Embong[2]

[1]Department of Management & Humanities, Universiti Teknologi Petronas, Bdr Seri Isknadar, Perak, Malaysia
[2]Department of Management & Humanities, Universiti Teknologi Petronas, Bdr Seri Isknadar, Perak, Malaysia

Abstract. The present study is an investigative work to examine learners' judgements on some proposed error correction methods, in connection with mispronounced English words. Learners are to evaluate the qualities of six different types of correction approaches, in terms of 'embarrassment', 'comprehensibility' and 'effectiveness' through a set of 36 error-correcting-act video clips representing the six correction types. A set of corresponding questionnaire survey in exploring respondents' insights was administered. Comparisons by their learning experience (year) and environment (institution) were carried out to study any effect on their judgements. The results of the survey conducted on 88 undergraduate students revealed that the respondents seem to share the same pattern of observations despite their learning experience and types of higher institutions. They have concern on the aspect of 'embarrassment' in relation to all corrective feedback types except for *Repetition* and *Clarification Requests* which are more communicative in their approach.

1 Introduction

Pronunciation skill has been the concern of many language learners, particularly to those who are learning a target language that shares very minimal similarity to the inventory of their first language (L1) sounds system. Being able to communicate in a target language, simply means that one is orally competent to speak and understand basic utterances of the language. The ability to produce the correct sound, pronounce meaningful words and phrases, and communicate for basic needs in the target language besides reading and writing, would boost learners' confidence to explore further and eventually master the language in question. Such description applies to all language learning scenario for whatever purposes. For one thing, learners and teachers of English as the Second Language (ESL) programs are concerned on the accuracy of the target pronunciation. Thus, attention on the guide to improve target pronunciation is paramount. However, the focus on pronunciation had been perceived as pointless by many ESL practitioners in the past (Chomsky, 1959; Krashen,1982,1985,1994). Pronunciation research became unpopular and was neglected due to the beliefs that learners will not be able to acquire native-like speech (Celce-Murcia, Brinton & Goodwind, 1996; Dalton & Seidlhofer, 2000). This aspect of language proficiency was neglected for teachers were skeptical of the right direction to follow in guiding their students. Researches on teaching pronunciation were close to none till the late 1990's. On the other hand, the interest in improving pronunciation is close to the hearts of the learners who inspire to speak the target language fluently (Abu Bakar & Abdullah, 2015). The strong concern on pronunciation training started to gain researchers' attention in early 2000's (Murphy & Baker, 2015) as well as Saito and Lyster (2012) compiled and studied several pronunciations training oriented classrooms to understand closely what take place during pronunciation-focused activities. Undoubtedly, there are numerous issues that requires experts' scrutiny in relation to enhance the impact of pronunciation training, in relation to teaching delivery, corrective feedback, attention, consciousness and awareness to name a few.

Similar to feedback on errors committed in other mechanics of a language, like grammar, vocabulary and writing, feedback on pronunciation errors is believed to be practical and applicable in assisting learners making adjustment towards the target accuracy (Neri, Cucchiari & Strick, 2001). In other words, it acts as a mechanism to raise learners' awareness on the target forms that they have been exposed to (Ellis, 2001; Ellis, Basturkmen &Lowen, 2001). Levelt (1989), in his model of speech production, theorizes that every learner is equipped with a support, what he termed as 'perceptual loop of monitoring'. The loop functions as a monitoring 'device' that evaluates the accuracy of a speaker's

*Corresponding author: zulqab@utp.edu.my

utterances. Upon self-realizing the mispronounced word, the speaker's knowledge on the inventory of the sound system will be triggered, the intended sound will be retrieved and the reformulation of the correct, intended sound takes place. The same principle can be applied in the pronunciation teaching and learning environment. However, instead of getting signals that triggered from the built-in parser which may not be readily functioning automatically as in the L1 system, an external assistance through corrective feedback can be utilized to raise learner's attention on the error committed. The learner will then retrieve the corresponding sound and make the correction accordingly (Abu Bakar, 2008). Lyster and Ranta (1997) who believe in the utilization of corrective feedback in response to a learner's errors, proposed six types of mechanisms to assist learners in adjusting toward the target forms of a language point in question. The six suggested corrective feedback types for the errors are *Explicit correction, Clarification Request, Verbal Metalinguistic Feedback, Elicitation, Repetition and Recast. Recast,* however has been perceived to have a very low tendency of learner uptake due to its ambiguity and subtleness to the extent that a learner may miss the corrective move (Lyster & Ranta, 1997). In the current study, *Recast* is not considered as a potential corrective move that may be practically applicable. Thus, it is not adopted as one of the corrective feedback types in the current study. Another form of corrective feedback used in many pronunciation workbooks is *Visual Cue* (Celce-Murcia, Brinton & Goodwind, 1996). This is another alternative form of corrective move that may be potentially applicable in the context of the current study. The following examples of the different corrective feedback types have been adopted and adapted to suit the application in the context of pronunciation training environment in the study as illustrated by Abu Bakar (2008).

Explicit Correction refers to the explicit provision of the target form. This is the type in which the instructor provides the correct form and clearly indicates the inaccuracy of a particular utterance.

Example:
Instructor: "…er not 'viskers', it's 'whiskers'…"

Here consciousness and attention are activated through auditory or external loop monitoring (Levelt, 1989). The auditory feedback gives the learner the opportunity to tune in to the whole pronunciation of the words and make a comparison between target and non-target pronunciation. As consciousness and attention are triggered, the target phoneme is retrieved, and the intended production is reformulated.

Clarification Requests indicates to the learner that the instructor has misunderstood his or her pronunciation due to the ill-formed pronunciation in some way and a reformulation is required.

Example:
Instructor: "…Sorry, what do you mean by flower base?" (as opposed to 'vase')

Here the non-target sound is highlighted to trigger the learner's consciousness and attention (Schmidt, 1990). The target phoneme is retrieved, and the word is reformulated.

Verbal metalinguistic feedback may contain comments, information or questions related to the well-formedness of the learner's utterance, without giving the explicit correct form of the pronunciation. Metalinguistic reference is employed to guide the learner by first implying that there is an error somewhere in the word or phrase.

Example:
Instructor: "…no...no…no, is that how you pronounce it?... should it be with the long or short vowel?"

In this type of feedback, the learner's consciousness and attention are drawn toward the metalinguistic knowledge related to the phoneme in focus. Upon realization of the discrepancy, the target phoneme is retrieved, and the target word is reformulated.

Elicitation refers to steps that the instructor employs to indirectly elicit the correct form of the pronunciation. First, the instructor elicits the completion of his or her own utterance by strategically pausing to allow learners to "fill in the blank".

Example:
Learner: "...because of the complica/ʒ/n...
everything went wrong"
Instructor: "No, not quite... How do you pronounce this word? (pointing to the word "she").

The target sound is /ʃ / in "complication". The elicit completion moves may be led by a metalinguistic comment e.g. "No, not voiced..." Next, the instructor uses a question, hinting at a clue that relates to the learner's prior knowledge e.g., "How do you pronounce this word (pointing to the word "she"). Then, the instructor may ask the learner to reformulate his or her utterance. In this type of feedback, consciousness and attention are triggered by the rising tone preceding the problematic pronunciation. The highlighting of a common word in which the target phoneme exists takes place and the target phoneme is then retrieved, and the production reformulated.

Repetition refers to the instructor's duplication in isolation of the learner's erroneous pronunciation. This may be done by adjusting the intonation to highlight the error.

Example:
Learner: "the /kɒmplɪkt/ (as opposed to 'conflict') is due to..."
Instructor: "the /kɒmplɪkt/?

The instructor repeats "the /kɒmplɪkt/?" with rising tone of voice indicating puzzlement. Here consciousness and attention are triggered by reinforcement of the non-target production. In response, based on the learner's prior knowledge, the target phoneme is retrieved and the target production reformulated.

Visual Cues refers to a type of feedback, based on the use of visual aids in pronunciation workbooks and software. The instructor directs the learner's attention to the visual representation of the segment in question.

Example:
Learner: ...they have booked the /hctel/ (as opposed to 'hotel').
Instructor: take a look at this (showing the cross-sectional view of the
articulatory system depicting the vowel trapezium at the same time
the target phoneme is uttered) .../həʊtel/

Error corrective feedback is one of the aspects in language training that has also been the interest of ESL researchers who believe that accuracy can be improved, given proper input and guidance in assisting learners to make adjustment (Lyster & Ranta, 1997; Ellis, 2001). However, the question is on how do learners perceive the qualities and judge the corrective feedback they receive in their experience, trying to enhance their performance.

A clear input in a lesson or instruction, naturally should be supported by an equally clear corrective move to facilitate learning. An error feedback will serve no purpose if it is not received and responded for the intended purposes accordingly. In other words, it should be provided, for one thing, with clear comprehensibility to ensure that learner understands the intervention and reacts with the appropriate reformulation (Ellis,2017; Price, Handley, Millar & O'Donovan, 2010; Poulos & Mahony, 2008). A learner needs to understand what has gone wrong before being able to provide the targeted output as desired. In the case for pronunciation training, the corrective move has to be clearly referred to the problematic sound as to avoid mistaken interpretation of the intervention. For example, a learner may mistakenly perceive a corrective feedback meant for a problematic sound in a target pronunciation to the choice of vocabulary used in the utterance. This will result in a mismatch in the response, contrary to the desired output.

Another aspect of corrective feedback that is considered vital as identified by Poulos and Mahony (2008) is effectiveness. This aspect of quality in corrective feedback deals with the way learners feel about the approach in the feedback provided and the believes in its workability in terms of the mannerism or how the corrective move is delivered to them. Naturally, a receiver of any corrective feedback would be able to experience if the assistance provided is in fact useful. Effective correction moves promote positive attitude towards improving one's targeted output (Vanosfadrani & Basturkment, 2009).

Face-saving and embarrassment are very vital to be considered in corrective feedback moves, particularly in the context of adult learning. A corrective feedback that is perceived to be disconcerting may not serve its purpose to assist learners making adjustment in the target output. It may deter learner's focus and confidence instead (Agudo & de Dios, 2013). A study conducted by Lasabagaster and Sierra (2005) on a big group of ESL learners revealed that learners may be disturbed by embarrassment, following a corrective feedback perceived to be jeopardizing learners' face.

Considering the aspect of qualities and judgements on the different corrective feedback types, the study seeks to accomplish the following objectives:

1. To elicit the learners' overall understandings of the different qualities in the different corrective feedback types
2. To elicit the learners' assessments on the qualities of the different corrective feedback types based on their learning experience
3. To elicit learners' assessments on the qualities of the different corrective feedback types based on and their learning environments
4. To elicit the learners' specific concern on the qualities of the corrective feedback types

Findings from these objectives will provide learners' perspectives on the qualities and judgements they have towards the different error correction types in questions.

2 Method

A set of survey questionnaire, corresponding to 36 corrective feedback video clips was employed to observe learners' assessment on 'Comprehensibility', 'Effectiveness' and 'Embarrassment', i.e., the qualities or the evaluation criteria in relation to the six different corrective feedback types. The matrix below summarizes the design of the questionnaire.

Table 1.0 Matrix of error correction types and evaluation criteria in relation to phonemes

Error Correction Type	Evaluation Criteria					
	Comprehensibility (X)		Embarrassment (Y)		Effectiveness (Z)	
Explicit Correction	*[I] C1X	[th] C33X	[o] C35Y	[zh] C26Y	[e] C30Z	[sh] C21Z
Visual Cues	[e] C25X	[sh] C34X	[I] C7Y	[th] C29Y	[o] C32Z	[zh] C2Z
Clarification Request	[o] C28X	[zh] C8X	[e] C3Y	[sh] C31Y	[I] C12Z	[th] C24Z
V. M. Feedback	[I] C16X	[th] C4X	[o] C23Y	[zh] C13Y	[e] C9Z	[sh] C27Z
Elicitation	[e] C14X	[sh] C22X	[I] C19Y	[th] C10Y	[o] C5Z	[zh] C17Z
Repetition	[o] C11X	[zh] C20X	[e] C18Y	[sh] C6Y	[I] C36Z	[th] C15Z

*NB. The phonetic equivalences are as the following: [I]=/ɪ/, [th] =/ð/, [o]=/əʊ/, [zh]=/ʒ/, [e]=/eɪ/, [sh]= /ʃ/

Table 1.0 also shows the labels used for each clip in the video compact disk (VCD); these correspond with the item statements in the questionnaire form, in which, for example, [th] represents the phoneme in question, "Z" represents the qualities or evaluation criteria addressed in the statement, in this case "effectiveness", and "C15" denotes clip number 15. There were 88 respondents of the first and final year students representing two institutions. One institution employs a totally English instruction in class deliveries, while the other is in mixed environment in which instructions are delivered in both English and the mother tongue (Malay language). The Cronbach Alpha coefficient for all items of the questionnaire survey is 0.801, a value of high reliability.

3 Findings and Discussions

A total of 3168 assessments were recorded as each of the 88 respondents evaluated 36 clips. 38 males and 50 female students from institution A (total English environment) and institution B (a mixed English-Malay environment) respectively, representing the first and final year students completed the evaluation in a lab provided. Learners' responses were analysed through descriptive statistics and the Ducan Post Hoc tests.

In response to the objectives of the study to assess the learners' judgements on the qualities of the different types of corrective feedback, the results revealed quite similar results at all levels.

Overall understanding on the different qualities

In response to first objective of the study, the analysis on the qualities or evaluation criteria show that learners seem to have the overall understanding on the qualities of all the corrective feedback types. The p-value for all the correction types are significant, suggesting that learners could see clear differences in the qualities for each of the corrective feedback types, except for *Repetition*. In other words, they have clear idea on how the different qualities function or work in the different manners of corrections. In the case of *Repetition*, the judgement on the qualities are somewhat equal with the p-value of 0.703, which is not significant. Though the ratings seem to be positive with relatively strong means, "embarrassment" is seen to be the most concerned qualities among the learners as depicted from all Duncan Post Hoc tests for all types of corrective feedback, except for *Clarification Request* with which "comprehensibility" is the major concern.

Table 1.1 summarizes the learners' judgments on the qualities or evaluation criteria in relation to all corrective feedback types, showing the comparison between "comprehensibility", "embarrassment" and "effectiveness".

Table 1.1 General mean scores, standard deviation and significant difference of respondents' assessment in relation to the evaluation criteria using one-way ANOVA

Correction Type	Evaluation Criteria	Mean	N	Std. Deviation	p-value
Visual cues	comprehensibility	1.87	176	.894	.013*
	embarrassment	2.13	176	.898	
	effectiveness	1.89	176	.897	
	Total	1.96	528	.902	
Explicit Correction	comprehensibility	1.90	176	.856	< 0.0001*
	embarrassment	2.18	176	1.094	
	effectiveness	1.63	176	.663	
	Total	1.90	528	.915	
Elicitation	comprehensibility	2.07	176	.950	.002*
	embarrassment	2.28	176	.919	
	effectiveness	1.94	176	.863	
	Total	2.10	528	.921	
Verbal Metalinguistic Feedback	comprehensibility	2.33	176	.971	<0.0001*
	embarrassment	2.66	176	.995	
	effectiveness	2.13	176	.812	
	Total	2.37	528	.954	
Repetition	comprehensibility	2.60	176	1.075	.703
	embarrassment	2.68	176	.998	
	effectiveness	2.68	176	1.117	
	Total	2.65	528	1.063	
Clarification Requests	comprehensibility	3.03	176	1.039	.033*
	embarrassment	2.76	176	1.008	
	effectiveness	2.80	176	1.071	
	Total	2.86	528	1.044	

* Significant at the 0.05 level
Note: 1 = strongly agree, 2 = agree, 3 = neutral, 4 = disagree, 5 = strongly disagree
A high mean indicates a less positive attitude and a low mean indicates a more positive attitude

Assessments on the qualities of the different corrective feedback types by learning experience

In relation to the learners' assessments on the qualities based on their learning experience, i.e., in terms of duration of time they have spent studying at tertiary level, results show that both, the first and final year students from the two institutions seem to share almost similar concerns. Contrasting with the overall results in which learners showed clear judgements on most of the corrective feedback with regards to the qualities, in comparison with regards to the duration on their learning experience, the results revealed that both sets of learners showed almost similar concerns on *Explicit Correction* and *Verbal Metalinguistics Feedback*. The final year students also seem to have clear judgement in relation to *Elicitation*. The results seem to suggest that both groups perceive all the qualities of the corrective feedback types with almost equally similar concerns in terms of their understanding for all the corrective feedback types except for *Explicit Correction* and *Verbal Metalinguistic Feedback*, with which "embarrassment" remains as the most concerned qualities for all corrective feedback types. Table 1.2 summarizes both the first and final year learners' judgements on the qualities of all the corrective feedback types.

Table 1.2 Mean scores, standard deviation and significance for correction type in relation to the evaluation criteria by year

First Year						Final Year					
Cor. Typ.	Evaluation Criteria	Mean	N	Std. Dev.	p-value	Cor. Typ.	Evaluation Criteria	Mean	N	Std. Dev.	p-value
VSC	comprehensibility	1.85	86	.901	.095	VSC	comprehensibility	1.89	90	.892	.131
	embarrassment	2.14	86	.922			embarrassment	2.11	90	.880	
	effectiveness	1.91	86	.953			effectiveness	1.88	90	.846	
	Total	1.97	258	.931			Total	1.96	270	.876	
EPT	comprehensibility	2.00	86	.894	.004 *	EPT	comprehensibility	1.80	90	.810	.000 *
	embarrassment	2.22	86	1.089			embarrassment	2.13	90	1.104	
	effectiveness	1.74	86	.785			effectiveness	1.52	90	.502	
	Total	1.99	258	.948			Total	1.82	270	.875	
ELT	comprehensibility	2.10	86	.970	.221	ELT	comprehensibility	2.04	90	.935	.004 *
	embarrassment	2.28	86	.916			embarrassment	2.29	90	.927	
	effectiveness	2.03	86	.951			effectiveness	1.84	90	.763	
	Total	2.14	258	.948			Total	2.06	270	.894	
VMF	comprehensibility	2.19	86	.964	.005 *	VMF	comprehensibility	2.47	90	.962	.000 *
	embarrassment	2.53	86	1.026			embarrassment	2.79	90	.954	
	effectiveness	2.09	86	.792			effectiveness	2.16	90	.833	
	Total	2.27	258	.948			Total	2.47	270	.951	
RPT	comprehensibility	2.69	86	1.043	.841	RPT	comprehensibility	2.51	90	1.104	.782
	embarrassment	2.78	86	1.011			embarrassment	2.59	90	.982	
	effectiveness	2.73	86	1.056			effectiveness	2.62	90	1.176	
	Total	2.73	258	1.034			Total	2.57	270	1.087	
CLR	comprehensibility	3.08	86	.997	.196	CLR	comprehensibility	2.98	90	1.081	.169
	embarrassment	2.83	86	.984			embarrassment	2.70	90	1.033	
	effectiveness	2.86	86	1.031			effectiveness	2.73	90	1.110	
	Total	2.92	258	1.007			Total	2.80	270	1.078	

* Significant at the 0.05 level
Note: 1 = strongly agree, 2 = agree, 3 = neutral, 4 = disagree, 5 = strongly disagree
A high mean indicates a less positive attitude and a low mean indicates a more positive attitude
Abbreviations: VSC stands for *Visual Cues*, EPT for *Explicit Correction*, ELT for *Elicitation*, VMF for *Verbal Metalinguistic Feedback*, RPT for *Repetition*, and CLR for *Clarification Requests*

Assessments on the qualities of the different corrective feedback types by learning environment

In comparison of learners' assessments on the qualities of the different corrective feedback types by learning environment or institutions, the results reveal that learners of Institutions A seem to have more concern on 'embarrassment' as compared to those of Institution B, at which the learners seem to have insignificant difference in their assessment of the qualities except for *Explicit Correction* and *Verbal Metalinguistic Feedback*. This seems to indicate that learners whose English Language awareness are high, are also much sensitive in terms of face-saving for all corrective feedback types. Such obvious sensitivity does not happen in Institution B, perhaps this is due to the state of mediocrity environment where learners are mediocre in the knowledge of English Language. Table 1.3 summarizes the results on comparison of learners' judgements on the qualities of all the corrective feedback types between Institution A and Institution B.

Table 1.3 Mean scores, standard deviation and significance for correction type in relation to reaction type for respondents of Institutions A and B

Respondents of Institution A						Respondents of Institution B					
Cor. Typ.	Evaluation Crit.	Mean	N	Std. Dev.	p-value	Cor. Typ.	Evaluation Crit.	Mean	N	Std. Dev.	p-value
VSC	comprehensibility	1.82	88	.824	.046 *	VSC	comprehensibility	1.92	88	.962	.187
	embarrassment	2.07	88	.814			embarrassment	2.18	88	.977	
	effectiveness	1.80	88	.775			effectiveness	1.99	88	1.000	
	Total	1.89	264	.811			Total	2.03	264	.982	
EPT	comprehensibility	1.91	88	.839	.004 *	EPT	comprehensibility	1.89	88	.877	.000 *
	embarrassment	2.15	88	1.088			embarrassment	2.20	88	1.105	
	effectiveness	1.69	88	.764			effectiveness	1.57	88	.542	
	Total	1.92	264	.923			Total	1.89	264	.907	
ELT	comprehensibility	2.12	88	.980	.015 *	ELT	comprehensibility	2.02	88	.922	.082
	embarrassment	2.42	88	.968			embarrassment	2.15	88	.851	
	effectiveness	2.02	88	.871			effectiveness	1.85	88	.851	
	Total	2.19	264	.952			Total	2.01	264	.881	
VMF	comprehensibility	2.49	88	.983	.002 *	VMF	comprehensibility	2.17	88	.937	.000 *
	embarrassment	2.65	88	.898			embarrassment	2.68	88	1.089	
	effectiveness	2.16	88	.843			effectiveness	2.09	88	.783	
	Total	2.43	264	.929			Total	2.31	264	.977	
RPT	comprehensibility	2.80	88	1.105	.878	RPT	comprehensibility	2.40	88	1.012	.675
	embarrassment	2.88	88	.969			embarrassment	2.49	88	.994	
	effectiveness	2.82	88	1.120			effectiveness	2.53	88	1.103	
	Total	2.83	264	1.063			Total	2.47	264	1.035	
CLR	comprehensibility	3.24	88	1.028	.036 *	CLR	comprehensibility	2.82	88	1.012	.491
	embarrassment	2.89	88	.915			embarrassment	2.64	88	1.085	
	effectiveness	2.91	88	1.079			effectiveness	2.68	88	1.056	
	Total	3.01	264	1.019			Total	2.71	264	1.050	

* Significant at the 0.05 level
Note: 1 = strongly agree, 2 = agree, 3 = neutral, 4 = disagree, 5 = strongly disagree
A high mean indicates a less positive attitude and a low mean indicates a more positive attitude
Abbreviations: VSC stands for *Visual Cues*, EPT for *Explicit Correction*, ELT for *Elicitation*, VMF for *Verbal Metalinguistic Feedback*, RPT for *Repetition*, and CLR for *Clarification Requests*

Specific concern on the qualities of the corrective feedback types

Based on all the results shown in Table 1.0, 1.1,1.2 and 1.3 it is obvious that the learners are very much aware of the different qualities or evaluation criteria for the different corrective feedback types. Although the assessment on the qualities are generally positive as depicted by the means for all the qualities in connection with all the corrective feedback types. The learners in the current study generally indicate that "embarrassment" is the most obvious quality of their concern compared to "comprehensibility" and "effectiveness", except for *Clarification Requests* in which "comprehensibility" is the most concerned qualities. For this type of corrective feedback, perhaps when a request to clarify the words uttered by a learner in the corrective move is addressed to him or her, it may be perceived as requesting for the meaning of the word. Thus, causes confusion.

4 Conclusion

Errors in pronunciation may be corrected through several corrective feedback types, such as *Explicit correction, Visual Cues, Clarification Request, Verbal Metalinguistic Feedback, Elicitation and Repetition*. Every corrective feedback type has its own qualities that may be workable or otherwise for a learner. The purpose of this study has been to gauge the qualities of these corrective feedback types in terms of "comprehensibility", "effectiveness" and "embarrassment". Despite the generally positive assessments on the qualities for all the corrective feedback types, there has been a quite consistent concern on "embarrassment" particularly with regards to *Explicit Correction* and *Verbal Metalinguistic Feedback* with which "embarrassment" is a great concern at all levels of comparison, followed by *Elicitation*. The findings in the current study seems to strongly suggest that the three corrective feedback types may need much extended study with bigger population to ascertain a much clearer state of these corrective feedback types in relation to learners' perceptions and the way forward.

5 References

Abu Bakar, Z & Abdullah, M. R. T. (2015). Importance of Correct Pronunciation in Spoken English: Dimension of Second Language Learners' Perspective. *Pertanika Journal of Social Science and Humanities*, 23(S), 143-158.

Agudo, M. & de Dios, J. (2013). An investigation into how EFL learners emotionally respond to teachers' oral corrective feedback. *Colombian Applied Linguistics Journal*, 15(2), 265-278.

Celce-Murcia, M., Brinton, D. M., & Goodwind, J. M. (1996). *Teaching pronunciation: reference for teachers of English to speakers of other languages*. Cambridge: Cambridge University Press.

Chomsky, N. (1959). Review of Verbal Behaviour by B.F.Skinner. *Language, 35*, 26-58.

Dalton, C., & Seidlhofer, B. (2000). Pronunciation. Oxford: Oxford University Press.

Ellis, R. (2001). Investigating form-focused instruction. *Language Learning, 51*(Supplement 1), 1-46.

Ellis, R., Basturkmen, H., & Loewen, S. (2001). Learner uptake in commnicative ESL lessons. *Language Learning, 51*(2), 281 - 318.

Ellis, R. (2017). Oral corrective feedback in L2 Classrooms: What We Know So Far. In N. Hossein & K. Eva (Eds.), *Corrective Feedback in Second Language Learning and Teaching* (pp.3-18). New York: Rouledge

Krashen, S. D. (1982). *Principles and practice in second langauge acquisition*. Englewood Clifts, NJ: Prentice-Hall.

Krashen, S. D. (1985). *The input hypothesis: issues and implications*. New York: Longman.

Krashen, S. D. (1994). The input hypothesis and its rivals. In N. Ellis (Ed.), *Implicit and explicit learning of languages* (pp. 45-77). London: Academic Press.

Levelt, W. J. M. (1989). Speaking: from intention to articulation. Cambridge, MA: MIP Press.

Lyster, R., & Ranta, L. (1997). Corrective feedback and learner uptake: negotiation of form in communicative classrooms. Studies in Second Language Acquisition, 19, 37-66.

Murphy, J & Baker, A.A. (2015). History of ESL pronunciation teaching. In M. Reed & J.M.Levis (Eds), The Handbook of English Pronunciation (pp.36-65), United Kingdom: Wiley-Blackwell.

Neri, A., Cucchiari, C., & Strick, H. (2001). *Effective feedback on L2 pronunciation in ASR-based CALL.* Proceedings of the workshop on Computer Assisted Language Learning, Artificial Intelligence in Education Conference. San Antonio, Texas.

Poulos, A. & Mahony, M. J. (2008). Effectiveness of feedback: the students' perspective. *Assessment & Evaluation in Higher Education,* 33(2), 143-154.

Price, M., Handley, K., Millar, J., & O'Donovan, B. (2010). Feedback: all that effort, but what is the effect? *Assessment & Evaluation in Higher Education*, 35(3), 277-289.

Saito, K. & Lyster, R. (2012). Effects of Form-Focused Instruction and Corrective Feedback on L2 Pronunciation Development of /r{turned}/ by Japanese Learners of English. *Language Learning,* 62(2),595-633.

Schmidt, R. (1990). The role of consciousness in second language learning. Applied Linguistics, 11, 129-158.

Vanosfadrani, A. D. & Basturkment, H. (2009). The Effectiveness of Implicit and Explicit Error Correction on Learner's Performance. System, 37(1),82-98.

Implementation of PBL to enhance the soft skills of engineering students

Hairuzila Idrus[1,*], and *Muhammad Ridhuan Tony Lim* Abdullah[1]

[1]Department of Management & Humanities, Universiti Teknologi PETRONAS, Perak, Malaysia.

Abstract. The objective of this paper is to look into the implementation of PBL in the teaching of engineering courses to enhance the students' ability in soft skills. The importance of soft skills has been acknowledged in most occupations and many organizations make selection and performance evaluation decisions based in part on employees' soft skills. Problem Based Learning (PBL) is a pedagogical approach that has the potential to harness students' ability in soft skills. Thus, this study looks into the implementation of PBL at a private university in Malaysia based on the lecturers' industry experience, teaching experience and the engineering program they are in. A set of questionnaire is given to 90 lecturers teaching engineering courses to gauge the extent to which PBL is employed by the lecturers. An independent-samples t-test was conducted to compare the mean scores of the two groups which were the lecturers with industry experience and the ones without industry experience on the teaching approaches employed by them. A one-way between-groups analysis of variance (ANOVA) is conducted to explore the teaching methods employed by lecturers of different years of teaching experience and engineering programs. The analysis showed that 77.7 % of the lecturers prefer to employ problem-based learning (PBL) in their teaching of engineering courses. The results also revealed that the preference stayed constant in relation to the lecturers' experience with the industry, years of teaching experience and the engineering program they are in. Thus, this illustrates that the lecturers prefer to employ student-cantered approach which lead to producing students with better soft skills like critical thinking. Similar patterns appeared in relations to the comparison by years of teaching experience, where lecturers with 6-10 years of teaching experience highly employed PBL. On the comparison by engineering program, lecturers from Chemical engineering (CE) program show the most frequently usage of PBL. The study also found that the soft skills that could be enhanced through PBL are critical thinking and problem solving skills, communication skills and teamwork skills.

1 Introduction

The main objectives of education are the holistic development of character and capabilities, the acquisition of specific skills, the realization of intellectual, physical and spiritual potential, and the training of human capital. Therefore, as noted by Hazadiah, et al. (2008), the Malaysian Ministry of Higher Education has produced a profile of the desired human capital based on three principles: knowledge, personal, and interpersonal attributes.

---

* Corresponding author: hairuzi@utp.edu.my

Higher education institutions must ensure that their teaching resources are in place, revamp programs and systems, and re-engineer learning processes to develop these attributes in all students. These initiatives signify the importance of teaching and learning in this transformation of education system in Malaysia.

In today's society and in the future soft skills are becoming more important especially baseline soft skills for instance communication and continuous learning among others. The importance of soft skills has been acknowledged in most occupations and many organizations make selection and performance evaluation decisions based in part on employees' soft skills. During economic uncertainties, graduates have the more reason to improve themselves in the soft skill areas and be ready for an upswing in the job market. This was because there were still job openings for entry-level positions for fresh graduates but the competition was tough, thus graduates who were equipped with substantial skills, knowledge and the right attitude to survive in any crisis would have high chances to get the jobs. This implied that graduates were able to secure jobs if they possessed good soft skills (Beagon & Niall, 2015; Hamidah, Yuriani & Palupi, 2016; Othman et al., 2017).

PBL is a learner focused educational approach where the student extends previous knowledge to new problems through self-directed reflection, research and practice in solving a problem (Khairiyah et al., 2007). PBL was adopted by many engineering programs as an approach to help students learn the skills required by the employers and to address many of the concerns that engineering graduates are lacking in of soft skills.

The objectives of this paper are first, to find out to what extend PBL is being implemented by the lecturers at this university in their teaching based on their experience with the industry, years of teaching and the engineering program they are in. And second to find out the soft skills that could be enhanced through the implementation of PBL in teaching engineering courses from the lecturers' perspectives.

2 Methodology

The respondents of this study were lecturers from four engineering programs, namely Mechanical Engineering (ME), Chemical Engineering (CE), Civil Engineering (CV) and Electrical and Electronics Engineering (EE), at a private university. The total number of lecturers from these program is 125 which consist of: ME 34, CV 26, EE 36 and CE 29. Since the number is not large, all of them were taken as the respondents in the study. Out of 125 questionnaires sent through the email and by hand, 90 lecturers responded. The 90 completed questionnaires that are used in the study represent a return rate of 72 percent. Table 1 below summarizes the lecturers' demographic information.

Table 1. Lecturers' demographic information

Demographic items	n	%
Gender		
Male	71	78.9
Female	19	21.1
Educational achievement		
PhD	62	68.9
Master	28	31.1
Engineering program		
EE	24	26.7
ME	24	26.7
CE	21	23.3
CV	21	23.3
Teaching experience		
1 – 5 years	30	33.3
6 – 10 years	20	22.2
11 – 15 years	18	20.0
16 – 20 years	6	6.7
More than 20 years	16	17.8
Industry experience		
Yes	60	66.7
No	30	33.3

The set of questionnaire for the lecturers focuses more on teaching approaches that the respondents employ in their teaching of technical courses. The items are generated based on the literature review carried out by the authors on the approaches taken to integrate soft skills in the teaching of technical courses in order to enhance the students' ability in soft skills. The teaching approaches included in the questionnaire are problem based learning, project oriented problem based learning, cooperative learning, teacher centred approach, project based approach. This section also asks on the understanding of the lecturers on the importance of soft skills by stating statements regarding benefits of soft skills to technical students, common perception pertaining to soft skills and the ability of students in soft skills.

3 Results and findings

This section is divided into 3 sections for discussion which are the respondents' teaching approaches, industry experience, teaching experience and engineering program.

3.1 Teaching approaches

This particular section assessed the teaching approaches employed by the lecturers in integrating soft skills in their teaching of technical courses. In the questionnaire, the authors placed both the traditional and non-traditional teaching methods because they did not want to assume that only non-traditional teaching methods can be used to integrate soft skills. The analysis showed that the teaching approach employed by the majority, which was about 69 (77.7 percent) of the lecturers was problem-based learning (PBL), followed by project-oriented problem-based learning approach (POPBL) employed by about 57 lecturers or 63.3 percent, project-based learning (PjBL) 60 percent or about 54 lecturers, cooperative learning was employed by about 53 or 58.9 percent and lastly the traditional approach, teacher-centred approach employed by 45.6 percent or 41 of the lecturers.

The overview revealed that the lecturers seemed to have a greater preference for problem-based learning, project oriented problem-based learning, project-based learning and cooperative learning, which were more student-centred as opposed to teacher-centred approach, a more traditional approach.

3.2 Industry experience

The reason why this matter is looked into was to find out if the lecturers' experience working with the industry affects their teaching approach. This was because the people from the industry were the ones who pointed out that many graduates are lacking in their soft skills, as such it would be interesting to find out whether the teaching approaches by those academics who were in the industry before were any different from those who were purely from academic and had no experience working with the industry. The numbers of respondents for the lecturers with industry experience and without industry experience were 60 and 30, respectively. The respondents were required to indicate their level of frequency in employing the teaching approaches listed in the questionnaire. They indicated their frequency in employing each teaching approach in a five-point Likert scales from '1' meaning never use the teaching approach, to '5' meaning always use the teaching approach. The interpretation of the mean score was adapted from Landell (1997) on the frequency of using the teaching methods as in Table 2 below:

Table 2. Interpretation of mean score

Range of mean score	Frequency
1.00 – 2.33	Low
2.34 – 3.67	Average
3.68 – 5.00	High

Based on their responses, the data were analyzed and the mean score of each teaching approach is shown in the table below. Table 3 below presents the summary of means for the teaching approaches employed by the lecturers in their teaching by making comparison between the mean score of the lecturers with industry experience to the ones without industry experience.

Table 3. Summary of mean score for teaching approach

Teaching approach	Industry experience	N	Mean score	Std. Dev.	Frequency
PBL	Yes	60	4.10	0.86	High
	No	30	3.97	0.96	High
POPBL	Yes	60	3.85	1.04	High
	No	30	3.53	1.36	Average
PjBL	Yes	60	3.82	0.91	High
	No	30	3.53	1.22	Average
CL	Yes	60	3.67	0.99	Average
	No	30	3.83	1.05	High
TCA	Yes	60	3.52	1.10	Average
	No	30	3.67	0.96	Average

From Table 3 above it was evident that most items have a mean score ranging between 3.53 and 4.10 implying that the lecturers no matter whether they have the industry experience or not did employ the stated teaching approaches. However, the frequency of using teaching approaches like POPBL, PBL and CL differed where the lecturers with industry experience used POPBL and PBL more regularly than lecturers without industry experience. In addition, the lecturers without industry experience frequently use CL as compared to the lecturers with industry approach. Thus, the analysis revealed that there were differences on the frequency in the teaching approaches employed by lecturers with industry experience and without the experience. Both groups frequently employed PBL in their teaching.

3.3 Teaching experience

The lecturers were divided into five groups which were those with teaching experience of 1 – 5 years, 6 – 10 years, 11 – 15 years, 16 – 20 years and more than 20 years. The realization on the importance of soft skills started in the 1980s, which was about 30 years ago. Thus, the aim was to find out if the years of teaching experience of the lecturers have any influence in the teaching approach that they employed in integrating soft skills in their teaching. This somehow could indicate the awareness of the lecturers on the teaching approaches that could be employed to enhance students' soft skills ability. The result tabulated in Table 4 represents the mean scores for the five different groups of lecturers on the teaching approaches addressed in this study. The following results are observed.

Table 4. Mean score

Teaching approach	Teaching Experience	N	Mean	Std. Dev.	Frequency
PBL	1-5 years	30	3.97	.85	High
	6-10 years	20	4.60	.68	High
	11-15 years	18	3.67	1.08	High
	16-20 years	6	3.50	.84	Average
	>20 years	16	4.19	.65	High
	Total	90	4.05	.89	High
CL	1-5 years	30	3.53	1.00	Average
	6-10 years	20	4.40	.82	High
	11-15 years	18	3.72	.96	High
	16-20 years	6	3.16	.98	Average
	>20 years	16	3.43	.96	Average
	Total	90	3.72	1.00	High
TC	1-5 years	30	3.43	.89	Average
	6-10 years	20	3.35	.69	Average
	11-15 years	18	3.78	1.06	High
	16-20 years	6	4.00	.89	High
	>20 years	16	3.69	1.48	High
	Total	90	3.57	1.50	High
PjBL	1-5 years	30	3.80	.87	High
	6-10 years	20	4.45	.69	High
	11-15 years	18	3.33	.97	Average
	16-20 years	6	2.67	.82	Average
	>20 years	16	3.50	1.21	Average
	Total	90	3.72	1.03	High
POPBL	1-5 years	30	4.03	.81	High
	6-10 years	20	4.20	1.28	High
	11-15 years	18	3.23	1.02	Average
	16-20 years	6	2.67	.82	Average
	>20 years	16	3.56	1.41	Average
	Total	90	3.74	1.16	High

With reference to Table 4 above, the analysis showed that the mean score for PBL for lecturers with 1-5, 6-10, 11-15, 16-20 and more than 20 years of experience ranged between 3.50 and 4.60, which was quite a high frequency.

The analyses have also shown that the group of lecturers with 16 – 20 years of teaching experience seem to obtain average mean scores for all four student-centred learning approaches as the mean scores were 3.50 for PBL, 3.16 for CL, 2.67 for PjBL and 2.67 for POPBL, but high frequency for teacher-centred approach with a mean score of 4.00 which signifies this group frequently employed the traditional approach.

Based on the mean scores, the results revealed that lecturers with teaching experience of 1-5 years and 6 – 10 years most of the time employed more of the student-centred teaching approaches like PBL with mean scores of 3.97 and 4.60 respectively, CL mean scores of 3.53 and 4.40, PBA with mean scores 3.80 and 4.45 and POPBL with mean scores of 4.03, 4.20, while the lecturers with 11 – 15 years, 16 – 20 years, more than 20 years teaching experience employed more of the teacher-centred approach with mean scores of 3.78, 4.00, and 3.68, respectively. To determine the statistically significant difference in the comparison between the lecturers of different years of teaching experience, a one-way between-groups analysis of variance (ANOVA) was used since the independent sample t-test was not applicable.

A one-way between-groups analysis of variance was conducted to explore the teaching methods employed by lecturers of different years of teaching experience in integrating soft skills in their teaching of technical courses, as measured by the questionnaire. Subjects were divided into five groups based on their years of teaching experience, namely, 1 – 5 years, 6 – 10 years, 11 – 15 years, 16 – 20 years and more than 20 years. The results of ANOVA test are summarized in Table 5 below.

Table 5. General mean scores, standard deviation and p-value of respondents' teaching approach based on years of teaching experience using one-way ANOVA

Teaching approach	Teaching Experience	Mean	N	Std. Dev.	p-value
PBL	1-5 years	3.97	30	.85	.006*
	6-10 years	4.60	20	.68	
	11-15 years	3.67	18	1.08	
	16-20 years	3.50	6	.84	
	>20 years	4.19	16	.65	
	Total	4.05	90	.89	
CL	1-5 years	3.53	30	1.00	.008*
	6-10 years	4.40	20	.82	
	11-15 years	3.72	18	.96	
	16-20 years	3.16	6	.98	
	>20 years	3.43	16	.96	
	Total	3.72	90	1.00	
TC	1-5 years	3.43	30	.89	.517
	6-10 years	3.35	20	.69	
	11-15 years	3.78	18	1.06	
	16-20 years	4.00	6	.89	
	>20 years	3.69	16	1.48	
	Total	3.57	90	1.50	
PjBL	1-5 years	3.80	30	.87	.000*
	6-10 years	4.45	20	.69	
	11-15 years	3.33	18	.97	
	16-20 years	2.67	6	.82	
	>20 years	3.50	16	1.21	
	Total	3.72	90	1.03	
POPBL	1-5 years	4.03	30	.81	.007*
	6-10 years	4.20	20	1.28	
	11-15 years	3.23	18	1.02	
	16-20 years	2.67	6	.82	
	>20 years	3.56	16	1.41	
	Total	3.74	90	1.16	

* Significant at the 0.05 level

The one-way ANOVA test was used through the manipulation of the data to test for a significant difference. The results of the one-way ANOVA in Table 5 indicated that there were significant differences (p-values) in all the teaching approaches with regard to the lecturers' years of teaching experience except for Teacher-centred Approach. The result found that there was a statistically significant difference at the p less than 0.05($p < .05$) level in the mean scores of teaching approaches employed by the lecturers with regard to PBL, CL, PjBL and POPBL in which the p-values were .006, .008, 0.000 and 0.007, respectively.

In order to find out where these differences laid, post-hoc test was used. The ANOVA for PBL revealed that there was a statistically significant difference in the means of the lecturers' years of teaching experience for PBL. With the p-value of 0.006, the result of the Tukey HSD post hoc test showed that the group with 6-10 years of teaching experience was significantly different from groups of 11-15 and 16-20 years of experience as illustrated in Table 6 below. However, this group did not differ significantly from the group of lecturers with 1-5 and more than 20 years of teaching experience.

Table 6. Tukey HSD post hoc test: Problem-based Learning

Teaching experience	N	Subset for alpha = .05	
		1	2
16 - 20 years	6	3.5000	
11 - 15 years	18	3.6667	
1 - 5 years	30	3.9667	3.9667
> 20 years	16	4.1875	4.1875
6 - 10 years	20		4.6000
Sig.		.214	.290

3.4 Engineering program

This sub-section deals with exploring the teaching approaches employed by the lecturers with regard to the engineering program they are teaching which are civil (CV), mechanical (ME), electrical and electronics (EE) and chemical (CE). The aim was to investigate whether the engineering program they were in has any influence on the teaching approaches they were using. In other words, the interest was to discover whether the engineering program they were teaching in differed over the teaching approaches they employed which may shed some light on the teaching approaches used in integrating soft skills with regard to engineering programs. The descriptive analysis of the assessment is followed by the ANOVA reports in which the test was performed to compare the results across programs. The mean score was analyzed and the frequency of employing teaching approaches like PBL, CL, TC, PjBL and POPBL were looked into. The summary of the results is presented in Table 7 below.

Table 7. Mean scores for engineering programs with regard to teaching approaches

Teaching approach	Engineering Program	N	Mean	Std. dev	Frequency
PBL	CV	21	3.76	.77	High
	ME	24	3.83	1.13	High
	EE	24	4.29	.69	High
	CE	21	4.33	.79	High
	Total	90	4.06	.89	High
CL	CV	21	3.43	1.07	Average
	ME	24	3.87	.99	High
	EE	24	3.58	.97	Average
	CE	21	4.00	.95	High
	Total	90	3.72	1.01	High
TC	CV	21	3.67	1.19	High
	ME	24	3.67	1.09	High
	EE	24	3.87	1.07	High
	CE	21	3.00	.55	Average
	Total	90	3.57	1.05	Average
PjBL	CV	21	3.52	.98	Average
	ME	24	3.75	.94	High
	EE	24	3.46	1.18	Average
	CE	21	4.19	.87	High
	Total	90	3.72	1.03	High
POPBL	CV	21	3.48	1.08	Average
	ME	24	3.63	1.28	Average
	EE	24	3.67	1.24	High
	CE	21	4.24	.89	High
	Total	90	3.74	1.16	High

With reference to Table 7, the analysis showed that the lecturers' means for PBL in relation to all the engineering programs ranged between 4.33 and 3.76, i.e. a high frequency. correspondingly. This has indicated that the usage of POPBL by the EE and CE lecturers in their teaching was highly frequent.

The analyses have also shown that lecturers from all four engineering programs seemed to employ all teaching approaches investigated in this study but by different degree of frequency. Interestingly, the lecturers in CE program obtained high frequency of usage for all four student-centred approaches addressed in this study which were PBL, CL, PjBL and POPBL with mean scores of 4.33, 4.00, 4.19 and 4.24, correspondingly. Nevertheless, for teacher-centred approach this group obtained lower means, 3.00 which signified average frequency. The group from CV, on the other hand, obtained average frequency on the use of student-centred approach like CL, PjBL and POPBL with mean score of 3.43, 3.52 and 3.48 respectively except for PBL which has slightly high mean score of 3.76. In contrast, the CV group obtained slightly high mean score for teacher-centred approach with the score of 3.67. This indicated that the lecturers in CV program prefer the traditional teacher-centred approach than student-centred approach.

ANOVA test was conducted to explore the teaching methods employed by lecturers of different engineering programs in integrating soft skills in their teaching of technical courses, as measured by the questionnaire. Subjects were divided into four groups according to the engineering programs they were in (Civil, Mechanical, Electrical & Electronics and Chemical). The analysis of the ANOVA test indicated that there was a statistically significant difference at the $p < .05$ level in the scores for teacher-centered approach (TC) where p=0.032. The other teaching approaches like PBL, CL, PjBL and POPBL obtained p-value of more than 0.05 which indicated no statistically significant difference in the scores. Table 8 shows the details of the results.

Table 8. General mean scores, standard deviation and p-value of respondents' teaching approach based on engineering programs using one-way ANOVA

Teaching approach	Program	Mean	N	Std. dev	p-value
PBL	CV	3.76	21	.77	.055
	ME	3.83	24	1.13	
	EE	4.29	24	.69	
	CE	4.33	21	.79	
	Total	4.06	90	.89	
CL	CV	3.43	21	1.07	.223
	ME	3.87	24	.99	
	EE	3.58	24	.97	
	CE	4.00	21	.95	
	Total	3.72	90	1.01	
TC	CV	3.67	21	1.19	.032*
	ME	3.67	24	1.09	
	EE	3.87	24	1.07	
	CE	3.00	21	.55	
	Total	3.57	90	1.05	
PjBL	CV	3.52	21	.98	.078
	ME	3.75	24	.94	
	EE	3.46	24	1.18	
	CE	4.19	21	.87	
	Total	3.72	90	1.03	
POPBL	CV	3.48	21	1.08	.149
	ME	3.63	24	1.28	
	EE	3.67	24	1.24	
	CE	4.24	21	.89	
	Total	3.74	90	1.16	

In examining the teaching approaches employed by the lecturers to integrate soft skills in their teaching of technical courses, it can be summarized that first, the descriptive analysis have shown that all lecturers always use PBL as their teaching approach while the least approach being used was teacher-centered teaching approach. Second, the t-test indicates that there was no statistically significant difference in the approaches employed by the lecturers with industry experience (prior to becoming full time lecturers at this university) and those without industry experience. Similar patterns appeared in relation to the comparison by years of teaching experience, although with much bigger differences in the significantly positive ratings for PBL,PjBL, CL and POPBLin lecturers with 6-10 years of teaching experience. This group highly employed the approaches as compared to the other groups. Although these significant differences did not represent a big contrast, they were significantly different within the higher range of the frequency in utilizing the approaches. The patterns were also similar in the comparison by engineering program, with the lecturers from Chemical engineering (CE) program showing the most frequently usage of the student-centered approaches, the PBL, PjBL, CL and POPBL approaches. These approaches are effective in integrating soft skills in the teaching and learning process.

4 Discussion

Problem-based learning (PBL) is an inductive learning, team-based approach that focuses on developing thinking and learning skills in students. It utilizes real world problems, not hypothetical case studies with neat, convergent outcomes. In PBL students learn how to learn in a systematic way. This approach places particular emphasis on learning through questioning, self-directed learning, and also learning with their peers. Several studies have also revealed that PBL could help enhance the soft skills ability of students. As a result, students develop communication skills, critical thinking skills and problem solving abilities as well as team working skills. Also, this approach could help students to pave their ways to enhancing their personal development, increasing their level of self-confidence and empowering the students to be active learners, excel in team working as well as able to accept challenges (Aleem & Jamaludin, 2007; Garcia & Garza, 2015; Othman et al., 2017). The findings of this study have also revealed the same result. The lecturers indicated that PBL was used to integrate critical thinking and problem solving skills as well as communication skills. However, team work skills were also integrated indirectly because in PBL students are required to work in a team.

Analysis of the questionnaire survey reveals that the lecturers preferred to use the student-centred approach namely the Problem-based Learning (PBL), Project-based learning (PjBL), Cooperative Learning (CL) and Project Oriented Problem-based Learning (POPBL), more than the Teacher-Centred (TCA) approach. The results also show that the preference stayed constant in relation to the lecturers' experience with the industry, years of teaching experience and the engineering program they are in. Both groups with and without industry experience had the same patterns of preferences with more preference to PBL approach and least preference to Teacher-centred approach. Thus, this illustrates that the lecturers at this University prefer to employ student-centred approach which lead to producing students with better soft skills like critical thinking. This is further supported by Nygaard and Holtham (2008), who argue that a shift from a traditional paradigm with syllabus-driven didactic teaching to a learning-centred paradigm in higher education benefits student learning and facilitates students' self-development of higher order thinking skills and transferable competencies.

It is also worth noting that the lecturers with industry experience embarked into the teaching line without prior pedagogical experience. They were hired based on their industry experience which the University believes that could impart knowledge to the students through their hands-on experience in the industry. It is also one of the ways to bridge the gap between academic and industry. This study found that the lecturers with industry experience employ similar teaching methods as the ones without industry experience. Thus, this illustrates that the University has taken a profound approach by hiring people from the industry as academic staff. No doubt that knowledge on pedagogy is important for educators to possess but they could gain the knowledge by attending trainings on teaching and learning after they have entered the teaching world. A combination of their industry experience and pedagogical knowledge could make them better lecturers because they know the needs of both academic and industry.

Similar patterns appeared in relations to the comparison by years of teaching experience, although with much bigger differences in the significantly positive ratings for PBL, PjBL, CL and POPBL in lecturers with 6-10 years of teaching experience. This group highly employed the approaches as compared to the groups of 1-5, 11-15, 16-20 and more than 20 years of teaching experience. The study reveals that lecturers with more teaching experience employ more of the teacher-centred approach as compared to the less experience. Although these significant differences did not represent a big contrast, they were significantly different within the higher range of the frequency in utilizing the approaches. This may be because the lecturers with less than 10 years of teaching experience are more open to new approaches since they are still at the learning stage. The lecturers with more teaching experience may be more complacent with their regular approach.

5 Conclusion

Problem-Based Learning is a problem-centered teaching method with exciting potential in engineering education for motivating and enhancing student learning. Implementation of PBL in engineering education has the potential to bridge the gap between theory and practice. PBL also was adopted by many engineering programs as an approach to help students learn the skills required by the employers and to address many of the concerns that engineering graduates are lacking in of soft skills. In acknowledging the potential of PBL to develop the engineering students' soft skills ability, most of the lecturers do use this approach in their teaching but the frequency slightly differs based on their teaching and working experience.

6 References

Aleem, S.A. & Jamaludin, M.Z. (2007). *Problem-based learning (PBL) in an engineering course from students perspective.* Proceedings of the Regional Conference on Engineering Education, Johor Bahru, 3-5 December 2007.

Beagon, U. & Niall, D. (2015). *Using problem-based learning to develop graduate attributes in first year engineering students.* Proceedings of 43rd Annual SEFI Conference, Orlean, France, June 29 – July 2, 2015.

Hamidah, S., Yuriani, S., & Palupi. (2015). *Integrated problem-based learning for improvement soft skill and high order thinking of vocational students.* Proceedings of International Conference on Innovation and Vocational Education 2015,

Hazadiah, M. D, Marzita, P., Sidhu, G., & Nor Aziah, A. (2008). Reengineering teaching and learning in higher education in the development of human capitol: The Malaysia initiatives. In C. Nygaard and C. Holtham (eds) *Understanding learning-centred higher education.* Denmark: Copenhagen Business School Press.

Khairiyah, M. Y., Syed Ahmad Helmi, S. H. and Zaidatun, T. (2007). *A problem-based learning (PBL) model for engineering and technical courses.* Proceeding for Regional Conference of Engineering Education, Johor Bahru, Malaysia, 3-5 December, 2007.

Landell, K. (1997). *Management by menu.* London: Wiley & Sons Inc.

Nygaard, C., & Holtham, C. (2008). The need for learning-centred higher education. In C.Nygaard & Holthan, C. (ED.), *Understanding learning-centred higher education* (pp. 11-29). Copenhagen, Denmark: Copenhagen Business School Press.

Othman, H., Mat Daud, K. A., Ewon, U., Mohd Salleh, B., Omar, N. H., Abd Baser, J., Ismail, M. E & Sulaiman, A. (2017). *Engineering students: Enhancing employability skills through PBL.* IOP Conference Series: Materials Science and Engineering 203 012024

How 'Straight' Has Developed Its Meanings — Based on a metaphysical theory

Kosuke Nakashima*

Hiroshima Institute of Technology, Faculty of Applied Information Science, 2-1-1 Miyake,Saeki-ku,Hiroshima, Japan

Abstract. This paper makes the development of the word 'straight' clear. Dictionaries define 'straight' as; 'without a curve or bend', 'going direct to the target', 'honest', 'in continuous succession' or 'neat'. As can be seen, 'straight' has a variety of meanings. Some connections between these meanings can be guessed but others cannot. The mechanism of how 'straight' has obtained each meaning will be explained logically in this paper. First, the most basic or central meaning of 'straight' is 'without a curve or bend'. So 'straight' modifies the verb 'stand' like the following: He stood up straight. 'Straight' has a strong conceptual connection with a standing posture. At the present time, 'straight' is the normal posture for human beings. This means 'straight' has a concept for 'normal' for human beings like the following: He grew up straight. This sentence does not mean that his posture is straight up. It means that he grew up honest. 'Straight' gives a normal image to English speakers both physically and metaphysically. Like an example above, this paper explains how 'straight' has developed its meanings through the human beings' way of thinking. And some comparisons between English expressions and Japanese ones, which is the author's mother tongue, are made in this paper as well.

1 Cognitive approach: A methodology for this paper

The linguistic method I have adopted here is a cognitive approach. Lakoff and Johnson (1999) mentioned that there are three major findings of cognitive science. One of them is that the mind is inherently embodied. The fact that we have our own bodies and minds is quite natural for us. Our bodies are concrete and we can see our bodies while our minds are metaphysical and cannot be seen. However, we are certain of having minds and they are understood through the filter of our bodies. Another finding of cognitive science is that thought is mostly unconscious. Our construal of outer world is unconsciously reflected in our verbal expressions. The last is that abstract concepts are largely metaphorical. Abstract things cannot be seen, so we need sources to understand them. The sources should be concrete things. Because of this, almost all abstract concepts are metaphorical.

2 Conceptualizations

One methodology from cognitive linguistics was adopted to unravel how 'straight' has developed its meaning. It is conceptualization. Regarding this idea, Lakoff and Johnson (1980) said the following:

> The concepts that govern our thought are not just matters of the intellect. They also govern our everyday functioning, down to the most mundane details. Our concepts structure what we perceive, how we get around in the world, and how we relate to other people. Our conceptual system thus plays a central role in defining our everyday realities. (p.3)

They said that we have concepts for each thing around our world, and the concepts are formed through our daily lives. And the process or result of forming concepts is called conceptualization. Some examples of conceptualizations are given.

2.1 Time is money

Abstractions are understood through the concrete. For example, how do we understand 'time'? We cannot see time, but we can say 'The time is coming', and we cannot touch time, but we can say 'I have some time for a lunch break.' or 'we have five minutes to leave.' In verbal expressions, English speakers handle time as if it were a concrete object. Furthermore, native English speakers have a more specific concept for time. Lakoff and Johnson (1980) founded the following:

* Corresponding author: kosuke-nakashima@outlook.jp

> TIME IS MONEY, TIME IS A LIMITED RESOURCE, and TIME IS A VALUABLE COMMODITY are all metaphorical concepts. They are metaphorical since we are using our everyday experiences with money, limited resources, and valuable commodities to conceptualize time. (pp.8-9)

They meant that time is not a mere object but has specific features similar to money, a limited resource and valuable commodity. You can see this concept of English speakers more clearly through the expressions below:

He <u>spent</u> a lot of time in doing his homework.
This task <u>costs</u> much time.
You should <u>budget</u> your time.
She <u>invested</u> a lot of time in studying.
I <u>lost</u> valuable time waiting for a train.
You <u>waste</u> a lot of time.

Through these verbal expressions, we can get a look into an English speakers' mind. They have conceptualized time as a limited and valuable thing through everyday experiences. This conceptualization is common with native Japanese speakers too. Japanese have a proverb that *toki-wa kane-nari*, that means time is money.

2.2 An organization is a human being

We conceptualize not only abstract objects but also concrete ones. For example, English speakers conceptualize an organization like a company as a human being*. See the sentences below:

She became <u>head</u> of the company.
The company plans to strengthen its <u>hand</u> in Asia.
I am working at the marketing <u>arm</u> of the company.
Our company is on its last <u>legs</u>.
The company needs new <u>bloods</u>.

As you can see, it is natural to treat a company as if it were a human being in English expressions. The structure of an organization is similar to that of a human being. For example, a human being can take action because its head, more precisely the brain, can make an order to each body part. This is the same as an organization. The head of an organization makes a decision and an order to each division of the organization, and it can take action.

2.3 Parallelism between English and Japanese

Some conceptualizations are held in common between different languages. Lakoff and Johnson (1980) said conceptualizations are tied to each culture, so some cultures have the same concept for things, but others have differences. It is up to the cultures. As mentioned above, for example, native Japanese speakers have the 'time is money' concept, and also have 'an organization is a human being' concept. This is because we sometimes have the same way of thinking as human beings even though our languages are different from each other. Other conceptualizations vary among different languages because each language has their own history, policy, culture, social system and religion. But the main target language of this paper is English, and some comparisons between English expressions and Japanese will be examined.

3 Polysemy of straight

Straight has many varied definitions in some dictionaries. For example, it has 12 definitions in *The Oxford Advanced Learner's Dictionary* 9th edition (2015) (=*OALD*), 10 definitions in *Longman Dictionary of Contemporary English* 6th edition (2014) and 6 definitions in *The New Oxford American Dictionary* Second edition (2005) in each adjective section. Each definition does not have an organic connection, but each definition does have a logical connection. Every word, including straight, must have had only one meaning when it emerged. Then, the words have developed their meanings one by one. So, to reveal the organic connections between the definitions of straight, we need to find the original meaning of the word.

3.1 The original meaning of straight

---

* An organization itself seems to be an abstract object, but the members or buildings of an organization can be seen physically. So I define an organization is a concrete object.

To make clear what the original meaning of straight might be, here is the entry from the *Oxford Dictionary of Word Origins* (2010):

straight [ME] The word straight is the old past form of Old English **stretch**, and originally meant 'extended at full length'. (p423)

In fact, according to *Oxford English Dictionary* Second edition on CD-ROM Version 4.0 (2009), straight appears for the first time in a written work around 1325:

Straight *a, sb,* and *adv.*
With a strught arme he keppit the caupe on his clene sheld.

This description can be translated into modern English as 'With a stretched arm, he kept the headpiece on his clean shield.'. These descriptions indicate that the original meaning of straight is 'physically extended at full length'.

3.2 Abstract meanings of straight

Though straight originally indicated a physical state of things, it has increased in meanings and can now indicate abstract meanings. Each abstract meaning is strongly connected to the original meaning. This paper focuses on some abstract meanings of straight such as honest, clearly, continuous, and neat.

3.2.1 Conceptual connection between straight and honest

Though the main purpose of this section is revealing how 'straight' has developed its original meaning to honest, I must mention the concept for standing posture first. This is because straight is strongly connected to standing posture. Human beings have realized tremendous evolution by taking 'standing posture'. To take the posture let us use our hands freely and we have learned to use tools. This has made big difference from other living things. Therefore, standing posture is the most important and basic posture for human beings. It is natural that we always try to keep the posture. This way of our thinking is reflected in verbal expressions, see the 2 examples from *OALD* below:

stand on your own (two) feet
to be independent and able to take care of yourself:
◆ When his parents died he had to learn to stand on his own two feet.
on your feet
completely well or in a normal state again after an illness or a time of trouble:
◆Sue's back <u>on her feet</u> again after her operation.
◆The new chairman hopes to get the company back <u>on its feet</u> within six months.

Of course, the expressions like 'stand on your own feet' or 'on your feet' can mean that someone takes a standing posture physically. But as you can see, these explanations refer to metaphysical meanings. It can be said that standing posture is a desirable posture for human beings and can also relate to the more metaphysical idea in a good condition.

When a human being takes a standing posture, the body gets straight. This fact connects straight and standing posture. Straight is physically normal for human beings, therefore it can also indicate the meaning of metaphysically normal for human beings. Look at the sentences below:

She grew up to be straight.
He is a bent cop.
He has a twisted personality.
She is a crooked business person.

The 'straight', 'bent', 'twisted' and 'crooked' in these sentences do not indicate physically state of each human being. They indicate mental situations. It is obvious that having an honest personality is normal or desirable for human beings, so straight can indicate the meaning of honest. On the contrary, not straight like bent, twisted and crooked, can indicate being dishonest.

Moreover, Japanese has the same concept as English speakers. In Japanese, *su-nao* means honest. *Su* means frankly and *nao* means straight.

3.2.2 Conceptual connection between straight and clearly

Straight can also indicate the meaning of clear/clearly like shown in the sentences below:

Let me just get this straight.
I'm going to ask you straight.

The connection between 'extended at full length', which is the original meaning of straight and clear/clearly is ambiguous. To make it clear, let me give some examples of the opposite words:

This is a knotty topic.
He has gotten tangled up with three women.
She was entangled with that man.

'Knotty', 'tangle' and 'entangle' originally meant the state of a thing being like thread or string. Complicated situations or problems can be an imagined like a tangle. And to undo a knot is connected to the image of solving those complicated situations, making them straight. Let me show you a visual image of this comparison below:

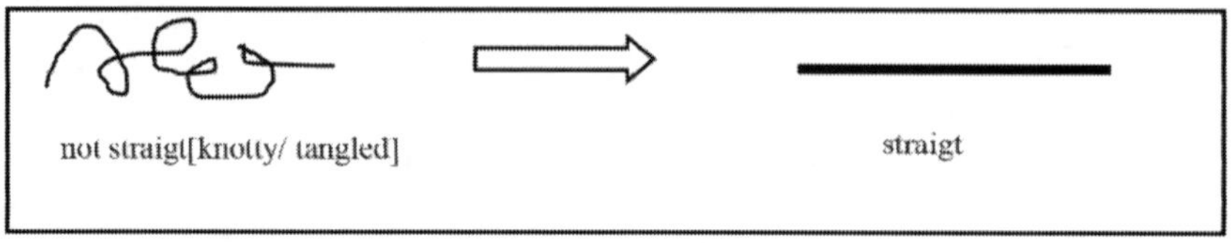

Fig. 1. Comparison between straight and not straight

As you know, things are usually more desirable when they are not knotty, tangled or entangled. And we can easily grasp the whole image of the thread when it is straight. In other words, we can see the straight clearly when it is straight but not when it is tangled. So straight can indicate the meaning of clear/clearly. And Japanese also has the same concept. In Japanese, we say *mondai-wo kaiketsu-suru.* This sentence means 'I solve the problem.' *Kaiketsu-suru* means to solve, but originally meant undo a knot.

3.2.3 Conceptual connection between straight and continuous

The fact that straight can indicate the meaning of continuous is obvious from the examples below:

He got straight A's at the end of this year.
Our team won five straight games.

To make clear the image of these sentences, let me give visual images below:

Mathematics	Science	Social Studies	Visual Arts	Music
A ——→	A ——→	A ——→	A ——→	A

Fig. 2. Image of straight A's

Game1	Game2	Game3	Game4	Game5
win ——→	win ——→	win ——→	win ——→	win

Fig. 3. Image of winning five straight games

As this image shows, to get straight A's can be imagined as grade As stretched from the start of the report card to the end. This chain of As gives English speakers the image of straight. In the same way, a chain of wins suggests the image of straight to English speakers. So straight can have the meaning of continuous.

3.2.4 Conceptual connection between straight and neat

As mentioned above, straight can create the image of continuity. Moreover, straight can indicate the continuity in number. As in the example below:

I tried to fill in a straight in poker.

As you know, a straight is a kind of hot hand of a poker game. Five cards must be in order like from ace to 5 or from 8 to queen. In the mind of native English speakers, this continuity in number also has the same image as straight. The number seems to stretch like ace to 2, 2 to 3, 3 to 4, and then 4 to 5. This image of stretching is strongly connected to straight. In other words, straight can indicate the meaning of in order. So, it is possible that straight is used in sentences like below:

You should put the room straight.

This straight does not mean not bent. It means that everything in the room is put where it should be, in the right place. In short, the room is neat. On the contrary, we can say 'you have put the room out of order', when the room is not neat.

4 Conclusion

The original meaning of straight is 'extended at full length.' From there, straight has come to indicate 'not bent'. And then, 'not bent' was strongly connected to the standing posture of human beings. Straight can indicate mentally not bent, or 'honest'. Furthermore, straight can indicate the meaning of 'clear/clearly' when we imagine of a piece of string. Also, the image of stretching something leads native English speakers to regard straight as continuous and in order. The visual image below shows you how straight has developed its meanings:

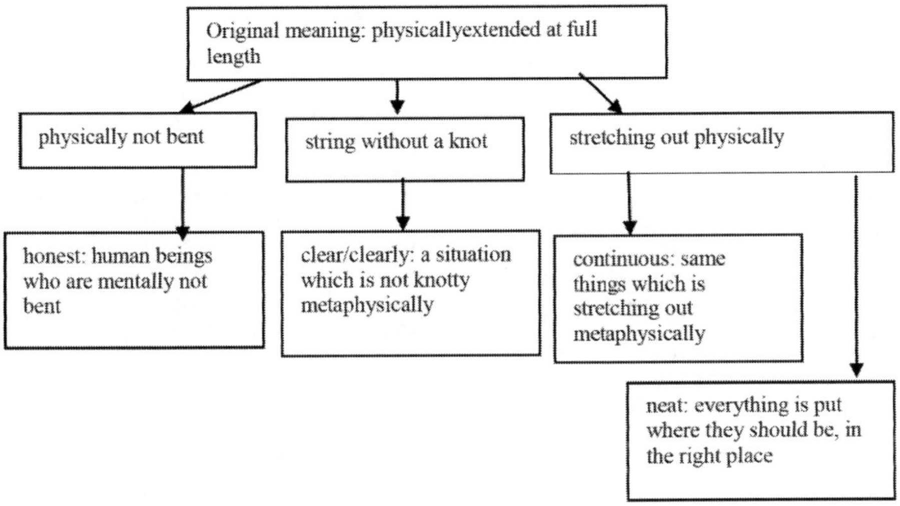

Fig. 4. How straight has spread its meanings

In this way, straight has developed in meaning from its original meaning. How human beings recognize things in the world is deeply connected to our linguistic activity. I hope this paper could reveal some parts of the mechanism of the magnificently intelligent function of human beings.

5 References

Cresswell, J. (ed.) (2010) *Oxford Dictionary of Word Origins*. London: Oxford University Press.

Fox, C. & Combley, R. et al (ed.) (2014) *Longman Dictionary of Contemporary English* 6th edition. London. Pearson Education Limited.

Hornby, A. S. et al. (ed.) (2015) *Oxford Advanced Learner's Dictionary. (=OALD)* London: Oxford University Press.

Lakoff, G. and M. Johnson. (1980) *Metaphors We Live By*. Chicago: University of Chicago Press.

Lakoff, G. and M. Johnson. (1999) *Philosophy in the Flesh. —The Embodiment Mind and its Challenge to Western Thought—*. New York: Basic Books.

Simpson, J. et al. (ed.) (2009) *Oxford English Dictionary* Second edition on CD-ROM Version 4.0. London: Oxford University Press.

Stevenson, A. et al. (ed.) (2005) *The New Oxford American Dictionary* Second edition London: Oxford University Press.

Muslim scholar's discourse on Buddhism: a literature on Buddha's position

Ahmad Faizuddin *Ramli*[1,*], Jaffary *Awang*[2], and Zaizul *Ab Rahman*[3]

[1]PhD Candidate, Centre for Akidah and Global Peace, Faculty of Islamic Studies, The National University of Malaysia, Selangor, Malaysia and Lecturer at Department of Social Sciences, Faculty of Humanities and Social Sciences, Nilai University. Negeri Sembilan, Malaysia.
[2]Assoc. Professor at Centre for Akidah and Global Peace, Faculty of Islamic Studies, The National University of Malaysia and Senior Fellow at The Institute of Islam Hadhari, The National University of Malaysia, Selangor Malaysia.
[3]Senior Lecturer at Centre for Akidah and Global Peace, Faculty of Islamic Studies, The National University of Malaysia, Selangor Malaysia.

Abstract. The coexistence of Islam and Buddhism relation took place in the middle of the 8th century. Although both religions originated from different sources, Muslims scholars were the first to study about Buddhism. Inspired by selected verses in the Quran, Muslims scholars recognized an element of similarity of Buddha's teaching with Islamic teaching. This paper examines the views of Muslim scholars on Buddha's position in the Islamic tradition in early and contemporarily literature. The method of this study is qualitative by emphasizing on contain analysis using three research design: philological, historical, and theological. Among the early Muslim scholars discussed in this paper are al-Baladhuri (d. 892), al-Biruni (973-1048 AD), Ibn al-Nadim (995 AD), al-Shahrastani (1086-1153 AD), al-Ṭabarī (838-923 AD), al-Hamadhani (1247-1318 AD), and the contemporary Muslim scholars is Muhammad Hamidullah (1908-2002), Hamid Abdul Qadir (1957), Hamza Yusuf (1958 –), Shah Reza Kazemi (1960 –), and Imtiyaz Yusuf. The study suggests the position of Buddha as a prophet in the Islamic tradition are justified based on selected terms in the Quran like *tīn*, *Dhu al-Kifl and Ṣābi'īn*. This paper concludes that there is a strong relation between Islam and Buddhism in early Islamic literature.

1 Introduction

Islam is a religion that recognizes the existence of various religions in the world. In the Quran, Allah s.w.t. states the names of world religions such as "Islam, the Jews, the Sabeans, the Christians, the Magians, and the polytheists" (22:17). There are religions that are not clearly mentioned in the Quran as Buddhism, although those religions existed thousands of years earlier before the birth of Prophet Muhammad (PBUH) Apart from the name of the religion, the Qur'an also classifies the people into three categories, namely *Ahl*

* Corresponding author: ahmadfaizuddin@outlook.com

al-Tawhīd (the people of Monotheism), which refers to a people who believed in the oneness of Allah s.w.t. and all prophets as His messenger; *Ahl al-Kitāb* (the People of the Book), which refers to the Jews and Christians, who are possessors of divine books, as distinguished to the People of Dubious from those whose religions are not based on divine revelations such as Zoroastrianism and pagans. Although each religion has its own theological and ritual system, the interaction between the believers is harmonious as demonstrated by the Prophet Muhammad (PBUH) in his series of dialogues especially with the *Ahl al-Kitāb* like the Jews and the Christians. The initial approach of interaction that is fully guided by the Qur'an and Hadith is manifested verbally and in writing, together with the good example - *qudwah al-ḥasanah*. In the 7th century, after the death of the Prophet Muhammad (PBUH), effort to convey the teachings of Islam was continued by the Companions and the Caliphs through expeditions to expand the Islamic empire that includes the Central Asian regions. Previously the interaction only involves the religions mentioned in the Quran, while the expansion of the Islamic empire in the Persian East, Transoxiana, Afghanistan and Sindh areas opens a new avenue through interaction with the Buddhists. Due to the lack of information about Buddhism, there are polemic among Muslims in determining Buddhist's position according to the Islamic jurisprudence, that is, whether to consider them same as the *dhimmi* like the People of the Book and vice versa. This has led to the in-depth study of scholars on Buddhist origins, doctrines and rituals of Buddhism based on historical and theological view in the form of narrative and comparative discourses.

2 Literature Review

The literature on Buddhism among early Muslim scholars is limited, due to the lack of a fundamental discussion in the major sources of Islam, the Quran and the Hadith. For example, the Quran only states about the six main religions, namely Islam, the Jews, the Christians, the Sabean, the Magians, and the polytheists (22:17) without mentioning any term concerning the Buddha, foremost in the early Quran exegesis as Ibn Abi Hatim (811-890 M) and al-Ṭabarī (838-923 AD). However, the implications of the power expansion policy of the seventh and eighth centuries, beginning in the Umayyad (661-684 AD) expansion to the Central Asia region, interaction with the Buddhist had occurred. In the early stages, Buddhists were categorized as People of the Book, and was subsequently considered as the People of the *Dhimmi* after reviewed by the caliph. Starting from the foundation that had been built in the early period, the Muslims-Buddhism interaction grew during the Abbasid (775-809 AD) period, foremost in intellectual aspects. Subsequently, the integration in civilization, science and culture between the Arab-Persian world, Central Asian and Western worlds was established through the translation works in various fields, for example in astronomy, mathematics and medicine. On the other hand, the Caliph also offered a special position to Indian scholars including Buddhist scientists to serve in Baghdad to develop the Baitul Hikmah as a distinguished scholarly institution. Its significance was continued through the initiative initiated by the Caliph al-Mansur (754-775 AD) and was continued by the caliph of Harun al-Rashid (786-809 AD), and thus recorded the first theological dialogue of Muslim-Buddhist (Elverskog, 2010). This scenario then inspired the writing of Buddhism through the work of *Bilawhar wa Budhasaf* by Abban al-Lahiki (750-815 AD), a leading poet in Baghdad. The work produced in Persian was later translated into Sanskrit, Arabic and other languages (Gibb, 1986). Another successful work is the *Book al-Budd* which is also translated from Sanskrit to Arabic and subsequently became the main reference in the study of Buddhism (Imtiyaz Yusuf, 2010).

The study of Buddhism during the period of Harun al-Rashid (786-809 AD) was increasingly gaining attention when the Caliph appointed the Barmakid family, that is Yahya bin Barmak, to hold important positions in the government administration. The Barmakid family, originated from Balkh, South Afghanistan held high positions in the Nava Vihara Temple, a Buddhist Center. Yahya bin Barmak (806 AD), the grandson of the Monk leader was also a tutor to Caliph Harun al-Rashid (Azad, 2017; Bladel, 2011; Berzin, 2010; Imtiyaz Yusuf, 2007; Sachau, 1910). In that position, he sent several of his officers to India to explore the ancient knowledge in medical (Levinson & Christensen 2002) as well as the religions that existed in India, including the Buddhas who had once been the religion of theirs. Yahya bin Barmak leads the translation of Sanskrit works and protects sources proving the Buddha's life (Azad 2017 & Bladel 2011), and even most of his early Arab and Persian literature sources were referred to by later Islamic scientists such as Ibn Khurradadbih (820-912 AD), al-Maqdisi (946-991 AD), al-Mawarzi (770-874 AD), Gardizi (1061) and al-Shahrastani (1086-1153 AD).

While in the ninth century, in the early stages, Muslim scholars focused on historical narratives (al-Baladhuri 1987; al-Maqdisi, 2004, al-Mas'udi, 1841, al-Ṭabarī, 1991) without discussing the character of the Buddha as Prophet. From theological aspect, the debate is limited to the result of the advent of Islamic scientists to the inhabited areas of the Buddha, giving certain characteristics to the worship houses, worshipers, and adherents of Buddhism (al-Maqdisi, 2004, al-Ṭabarī, 1991, al-Baladhuri, 1987) and there is an element of confusion when the character of God the Buddha is likened to the Lord of Hinduism (al-Maqdisi, 2004). Only in the tenth century, Islamic scientists raised talks about the position of the Prophet for instance under the topic of *Aṣhab Madhāhib al-Hind* by referring to the term *al-budd* to the character of *Budhasaf/Yudhasaf* (Ibn al-Nadim 2014, al-Biruni's 2008, al-Mas'udi 1841) or a Prophet (al-Shahrastani, 1992 & Sachau, 1910). The al-Biruni's account of India is more focused on Hinduism, and it is difficult to find Buddhism's debate (Sachau, 1910). This was justified by Sachau (1910) that al-Biruni did not know much about Buddhism because his writing was theoretical and never arrived in India. This impression is not true because there are many records of al-Biruni's visit to the Buddhist community including India. For example, when Mahmud Ghazni dominates Bukhara, al-Biruni has spent his life serving him and his son, Mas'ud. The domination of South India by Mahmud Ghazni succeeded with the help of al-Biruni (Sparavigna, 2013; Bigelow, 2013; Berzin, 2008). Moreover, al-Biruni (2008) in his work *al-Āthār al-Bāqiyah min al-Qarūn al-Khāliyah*, does mentioned the character of *Yudhasaf* with the Sabean religion which later supported by Tabrizi (2012) and Tabataba'i (1997) as the Buddhism in the Quran.

Regarding Buddhism, al-Biruni in the *Book of al-Hind* explains the basis of Buddhist traditions and beliefs, as well as the perception of the Buddhist community as a Prophet. Nevertheless, it does not mean that he proposes to the Muslims to recognize the position of Buddha as Prophet, but he understands that Buddhists do not consider Shakyamuni's character as God (Berzin, 2008). The al-Biruni's legation was further developed by al-Shahrastani (1086-1153 AD) who also visited India in the eleven and twelve centuries, regarding Buddhism's standing in India (Berzin, 1994).

Entering the 13th century, the Mongols' attack on Baghdad led to the destruction of the Muslims Scholar's works on Buddhism, which were either burned or partly thrown into the Tigris river until the ink transform the river's water into black color. The implications paralyzed the Islamic world which was centered in Baghdad. The only surviving work that brings together a significant number of important Islamic-Buddhist relation in the Central Asian region is *Jāmi' al-Tawārikh's* by Rashid al-Din al-Hamadhani (1247-1318 M). The work of Rashid al-Din al-Hamadhani on Buddhism derived from Buddhist material from Kashmiri Buddhist monk (bakhshī) Kamāla Śrī, besides from Central Asian Buddhism

(Jackson, 2017 & Canby, 1993). Perhaps the last was by Ibn Battuta (1304-1369), but works on Buddhism or the Buddhist position were no longer discussed among Muslims traditions until the twentieth century, where Muslims scholars from Central Asia like Muhammad Hamidullah (1908-2002) and Hamid Abdul Qadir (1957) started to stress the theme in their works. The same premise was later followed by other prominent Muslim scholars like Hamza Yusuf (1958 –), Shah Reza Kazemi (1960 –) and Imtiyaz Yusuf who discussed the significant relation of Muslim-Buddhist relations in contemporary world. Discussions among prominent Muslim scholars like Imtiyaz were supported by philological arguments, instead of theology and history. However, the premise is open to debate due to lack of support from early Muslims *mufassirin* (exegesis) on the selected terms in the Quran which is *tīn, Dhu al-Kifl and Ṣābi'īn* (Sabean). However, this argument was supported by prominent Muslims *mufassirīn* (exegesis) al-Qasimi (d. 1914) and Tabataba'i d. 1981) regarding the Buddha as one of the prophets.

3 Result and Discussion

The Muslim scholars' discourse on Buddha position was analysed from three arguments; *first*, philology; *second*, historical; *third*; theological. The philology argument refers to the special terms in the Quran that refers to the Buddha indirectly like the other Prophets – Adam, Noah, Moses, Jesus, etc., while the historical argument is based on the early narrative on Buddhism by focusing on the general concept of prophet, while on the theological, the argument stress on the similarity element of Buddhism with Islamic teaching.

3.1 Philological Argument

In terms of philology, it is based on three verses in the Holy Quran, the first two verses refer Siddhartha as the prophet Dhu al-Kifl:

وَإِسۡمَٰعِيلَ وَإِدۡرِيسَ وَذَا ٱلۡكِفۡلِۖ كُلّٞ مِّنَ ٱلصَّٰبِرِينَ ٨٥ [الأنبياء:85-85]

"And (mention) Ishmael, and Idris, and Dhu al-Kifl. All were of the steadfast." (85)
[Quran al-Anbiyā' 21: 85]

وَٱذۡكُرۡ إِسۡمَٰعِيلَ وَٱلۡيَسَعَ وَذَا ٱلۡكِفۡلِۖ وَكُلّٞ مِّنَ ٱلۡأَخۡيَارِ ٤٨ [ص:48-48]

"And make mention of Ishmael and Elisha and Dhu al-Kifl. All are of the chosen" (48)
[Quran Ṣad 38: 48]

As a guide to all mankind, the Quran describes the Prophets, either directly or indirectly. There are 24,000 of prophets sent by Allah to all mankind, and only 25 of them were selected as an apostle, while five of them were selected as the *ulul azmi* – the higher ranking of apostle. Among the apostle is prophet Dhu al-Kifl as stated in two verses above. The history of prophets Dhu al-Kifl is not discussed in many Islamic narrations. With regards to the founder of Buddhism, Siddharta Gautama, Muslims scholar like Muhammad Hamidullah (1974) and Hamid Abdul Qadir (1957) stated that the Buddha was Prophet Dhu al-Kifl in the Qur'an. Justification for this was made based on the word *kifli*, which means 'someone from Kifli'. The word of *kifl* is Arabicized from the *Kapila* phrase, is the short name for Kapilavastu. Kapilavastu is the birthplace of the Buddha, hence it is named by

Dhu al-Kifl (Berzin, 1994; Imtiyaz Yusuf, 2013; Obuse, 2010; Perry Schmidt-Leukel, 2010).

Instead the terms of *kifl*, the word of *tīn* was regarded by Muhammad Hamidullah and supported by Hamza Yusuf (2010), Reza Kazemi (2010), Imtiyaz Yusuf (2003), al-Qasimi (2002) as a Bodhi tree where the Siddhartha mediated and attained the enlightenment (*nirvana*). Thus, only Gautama Buddha is the only closest figure in the Quran with the fig (*tīn*) tree. Allah said in the Holy Quran:

وَٱلتِّينِ وَٱلزَّيْتُونِ ١ وَطُورِ سِينِينَ ٢ وَهَٰذَا ٱلْبَلَدِ ٱلْأَمِينِ ٣ [التين:1-3]

"I swear by the fig and the olive (1), And mount Sinai (2), And this city made secure (3)"
[Quran al-Tīn 95: 1-3]

The four symbols in the surah at-Tīn are ironically symbolic to the Prophet. Muslim scholars have different views on the *tīn* (fig) while they agreed with other terms such as *al-Zaytūn* as a symbol of Jerusalem the birthplace of the Prophet Isa., *Sinīn* (Mount Thursina) as a symbol for the place of Moses a.s., and *al-Balad al-Amīn* is symbolized as the Prophet Muhammad (PBUH). Early Muslims Exegesis scholars in the like of Ibn Abi Hatim (811-890 M) quotes the view of Qatadah that the *tīn* is the name of a hill in Damascus, the blessed hill in Sham and contained a tree, while Muhammad bin Ka'ab interpreted the *tīn* as the Companions of the Cave (*Aṣḥabul Kahfi*) as Al-Qurtubi. Ibn Abbas and Mujahid hold same view by regarding it as a fruit that is ate by people (Ibn Abi Hatim, 1997), later followed by al-Tabari (2001), Ibn Kathir (2000) and Abdul Razzaq (1999) quoted Ibn Abbas views, it is a mosque built by Noah on the Mount of Judi besides repeating the al-Qurtubi's view.

If observed, Muslims *mufassirin* differ in their views on the true meaning of the word *tiin*. Some interpret it as a fruit (Ibn Abbas and Mujahid), while some interpret it symbolical to some places (Qatadah & Muhammad bin Ka'ab). However, by using philological argument, Imtiyaz Yusuf (2017) suggests the precise interpretation is symbolical for a place in the Middle East. Using scientific argument, Imtiyaz Yusuf (2017) suggests the *tīn* (*ficus religiose*) refers to the Bodhi tree, due to the tree growing much in the Indian subcontinent and in Indo-China rather in the Middle east. Contrarily, the species of the Bodhi tree is from the *ficus carica,* not *ficus religiose* which only grows in the Middle East and West Asia. But the argument can be accepted since both are originated from the same family of *Moraceae*. Moreover, the Bodhi tree also known with seven names in Sanskrit tradition as *Patala* (*Bignonia*), *Pundrika* (*Lotus*), *Sala* (*Shoria Rabusta*), *Saresha* (*Accaciassirisa*), *Udambra, Nyagrodha,* and *Asvatha* (*ficus religiousa*). On the other hand, the Buddhist scholars define the Bodhi tree with different species like Dragon Flower tree, Champac tree, Dragon tree, Kesser Dragon tree, and Iron wood tree. In fact, not just in Buddhism, the species of *ficus religiose* is also considered as a sacred tree in the texts of Hindu and Jain, as in the Bhagavad Gita (Abdul Haq Vidyarthi & Abdul Ahad Dawud 2013). The claim to regard Buddha as a prophet, eventually recognized by contemporary Muslims *mufassirin* as one of the views (Al-Qasimi, 2002).

The Bodhi tree that related with Siddhartha's history, on the other hand, was compared by H.O.K Rahmat (1984) with the story of Prophet Moses during receiving revelation. When Siddhartha was under the Bodhi tree, he decided not to move until he had the great truth and knowledge. For forty-nine days he meditated and eventually the 'light' came to him. While Moses was mentioned in the Qur'an (Ṭāha 20: 9-13): "And has come to you the story of Moses. When he saw the fire then said to his family: Wait! I have seen fire, may I bring you a flame from it, or I can guide you from it. When he came to him, he was summoned: O Moses! Behold, I am your Lord, open your two shoes, indeed you are now in the valley of *Ṭuwa*; and I have chosen you, then consider what is revealed". So the 'light' for Siddhartha also can be symbolized as a 'fire' to Prophet Moses.

3.2 Historical Argument

From the historical points of view, al-Maqdisi (1907), divided the Buddhists into two categories, which one of it believed the Buddha is a prophet charged with a mission, and others believed he is divine incarnation. While a Muslim religious scholar and historian, al-Ṭabarī (839–923 CE), prefers not to involve with the polemic, instead describing what he experienced in general. He reported that Buddhist idols were brought from Kabul, Afghanistan to Baghdad in the ninth century. It is also reported that Buddhist idols were sold in Buddhist temple next to the Makh mosque in the market of the city of Bukhara in present Uzbekistan (Imtiyaz Yusuf 2007). Parallel with al-Maqdisi, for the Persian historian, al-Biruni wrote a book about India (*Kitab al-Hind*) and described the basic Buddhist customs and beliefs and noted that the Indians regarded Buddha as a prophet. Similarly, on his work *al-Āthār al-Bāqiyah min al-Qarūn al-Khāliyah*, al-Biruni (2008) did mentioned the Buddhsaf was among the psedo-prophet who came forward in India after the 1st year of Tahmurath. He introduced the Persian writing and called people to the religion of the Sabeans.

However, for Berzin (2008), that does not mean, of course, that he was suggesting that Muslims accept Buddha as a prophet of Allah, but it does indicate that he understood that Buddhists do not assert Shakyamuni as their God. The works of al-Biruni on Buddhism was repeated by Ibn Nadim (2014), in *al-Fihrist*. He delineates the different scholarly views of the Buddha, some believed he was the divine incarnate, some believed he was the Apostle from the kind of Angels, some believed he was the *Ifrit* (kind of jinn), while others believed him as an apostle from God in the form of human race. Ibn al-Nadim then stated:

> These people [Buddhists of Khurasan] are the most generous of all the inhabitants of the earth and of all the religionists. This is because their prophet Budhasaf [Bodhisattva] has taught them that the greatest sin, which should never be thought of or committed, is the utterance of "No." Hence they act upon this advice; they regard the uttering of "No" as an act of Satan. And it is their very religion to banish Satan.

By 12th century, al-Shahrastani also repeated al-Biruni's account of Buddhism in his book entitled The Book of Religions and Creeds (*Kitab al-Milal wa al-Nihal*). Al-Shahrastani (1992) placed the Buddha's position as Prophet Khidir which can be summarized as follows:

> Buddha was an ideal human. He was not born, did not get old, and even did not die. He avoided the pleasure of eating, drinking and remained celibate all his life. He was a perfect ascetic and an exemplary model for his disciples. Accordingly, a Buddhist can be defined as a seeker of truth, a practitioner of various ascetic works, and a holder of the deepest sense of inner morality. Buddhist ethical practices include having patience, giving alms, renouncing this world, abstaining from greed and worldly pleasure, and showing compassion for all creatures. Thus, a Buddhist is required to live an ethically correct life. In addition to practicing these ascetic disciplines, a Buddhist is encouraged to pursue spiritual perfection. He must enhance his intelligence and seek a higher existence. By doing so, he can liberate himself from this lower world that is destined to decline, live in the eternal world, and eventually be led to the presence of divine beings. Buddhism places importance on ascetic disciplines and spiritual endeavor. At the end of this description, Shahrastani states that Buddha in the Buddhist context can be compared to Khaḍir (Khiḍr), a legendary Islamic saint whose existence is taken for granted by Muslims.

Thus, from historical argument, Muslim scholars have different views on Buddha. al-Maqdisi, al-Biruni and Ibn Nadim agree with the position of Buddha as a prophet, without naming it, while al-Tabari did not discuss it except gave some general narratives on Buddhism. Only al-Shahrastani begins to compare it to the prophet Khaḍir (Khiḍr).

However, the timeline of Khadir that exist at the same period of Moses (d. 1407 BCE) contradicts to the period of Siddhartha (480 – 400 BCE) which exist thousand years earlier. Since there is no established view of his identity, Muslim scholars, mostly those with a South Asian background, like Muhammad Hamidullah (1974) followed by Hamid Abdul Qadir (1957) regards the Buddha as a prophet named Dhu al-Kifl. However, the character of Dhu al-Kifl was identified by some scholars with Ezekiel (622-570 BCE), born at Babylon that existed around hundred years before Siddhartha. The character of Ezekiel is an important figure in Abrahamic Faith – Islam, Christianity and Judaism. Hence, according to the history timeline, Dhu al-Kifl if referred to the prophet Ezekiel nor the Khaḍir was contradict. The only justification by 21st century Muslim scholars like Hamza Yusuf, Shah Reza Kazemi, and Imtiyaz Yusuf on characterization of Dhu al-Kifl with Siddhartha as prophets was based on the premise 'the unmentioned name of prophets in the Quran.' In this regard, the unmentioned name of Buddha in the Quran does not deny the existence of other verses that stated him indirectly. It is known that the number of the Prophets is 124,000 while the messengers are known to be 25. To strengthen the position of Buddha as Prophet, Reza Shah Kazemi (2010) outlines six premises based on the verses of the Quran; *First*, every community has a special Messenger (Quran 10:47); *secondly*, there are names of Prophets and messengers that are not narrated (Quran 4: 164); *thirdly*, the existence of religious and Sharia rules for every people (Quran 5:48); *fourth*, every messenger who was pronounced in the language of their people (Quran 14: 4); *fifth*, every prophet who was sent was revealed to reveal the teachings of Tawhid (Quran 21:25); the sixth, the challenge faced by the Prophet Muhammad (PBUH), exactly the challenge of the previous Prophet (Quran 41:43). So, these premises justify the position of the Buddha as a Prophet.

3.3 Theological Argument

The debate that described Buddha as a Prophet brought the teaching of tawhid backed up with the existence of several elements in parallel with the teachings of Islam. *Firstly*, the doctrine of God; *secondly*, universal values; *thirdly*, spiritual wisdom. For the doctrine of God, Muslim scholars have different view with regards the Buddhism's doctrine, whether as non-theistic or theistic. For al-Baladhuri, al-Ṭabarī, al-Maqdisi, they described Buddha's with theistic, which is apart from Islamic teaching. The same goes to al-Biruni, followed by al-Sharastani, that Buddhists have their own concept of idolatry and not worshipping Allah. However, for the twentieth century Muslims scholars, by supporting the idea of Buddha as a prophet, the Islamic worldview on Buddhism are open to debate. Like other prophets, it was Buddha's mission to correct the religious inclination of his time, otherwise the tradition subsequently became corrupted (Obuse 2010; Abul Kalam Azad 1981).

While justifying the Buddhism as non-theistic religion, Imran Nazar Hosein (1976) asserts, it is the original teaching of Buddha, and in the Hinayana sect, there is no God. However, this does not connote atheism (denial of existence of God), rather it affirms the existence of any Deity. Apparently, Buddha was once asked by a disciple whether God existed. He refused to reply. When pressed, he remarked to the effect that if you are suffering from a stomach ache, would you be concerned with the relief of the pain or with studying the prescription of the physician. "It is not my business or yours to find out whether there is a God – our business is to remove the suffering of the world. Hence, by providing the doctrine of Dhamma or the 'Impersonal Law', in place of God.

For Reza Kazemi (2010), he agrees with the non-theistic of Buddha's doctrine. In other words, it shows there is no Personal divinity playing the role of Creator, Revealer, Judge in Buddhism. But to assert that the Buddha's doctrine is 'atheistic' would be to attribute to him an explicit denial and negation of the Absolute—which one does not find anywhere in his teachings. Even regarded as non-theistic, the concept is close to the concept of *shūnya* (denial) in the Buddha and *shahādah* (testimony) in Islam. The concept of *shūnya* (denial) in Buddhism denies the existence of God, while the concept of *shahādah* (testimony) in Islam contains the pronouncement of denial (*la ilāh*) and pronouncement of confirmation (but Allah). The word rash is by denying the existence of God, in the sense that no God exists. Then followed by the word by affirming that only God is the only God who exists. Hence the disclaimer of the Buddha's teachings to the existence of God is exactly the first pronouncement in *shahādah* (testimony) of Islam (Imtiyaz Yusuf, 2003 & Reza Shah Kazemi, 2010). For the universal values that promoted by Islam-Buddhist, Hamza Yusuf (2010) suggests; Islam is based on six pillars of Faith and its five pillars of Islam, while the Buddha is subject to the Four Noble Truths and the Eightfold Paths. Both teachings also share the value of kindness, patience and love as the core of human nature. Whereas in terms of spiritual wisdom as asceticism, Buddhist teachings use the term *anicca* while Islam applies the terms of *Zuhd*. *Anicca* as in the Buddha refers to something that is immortal which becomes a barrier to one's to see the true nature of the world. Parallel with Islam, life in the world (*al-hayāt al-dunyā*) is not eternal life (Reza Shah Kazemi, 2010). Thus, in the concept of *zuhd*, one separates itself from worldly things, in the sense that life in the hereafter is eternal. Both terms though have different concepts but share same connotations (Imtiyaz Yusuf, 2003 & Reza Shah Kazemi, 2010).

Ironically, the theological arguments regarding the position of Buddha is more complex than the philology and historical arguments. Furthermore, there are three main schools in Buddhism – Theravada, Mahayana, Hinayana which originated from different background and have different concepts of theology. However, by relating the theological concept in Islam teaching with general doctrine in Buddhism like *sunya* with *shahādah* and *anicca* with *zuhd*, indirectly it can consider as coming from the same source which is from the divine. In short, the three of arguments – philology, historical and theological can be summarize as followed:

Table 1 Type of arguments on Buddha's position

Phase (Century)	Muslim Scholars	Views of Buddha as Prophet		
		Philological	**Historical**	**Theological**
9th	al-Ṭabarī (838-923 AD)		X	
	al-Maqdisi (974 AD)		X	X
	al-Baladhuri (m.892)		X	X
10th	al-Biruni (973-1048 AD)		X	X
	Ibn al-Nadim (995 AD)		X	X
11th	al-Shahrastani (1086-1153 AD)		X	X
13th	al-Hamadani (1247-1318 AD)	X	X	X
20th	Al-Qasimi (1866-1914)	X	X	X
	Muhammad Hamidullah (1908-2002)	X	X	X
	Imran Nazar Hosein (1942 –)		X	X
	Hamid Abdul Qadir (1957)	X	X	X
	Hamza Yusuf (1958 –)		X	X
	Shah Reza Kazemi (1960 –)		X	X
	Imtiyaz Yusuf (1900s)	X	X	X

4 Conclusion

In Islamic teaching, the religion of tawhid begins with the sending of Adam and followed by hundred thousand of prophets and messengers. Either named or characterized with a figure, vis versa, Muslims need to believe on their prophethood as to bring mankind to the truth. The Quran states, every nation has their own prophet, thus Muslims believe no nation will be abandoned without a prophet or messenger to guide them. As Islam is a propagation religion, Muslims need to deliver the message to all mankind. Furthermore, the Quran has classified religion into three categories – Islam, the people of book, and the people of dubious. The Quran also refutes the false doctrine of other religions to call them to the truth. Until the seventh century, by the expansion of Muslims empire to a region that was dominated by unmentioned religion like Buddhism, Muslims starts to learn about it, however with limited sources. From the ninth to eleventh century, Muslims scholars agree to the position of Buddha as a prophet, from theological and historical argument. These arguments include the refutation and criticism on the Buddhist doctrine which are apart from Islamic teaching. While, from the thirteen to twentieth century, Muslims scholars particularly originated from Central Asia, were keen to regard the Buddha as prophets, not just from theological and historical argument, but from the philological argument as well. By comparing the Buddha with Islamic teaching, there are proofs that Muslim Scholars are keen to be close with other religions – in particular Buddhism, by identifying the similarity elements of both religions. In additions, it is shown that Islam is always concern in promoting good relationship with other religions using the best approaches.

This study received the support and funding from the Universiti Kebangsaan Malaysia through the research code UKM Premier Challenge Grants DCP-2017-006/4 and FRGS/1/2011/SS/UKM/03/3. Appreciation is also conveyed to the Centre of Aqidah and Global Peace, Faculty of Islamic Studies, UKM, Faculty of Engineering and Built Environment, Faculty of Education, Faculty of Medicine, Faculty of Law, Faculty of Economy and Faculty of Social Science and Humanities, UKM as well as all those who have made this research a success.

5 References

Abdul Haq Vidyarthi and Abdul Ahad Dawud. (2013). *Ramalan tentang Muhammad dalam Kitab Suci Agama Zoroaster, Hindu, Buddha, dan Kristen*. Bandung: Noura Books.

Azad, A. (2017). The Beginnings of Islam in Afghanistan: Conquest, Acculturation and Islamization. In N. Green (Ed.), *Afghanistan's Islam*. Los Angeles: University of California Press.

al-Baladhuri, Abi Abbas Ahmad bin Yahya. (1987). *Futūh al-Buldān*. Beirut: Muassasah al-Maᶜarif.

al-Biruni, Abu Rayhan Muhammad bin Ahmad. (2008). *al-Āthār al-Bāqiyah min al-Qarūn al-Khāliyah*. Kaherah: Maktabah al-Thaqafah al-Diniyyah.

Berzin, A. (2008). Buddhist-Muslim Doctrinal Relations. In P. Schmidt-Leukel (Ed.), *Buddhist Attitudes toward Other Religions* (pp. 212–236). St. Ottilien: EOS Verlag.

Berzin, A. (1994). Islamic-Buddhist Dialogue. Retrieved September 13, 2017, from https://studybuddhism.com/en/advanced-studies/history-culture/buddhism-islam/islamic- buddhist-dialogue.

Berzin, A. (2010). Historical survey of the Buddhist and Muslim worlds' knowledge of each other's customs and teachings. *Muslim World*, *100*(2–3), 187–203.

Bigelow, A. (2013). Muslim-Hindu Dialogue. In C. Cornille (Ed.), *The Wiley-Blackwell Companion to Inter-Religious Dialogue*. West Sussex: John Wiley & Sons, Ltd.

Bladel, K. Van. (2011). The Bactrian Background of the Barmakids. In Islam and Tibet: Interactions along the Musk Routes. (Ed. Anna Akasoy, Charles Burnett, and Ronit Yoeli-Tlalim). Surrey: Ashgate.

Canby, S.R. (1993). Depictions of Buddha Sakyamuni in the Jami' al-Tavarikh and the Majma' al-Tavarikh. *Muqarnas*. 10: 299 – 310.

Elverskog, J. (2010). Buddhism and Islam on the Silk Road. American Historical Review, Philadelphia: University of Pennsylvania Press.

Gibb, H.A.R., et al (1986). Bilawhar wa-Yudasaf. In *The Encyclopaedia of Islam*. Vol. 1, pp. 1215-1217) Leiden: E.J. Brill.

Hamza Yusuf. (2010). Buddha in the Quran. *Common Ground Between Islam and Buddhism*. Louisville: Fons Vitae.

H.O.K Rahmat S.H. (1984). *Dari Adam sampai Muhammad: Sebuah Kajian Mengenai Agama-agama*. Kota Baharu: Pustakan Aman Press.

Ibn Abi Hatim, Abdul Rahman bin Muhammad ibn Idris al-Razi. (1997). *Tafsīr al-Qur'ān al-ᶜAdhīm*: Riyadh: Maktabah Nizar al-Baz.

Ibn al-Nadim. (2014). *Kitab al-Fihrist*. London: Al-Furqan Islamic Heritage Foundation.

Ibn Kathir, Imād al-Dīn Abi Fidā' Ismaᶜīl bin Kathīr al-Dimasyq. (2000). *Tafsīr al-Qur'ān al-ᶜAdzīm*. Kaherah: Mu'assasah Qurṭubah.

Imran Nazar Hosein. (1976). *Islam and Buddhism in the Modern World*. Karachi: World Federation of Islamic Missions.

Imtiyaz Yusuf. (2003). Religious Diversity in a Buddhist Majority Country: The Case of Islam in Thailand. *International Journal of Buddhist Thought and Culture*, 3(September), 131–43.

Imtiyaz Yusuf. (2017). *Muslims Don't Study Buddhism Enough*. Retrieved from http://crcs.ugm.ac.id/news/11079/muslims-dont-study-buddhism-enough-an-interview-with-prof-imtiyaz-yusuf-part-1.html.

Imtiyaz Yusuf. (2007). The Southern Thailand Conflict and the Muslim World. *Journal of Muslim Minority Affairs* 27(2): 319–339.

Imtiyaz Yusuf. (2013). Islam and Buddhism. In C. Cornille (Ed.), *The Wiley-Blackwell Companion to Inter-Religious Dialogue*. John Wiley & Sons, Ltd.

Jackson, P. (2017). *The Mongols and the Islamic World from Conquest to Conversion*. New Haven: Yale University Press.

Levinson, D., & Christensen, K. (2002). *Encyclopedia of Modern Asia*. New York: Charles Scribner's Sons.

al-Maqdisi, Mutahhir bin Tahir. (2004). *Kitāb al-Bad' wa al-Tārīkh*. Maktabah al-Thaqafah al-Diniyyah.

al-Mas'udi. (1841). *Meadows of Gold and Mines of Germs*. (Tras. Alloy Sprenger, M.D). London: Harrison and Co., Printers.

Muhammad Hamidullah (1974). Muhammad Rasullah. Hyderabad: Habib & Co.

al-Mujahid, Abu al-Hujjaj. (1989). *Tafsīr al-Mujāhid*. Misr: Dar al-Fikr al-Islami al-Hadithah.

Obuse, K. (2010). The Muslim doctrine of prophethood in the context of Buddhist-Muslim relations in Japan: Is the Buddha a prophet? *Muslim World* 100(2–3): 215–232.

al-Qasimi, Muhammad Jamal al-Din. (2002). *Tafsīr al-Qāsimi al-Musamma Maḥāsin al-Ta'wīl*. Beirut: Dar al-Kutub al-ᶜIlmiyyah.

al-Qurtubi, Abu Abdillah Muhammad bin Ahmad. (1964). *Al-Jamiᶜ al-Ahkām al-Qur'ān*. Kaherah: Dar al-Kutub al-Misriyyah.

al-Razzaq, Abu Bakr Abd. (1999). *Tafsir ᶜAbd al-Razzaq*. Beirut: Dar al-Kutub al-ᶜIlmiyyah.

Reza Shah Kazemi. (2010). *Common Ground Between Islam and Buddhism*, Louisville: Fons Vitae.

Sachau, E. C. (1910). *Alberuni's India; an account of the religion, philosophy, literature, geography, chronology, astronomy, customs, laws and astrology of India, about A.D. 1030. Trubner's Oriental series*. Edinburgh: Ballantyne, Hanson & Co.

Schmidt-Leukel, P. (2010). Buddhist-Muslim dialogue. Observations and suggestions from a Christian perspective. *Muslim World* 100(2–3): 349–363.

al-Shahrastani. (1992). *Al-Milāl wa al-Nihāl*. Beirut: Dar al-ᶜIlmiyyah.

Sparavigna, Carolina, A. (2013). The Science of al-Biruni. *International Journal of Sciences*, 1(12), 52–60.

al-Ṭabari, Abi Jafar Muhammad bin Jarir. 2001. *Tafsīr al-Ṭabari: Jāmiᶜ al-Bayān ᶜan Ta'wīl āya al-Qur'ān*. Kaherah: Dar Hijr.

al-Tabari. (1991). *The History of al-Tabari: Strom and Stress along the Northern Frontiers of the Abbasid Caliphate*. C.E. Bosworth (Ed.). New York: The State University of New York Press.

Tabataba'i, Sayyid Muhammad Hussayn. (1997). *Al-Mīzān fī Tafsīr al-Qur'ān*. Beirut: Mu'assasah al-ᶜAlami lil Matbuᶜah.

Tabrizi, T. G. (2012). Ritual purity and Buddhists in modern Twelver Shi'a exegesis and law. *Journal of Shi'a Islamic Studies*, *5*(4), 455–471.

The Holy Quran. (1987). *English Translation of the Meanings and Commentary*. Al-Madinah al-Munawarah: King Fahd Holy Quran Printing Complex.

Freedom of belief in the light of Maqasid al-shariah: Special reference to Sri Lanka

Najimudeen Mohamed Rishard[1]*, *Mohamed* Mohideen Nayeem[1], *Muhaimin* Sulam[1]

[1] Department of Management and Humanities, Universiti Teknologi PETRONAS, 32610 Bandar Seri Iskandar, Perak, Malaysia

Abstract. Concept of maqasid al-shariah (purposes of Shariah) is being re-interpreted in order to confront modern challenges and questions. This research analyses how the predominant segment of maqasid al-shariah, Protection of Religion, has been interpreted by modern scholars in a conflicting terminology as 'freedom of belief' and how they discuss the misapprehended texts which relate to the topic. Research focuses on applying the modern concept of freedom of belief which is one of the universal values, into Muslim majority countries as well as Sri Lanka where Muslims are a minority. A Special consideration has been specified on freedom of belief in Sri Lanka's constitution and its implementation. Qualitative data, mostly secondary data, has been used in this study. Data were collected from literary books, journal articles and internet sources. This research finds out that Sri Lanka in particular, a number of Muslim minority countries in general are healthier than several Muslim majority countries, especially Middle East countries, in practicing freedom of belief.

Key Words: Freedom of belief, Protection of religion, Maqasid al-shariah, Sri Lanka

1 Introduction

Maqasid al-shariah is a philosophy and it was implemented by the last messenger and his companions. A Number of prophetic narrations indicate that the companions of the messenger did not apply the 'apparent meaning' of his instructions (Auda, 2008). A famous incident of 'Banu Quraiza' was one of the transparent exemplifications for the above argument. The Second Khalifa Omar preferred to apply certain intent or purpose of text in his various decisions and judgments. After every battle he decided to distribute only one fifth of the conquered lands among the soldiers intending to enrich the public trust and achieve fairness amongst soldiers (Ibn Rushd, 2000).

After the era of the companions, traditional Imams applied some reasoning methods like Al-qiyas, Al-istihsan and Al-maslaha. Particularly, Abu Hanifah and Malik in their various legal opinions went beyond apparent meaning and gave their greatest possible effort to find out the cause or purpose of the text. But the maqasid terminology had not been used by Imams until the final phase of the 3rd Islamic century. In between 3rd to 5th Islamic centuries, some scholars paid their attention to maqasid al-shariah and they also had some publications in this regard (Raisuni, 2013). Between fifth to eight centuries, maqasid was drafted as a philosophy. The maqasid concept further developed and reached its matured stage in the 8th Islamic century.

After al-Shathibi, no proper efforts were put forward to enrich it until the modern Islamic renewal (Raisuni, 2013). Fortunately, Ibn al-Ashur was considered to be the first modern scholar to reopen the door for maqasid concept, who understood the significance of the maqasid in general and significance of al-Shathibi's tremendous and dynamic role in this subject in particular. He started to teach it in universities and academic forums. Furthermore, he was able to renew the maqasid thought in order to find solutions for modern problems and challenges. He included values such as equality and freedom in maqasid al-shariah (Raisuni, 2013).

Some other modern scholars tried to broaden the scope of the five traditional maqasid. For instance, Rashid al-Rila suggested to include reform and women's right in the theory of maqasid (Rila, 2010), Muhammad Al-Gazaly added justice and freedom to the pre-modern five maqasid (Atiya, 2008), and Taha Jabir al-Alwani added the concept of developing civilization on earth (Alwani, 2010).

The Maqasid philosophy is being reformed by contemporary Islamic scholars currently. They are completely involved in moving the philosophy of maqasid from a theoretical frame to a practical solution. Moreover, some

* Corresponding author: nm.rishad1985@gmail.com

activists and politicians try to employ concept of maqasid in their social and political reformist projects (Rane, 2015). Their strategies and approaches in the nation building process using maqasid al-shariah are highly welcomed and appreciated by Muslim majority countries as well as the western countries. Furthermore, maqasid al-shariah is applied in other fields like economic, health, environment and social development (Deuraseh, 2012).

2 Materials and Methods

This study is descriptive and interpretive and mainly uses secondary data collected from sources such as research journals, reports from research centres and online resources. The data analysis used in this study is carried out qualitatively based on text. This research tries to apply modern interpretation of first segment of maqasid into contemporary world. Special attention has been paid on Sri Lanka's constitution and its application.

3 Literature Review

3.1 Modern interpretation on Protection of Religion

Traditional Islamic scholars used 'preservation' and 'protection' in their writings when they described maqasid al-shariah. In contrast, contemporary scholars have developed traditional maqasid terminologies according to modern challenges and problems. They use new terminologies like 'Rights' and 'Development' despite the opposition of few other jurists regarding these terminologies (Auda, 2007).

Ibn al-Ashur was one of the first most renowned modern-era scholars to open the door for re-interpreting the 'Theory of maqasid' in general and 'Theory of protection of religion' in particular. In previous periods scholars like Al-Juwaini used the terminology 'protection of Religion'. In modern days Ibn al-Ashur put forward a dramatically different concept for the word 'Hifl al-deen'. He preferred to use different terminologies such as 'Freedom of faith' and 'Freedom of belief' instead of 'protection of religion' (Ibn al-Ashur, 2011). After Ibn al-Ashur, presenters of this perception frequently quote the following Quran verse in their writings: 'No compulsion in matters of religion' (Al-Baqara 256). They also have a different kind of interpretation for the punishment of apostasy (Had al-riddah) and their arguments will be analysed briefly in the following topic.

Ibn al-Ashur emphasized in his book 'Maqasid al-shariah' on re-interpreting some words like freedom. For instance, the meaning of freedom which was proposed by him is different from freedom which was mentioned by traditional jurists. Ibn al-Ashur included freedom of thought, expression, belief and action in this particular terminology (Ibn al-Ashur, 2011). After Ibn al-Ashur, a number of scholars tried to expand the theory of freedom and paid special attention to freedom of belief. Rashid al-Ghanuchi had a deep discussion on how we could interpret the terminology 'Freedom of belief' (Ghanuchi, 1993). Furthermore, he has been playing a vital role in applying freedom of belief in Tunisia's context since the Arab spring although his approach is being criticized by few scholars.

3.2 Misunderstood texts related to freedom of belief

Number of contemporary scholars differ from classical scholars in understanding and interpreting the texts related to apostasy and its punishments. Both classic and modern scholars approach this critical issue with their political and historical factors. There were different opinions among traditional scholars themselves. Most of the traditional scholars argued that the death penalty is the punishment for apostasy regardless of the difference of opinions among them with regard to the waiting period while in incarceration to repent and accept Islam again. In contrast, a few other scholars refused and did not agree with particular punishment (Alwani, 2012).

Firstly, Modern scholars try to take a holistic view on freedom of belief by focusing on al-Quran verses. They frequently quote the following verses in order to strengthen their argument that Islam provides freedom of belief to whole human being:

"There is no compulsion in religion. The right way is indeed clearly distinct from error" (Al-Baqara 256)

"Say: (It is) the truth from the Lord of you (all). Then whosoever will, let him believe, and whosoever will, let him disbelieve. Lo! We have prepared for disbelievers Fire. Its tent enclosed them. If they ask for showers, they will be showered with water like to molten lead which burneth the faces. Calamitous the drink and ill the resting-place!" (Al-Kahf 29)

"We have truly shown him the way; he may be thankful or unthankful" (Al-Insan 3)

"Clear proofs have indeed come to you from your Lord: so whoever sees, it is for his own good; and whoever is blind, it is to his own harm. And I am not a keeper over you" (Al-An'am 104)

"If they accept Islam, then indeed they follow the right way; and if they turn back, your duty (O Prophet) is only to deliver the message" (A'l Imran 20)

"And obey Allah and obey the Messenger; but if you turn away, the duty of Our Messenger is only to deliver the message clearly" (Al-Taghabun: 12; see also Al-Ma'idha 92)

Secondly, modern scholars refer to the Hadeeth where Prophet Muhammad emphasises the punishment for apostasy. There are number of narrations of the prophet that indicate the person who renounces Islam should be killed. Following two are among them:

"Whoever changes his religion, kill him" (Bukhari)

"The blood of a Muslim who confesses that none has the right to be worshipped but Allah and that I am His Apostle, cannot be shed except in three cases: In Qisas for murder, a married person who commits illegal sexual intercourse and the one who reverts from Islam (apostate) and leaves the Muslims." (Bukhari)

Most of the modern scholars argue that the background of above mentioned Hadeeth should be considered while interpreting. From a holistic view of the above mentioned and some other texts which are related to this issue, scholars come to the following conclusions:

1. Islam allows a person to embrace and practice any theology or religion which he believes in. He has the freedom and right to follow the instruction of his religion among the Muslims (Saeed, 2017). On the other hand, no one from the Muslims, even leader of the country, has the right to compel him to embrace Islam or leave his religion. Furthermore, Muslims should not disturb him during his prayers and while he is practicing his religious acts. The whole history from the prophetic era until the present would be the obvious evidence for this argument.
2. Al-Quran did not mention any punishment for apostasy in this world when describes about the people who returned back to their former religions and beliefs after embracing Islam. Following verse is a clear example for above mentioned argument: "And whoever of you reverts from his religion [to disbelief] and dies while he is a disbeliever - for those, their deeds have become worthless in this world and the Hereafter, and those are the companions of the Fire, they will abide therein eternally" (Al-Baqara 217). It is very interesting to make a note here that the writer Willi Heffening mentions in the Encyclopaedia of Islam: 'In the Quran the apostate is threatened with punishment in the next world only' (Encyclopaedia of Islam, vol. 3, p. 736).
3. Some scholars argue that the only apostates who did other crimes such as insubordination and plotting against the government, could be punished with death penalty. Committing crimes could be the only reason for which he deserves the punishment, and it has nothing to do with changing one's religion or belief (Nandwa, 2016). This was exactly what happened in fathhu Makka and some other occasions.
4. With a deep erudition into Madenian period, we could realize that some Jews changed their belief very frequently. Their intention was to make a mockery of purity of Islam (Raisuni, 2013). They tried to fabricate a public opinion that Islam is not a religion worth having. "Those who believe, then reject faith, then believe (again) and (again) reject faith, and go on increasing in unbelief, Allah will not forgive them nor guide them nor guide them on the way" (Al-Nisa 137). The Quran did not ask the prophet to punish them even having mockery of purity of Islam.
5. The misconception regarding apostasy appears to have arisen from the people who joined the enemies of Muslims after becoming apostates. However, they were treated as enemies and were killed during the war; not for changing their religion, but for committing crimes and murders (Nandwa, 2016). The Quran points out a few cases where some people left Islam and merged with the opposition during war. It further guides the prophet and his companions on how to deal with this kind of people (Al-Nisa 88, 91).
6. Some other scholars deal with this issue in a different way. They argue that understanding the socio-political context of the prophetic period is very important. During that time apostasy was not considered as merely changing belief or faith. It was an act declaring war on the former (Alwani, 2012). According to this argument, apostasy was a political crime not a religious one.

As we emphasised earlier, these interpretations provide a clear conclusion that Islam gives complete freedom to every person to follow and practice a religion he wishes. Moreover, it guides Muslims not to punish or insult the one who leaves and shifts to another religion.

4. Discussion and Results

4.1 Applying 'freedom of belief' in contemporary world

By understanding the above-mentioned paragraphs deeply, we will be able to measure the countries which provide religious freedom, and which restrict the freedom; which is the first and foremost objective of Islamic shariah.

Pew Research Centre which is located in USA conducted a research on religious freedom. Pew research categorized data from 18 published cross-national reliable sources and it referred to the sources only for well-documented, specific facts, not for commentaries and opinions. The staff of the forum used extensive data-verification checks and inter -rated reliability assessments in order to come out with genuine and scientific results (Pew Research Centre, 2011). Following are main results of the research:

1. Research finds that seven out of the ten countries with the worst grades for government restrictions on religion are Islamic states (Pew Research, 2011).

2. Geographically, the Middle East/North Africa region boasted the largest proportion of countries – 30 percent – where official restrictions on religion increased over the three-year period (Pew Research, 2011).

3. Seven of the ten countries with the highest – that is, worst – grades when it comes to government restrictions on religion were OIC countries. The other three were China, Burma and Eritrea (Pew Research, 2011).

4. In an index measuring official interference with religious practice, 18 out of 26 countries (69 percent) whose government "prohibits worship or religious practices of one or more religious groups as a general policy," were OIC members (Pew Research, 2011).

5. A grading of countries where conversion from one religion to another is restricted was also dominated by Islamic states, accounting for 25 out of 29 countries (Pew Research, 2011).

6. Taking the conversion issue a step further, among 13 countries where there were incidents of physical violence over conversions from one religion to another, 10 (77 percent) were Muslim majority countries (Pew Research, 2011).

7. A study conducted by Human Rights First and it documented more than 70 cases in 15 countries where the enforcement of blasphemy laws resulted in problems of various kinds since 2007. Of the 70 cases, only four were not in Muslim countries. By contrast, the vast majority of the cases documented in the report took place in Muslim majority countries and included lengthy prison terms and the imposition of the death penalty, as well as extrajudicial retribution such as mob attacks and killings (Human Rights First, 2012).

Number of contemporary Muslim scholars accept above mentioned facts and they keep on repeating that the Muslim majority countries should reduce religious restrictions and should provide religious freedom for their citizens (Alwani, 2012). Tariq Ramadan condemns the religious restrictions imposed by Muslim majority countries and he argues that all citizens have their freedom equally and the religious freedom should be guaranteed to the citizens (Ramadan, 2009). Rashid al-Ghanuchi criticizes the Muslim majority countries as well as the Islamic movements for their rigid approach to the religious freedom (Ghanuchi, 1993).

In summary, most of the Muslim majority countries neglect and ignore the first and foremost part of maqasid al-shariah as elaborated by contemporary Islamic scholars, which is freedom of belief. On the other hand, number of non-Muslim countries offer freedom of belief or religious freedom to their citizens in satisfactory manner. Freedom of belief is one of the universal values that can be applied in every countries regardless of religions and ethnics (Ramadan, 1999). In some countries where Muslims are minority, freedom of belief is granted to the citizens in a healthy way.

4.2 Freedom of belief in Sri Lanka: Constitution and its implementation

Sri Lanka is an island with a population of 20 million consisting of a mixture of Sinhalese, Tamils, Muslims and others. The Sinhalese community is the majority while Tamils and Muslims are the minority respectively.

As far as the Sri Lankan constitution is concerned, there are two chapters which are closely related to religious freedom. Chapter 2 talks about Buddhism while chapter 3 speaks about fundamental rights. In the following paragraphs we will analyse the articles of these chapters which have specific provisions to safeguard the religious freedom in the country.

Chapter 03 includes some articles which provide religious freedom and rights to minority community:

Article 10: Every person is entitled to freedom of thought, conscience and religion, including the freedom to have or to adopt a religion or belief of his choice.

Article 12:

(1) All persons are equal before the law and are entitled to the equal protection of the law.

(2) No citizen shall be discriminated against on the grounds of race, religion, language, caste, sex, political opinion, place of birth or any such grounds

(3) No person shall, on the grounds of race, religion, language, caste, sex or any one such grounds, be subject to any disability, liability, restriction or condition with regard to access to shops, public restaurants, hotels, places of public entertainment and places of public worship of his own religion.

Article 14. (1) Every citizen is entitled to

(a) The freedom of speech and expression including publication;

(b) The freedom of peaceful assembly;

(c) The freedom of association;

(d) The freedom to form and join a trade union;

(e) The freedom, either by himself or in association with others, and either in public or in private, to manifest his religion or belief in worship, observance, practice or teaching;

(f) The freedom by himself or in association with others to enjoy and promote his own culture and to use his own language;

(g) The freedom to engage by himself or in association with others in any lawful occupation, profession, trade, business or enterprise;

(h) The freedom of movement and of choosing his residence within Sri Lanka; and

(i) The freedom to return to Sri Lanka

Above mentioned chapter with its articles illustrate that constitution of Sri Lanka provides special protection to minority religions against unequal treatment before the law and it ensures equal protection for them. Minority communities were given the freedom to draft the matters related to personal law like divorce, child custody, and inheritance according to their religious principles (Khan and Rahman, 2009). For instance, the minimum age level for marriage is 18 except for the Muslims. In day to day life activities the minority communities are given rights equal to the Buddhist majority.

In Sri Lanka, minority communities are given admission to state schools without any discrimination, they are provided employment in state sector, they have the right to build mosque or temple to conduct their worship, and they have the right of freedom to travel throughout the country (Godahewa, 2017). This indicates that Sri Lanka provides more religious freedom and accepts of religious freedom than many Muslim majority countries.

However, a number of complaints have been filed before the Supreme Court with regard to the violation of fundamental human rights (Roshani, 2016). Moreover, a report on a series of human rights violations is to be released by the Human Rights Commission as well. In comparison between Sri Lanka and Muslim majority countries, the former is very high and stable in religious freedom even though there were few unpleasant incidents reported against the minority communities in the past. For instance, Kandy riot against Muslims that took place on 29th of February 2018 is one of the transparent examples for unethical violations. But the Sri Lankan government took necessary steps as much as to its possibility in order to curb and stop the riot. Nationwide state of emergency was imposed by the government to control the violations. Finally, Sri Lankan government arrested Amith Weerasingha who was alleged to be one of the masterminds of the communal unrest, along with 08 others. The rule of the law should be implemented to avoid these kinds of violations. Additionally, reasons for the breakdown of the rule of law should be identified and fulfilled (James, Craig, 2010).

There is a heated debate among political scholars regarding chapter 2 of the constitution. Following is the one and only article in this chapter:

The Republic of Sri Lanka shall give to Buddhism the foremost place and accordingly it shall be the duty of the State to protect and foster the Buddha Sasana, while assuring to all religions the rights granted by Articles 10 and 14(1)(e).

This preferential statement to Buddhism is highly controversial among scholars and researchers. Some argue that this article gives Buddhism the 'foremost place' and asks the government to protect it. The first meaning of that is, the minorities are second class citizens and they should be treated as second class not equal to majority Buddhist. Some activists argue that the only one solution for this kind of classification is to amend the constitution and make Sri Lanka as a 'Secular state' (Public Representations Committee on Constitutional Reform, 2016). They argue that secular state is a very positive one. History tells us that no religion or culture will automatically be protected merely by having state sponsorship. The best example would be Catholicism itself. As revealed by history, the decadence of the Catholic Church in Europe began with State sponsorship. Further they emphasize that a secular state does not mean being irreligious. But an equal freedom for people to pursue their spirituality and faith without the obstacle of a constitution that guarantees only one religion which can sometimes assault others.

On the other hand, some others emphasize that it does not recognize it as the state religion. Additionally, the constitution states that every person is "entitled to freedom of thought, conscience, and religion, including the freedom to have or to adopt a religion or belief of his choice." As we mentioned earlier, the constitution gives the priority and foremost place according to its overwhelming majority (76%) (Public Representations Committee on Constitutional Reform, 2016). Meanwhile there are no obstacles, or discriminations against minorities except a few incidents reported here and there. Minorities have been given separate religious affairs departments and they are provided the freedom to conduct their religious education outside the school system. National holidays have been declared for the major religious festivals of other religious groups. Moreover, they argue that a non-religious state is a leftist idea. Buddhism is accepted as the major religion by tradition and accepted as a characteristic of democratic society. A country cannot exist without a religion.

In this controversial issue, it is very difficult to make changes in the constitution that Sri Lanka as a secular state and it would be an impossible task in the near future. According to modern activists, the most important thing is the community should be educated and trained to respect others and to live with religious tolerance and harmony. Finally, all academicians without any exception emphasize that the rule of law should be guaranteed as well as applied regardless of race, ethnicity and religion.

5 Conclusion

Freedom of belief and apostasy are two controversial terms which have made for heated debate among modern Muslim scholars. A number of them re-interpret the texts related to apostasy in a different way to classical scholars and they come to some conclusions. They argue that Islam considers freedom of belief as one of the universal values and it provides freedom of belief to all human beings. Further, they reiterate that Muslims, either in Muslim majority countries or Muslim minority countries, should contribute in guaranteeing freedom of belief to their fellow citizens. Unfortunately, several Muslim majority countries, especially Middle East countries, have strict rules and regulation as well as restriction on religious freedom. Moreover, official interference with religious practices and physical violence over conversions from one religion to another have been recorded there. It is clear that a number of Muslim minority countries are healthier than several Muslim majority countries in practicing freedom of belief, first segment of maqasid al-shariah. The Constitution of Sri Lanka and its application are transparent evidences for above mentioned argument. For instance, Sri Lankan government applied the rule of law without any religious discrimination during the Kandy riot that took place by imposing nationwide state of emergency and arresting the person who was alleged to be one of the mastermind of the riot along with other violators.

6 References

Alwani, T. (2010). Afala yathadabbaruna al-Qur'an (pp. 62). Cairo: Dar al-Salam.

Alwani, T. (2012). Apostasy in Islam A Historical and scriptural analysis (pp. 03). USA: International Institute of Islamic Thought.

Atiya, J.D. (2008). Nahwa thaf'ili maqasid al-shariah (pp. 98). USA: IIIT.

Jasir, A. (2008). Maqasid al-shariah a beginner's Guide (pp. 11). USA: The international institute of Islamic thought.

Jasir, A. (2007). Maqasid al-shariah as Philosophy of Islamic Law A systems approach (pp. 21). USA: The International Institute of Islamic Thought.

Deuraseh, N. (2012). New essential values of Daruriyyah of the objectives of Islamic Law, *Journal Hadhari*, 4(2), 107.

Ghanuchi, R. (1993). al-Huriya al-Aamma fi al-dawla al-Islamiya (pp. 44). Beirut: Markaz dirasath al-Wahda al-Arabiya.

Godahewa, N. (2017). Discrimination in Sri Lanka: The reality, *Speech delivered at The 36th Session of the UNHCR* in Geneva On 27 September 2017.

Human Rights First. (2012). Blasphemy Laws Exposed: The Consequences of Criminalizing "Defamation of Religions".

Ibn al-Ashur, T. (2011). Maqasid al-shariah al-Islamiah (pp. 233). Cairo: Dar al-kithab al-Misri.

Ibn Rushd, W. (2000). Bidayah al-Mujthahid wa Nihaya al-Muqthasid (pp. 291). Dar al-Kutub al-Islamiyah.

James, Y, Craig, S. (2010). The Breakdown of the Rule of Law in Sri Lanka: An Overview (pp. 03). Sri Lanka Campaign for Peace and Justice.

Khan, B., Muhbubur R. (2009). Protection of Minorities: A South Asian Discourse (pp. 82). Dhaka: EURASIA NET.

Nandwa, W. (2016). Plurality and Religious Tolerance in Islam, *European Scientific Journal*, vol.12, No.32, 320

Pew Research Center's Forum on Religion and Public life, August, 2011.

Raisuni, A. (2013). Muhalarath fi maqasid al-shariah (pp. 62). 2nd Ed, Dar al-Kalima.

Raisuni, A. (2013). Al-fikr al-Islami wa kalayana al-siyasa al-muasira (pp. 90). 1st Ed, Dar al-Kalima.

Ramadan, T. (2009). Islam, the West and the challenges of Modernity (pp. 76). 4th Ed, Nigeria: The Islamic Foundation.

Ramadan, T. (1999). To be a European Muslim (pp.131). London: The Islamic Foundation.

Rane, H. (2013). The relevance of a maqasid Approach for political Islam post Arab Revolutions, *Journal of Law and Religion*, Vol.28, 490.

Report on public representations on Constitutional reform, 2016.

Rila, R. (2010). Al-wahyu al-Muhamadiyu (pp. 333). Al-Haia al-a'mma li kusur al-sakafa.

https://doi.org/10.1051/shsconf/20185304002

Roshani, W. (2016). Confronting intolerance: Continued violations against religious minorities in Sri Lanka (pp. 08). Minority Rights Group International.

The Paradigms of Consciousness: A Discourse

Khalidah Khalid Ali[1]*Muhaimin bin Sulam[2]

[1,2] Department of Management and Humanities, Universiti Teknologi PETRONAS, 32610, Seri Iskandar, Perak Malaysia

Abstract

Consciousness is deep in meaning and utilised in many different ways. It is often used to scientifically explain one's ability to discriminate stimuli, report information, monitor internal states or control behaviour. This paper investigates the concepts and paradigms of consciousness from scientific, Western and religious perspectives with focus on Islamic dimensions as a discourse. A library research is conducted and online resources are utilised to review literatures from multi-disciplines before arriving at a conclusion. Discursive paradigms of consciousness recognise it as an abstract concept. Consciousness is beyond cognition. While cognitive thoughts, theories and ideologies relate consciousness to the five senses of humans, the concepts of 'will', awareness (i.e. being aware of things happening around a human being) and having a sense of purpose or intent in undertaking actions are relevant in the study of consciousness. A conscious person will use his/her senses to conduct life as civilised beings. However, by linking the body and the soul in explaining consciousness from Islamic viewpoints, consciousness is the awareness of oneself driven by his/her commitment and conviction as a servant of God and total submission to the Creator. Consciousness from Islamic contexts indeed relates to *iman* (faith) and *taqwa* (God consciousness).

Keywords: Consciousness, Multi-disciplines, Islamic perspectives, Discourse.

* Corresponding author: khalidah_kh@utp.edu.my

1 Introduction

This paper is part of a doctoral work to study the consciousness of ethics among the Malay ethnic group in Malaysia embracing Islam as a religion. As a discourse, it will firstly investigate the concepts and paradigms of consciousness from scientific, Western and religious perspectives before focussing on Islamic dimensions in order to put in context with the above-mentioned full research study.

The word 'consciousness' is deep in meaning and utilised in many different ways. It is often used to scientifically explain one's ability to discriminate stimuli, report information, monitor internal states or control behaviour (Chalmers, 2002). In definition, it is the state of being awake and aware of one's surroundings (Oxford Dictionaries online, 2016). Consciousness is also defined as the quality or state of being aware especially of something within oneself (Merriam Webster Online Dictionary, 2017). Rationally, consciousness is a process in itself. As such, it is the sum of all system interaction, that is, the processes that finally may lead to a conclusion about the awareness of the surrounding or self. Consciousness, being defined as awareness, however, does not necessarily require reason, reasoning or rationality. Even an unreasonable and irrational man has consciousness from relativistic contexts.

Following a medical definition which often emphasises on objectivity and scientific viewpoints, consciousness is being aware of the stimulants through the five human senses, consistently alert and controlling one's behaviour and attitudes. In other words, what is meant by consciousness is one's ability to perceive through senses.

By examining this concept within philosophy, one form that definitions of consciousness may take is in the description of the phenomenal, experiential state in which a person is (Kriegel, 2006). Complementing earlier definitions, consciousness is the subjective experiencing of a stimulus or mental state. This state has been referred to as *qualia* by William James (1890) and has received a significant amount of attention from philosophers (Chalmers, 2002) who often use the above term (singular 'quale') to refer to the introspectively accessible, phenomenal aspects of our mental lives. In this broad sense of the term, it is difficult to deny that there are *qualia*. Disagreement typically centres on which mental states have *qualia*, whether *qualia* are intrinsic qualities of their bearers, and how *qualia* relate to the physical world both inside and outside the head. The status of *qualia* is hotly debated in philosophy largely because it is central to a proper understanding of the nature of consciousness. *Qualia* are at the very heart of the mind-body problem (For more elaborations, refer to http://plato.stanford.edu/entries/qualia/, retrieved online on 6 September, 2015).

While *qualia* have been discussed above, the concept of philosophical consciousness for man means 'being aware of his existence'. There exists, to a certain extent, willpower in consciousness; that influences one's decisions (Chalmers, 2002). After a human perceives things with his or her five senses, the brain processes the information that is sent to it and this consequently results in the direction of the decisions it (the brain) makes. In consciousness, to be able to conduct 'the behaviour directed toward the purpose' is in question.

To think about thought, which is the activity of high consciousness, is a characteristic of man (Chalmers, 2002). Consciousness, which is expressed through the word 'cognition', therefore is the mental activity and high brain functions that concern the processes of thought, decision-making and attention of a person. It is consciousness that directs the processes which can be summarised as 'quest for meaning with the perception of time and existence'. Man is a special being with intelligence and ability to distinguish between right and wrong; good and bad by relating to situations. This consciousness, which makes man 'real man', has high functions and it means being able to discern and differentiate abstract concepts.

2 Theories of Consciousness: An Overview

Consciousness theories are often divided into physicalist and anti-physicalist theories (Kriegal, 2006). Physicalist theories attempt to account for phenomenal consciousness in (micro) physical terms whereas anti-physicalist theories claim that it is impossible to account for phenomenal consciousness. A scholar, Uriah Kriegel (2006a), views that a more fundamental division for theories of consciousness is reductive and non-reductive theories. Reductive theories attempt to account for phenomenal character in non-phenomenal terms whereas non-reductive ones do not. Whether the relevant terms can in turn be accounted for in (micro) physical terms or not is left as an open question at this stage (Kriegal, 2006b).

Although most philosophical works have concentrated on reductive theories, some scholars have devoted to non-reductive ones as well. By claiming that phenomenal consciousness cannot be reductively accounted for in non-phenomenal terms, a non-reductive theory treats phenomenal consciousness as a pertinent feature of the world. It thus considers that there are (at least) two fundamental kinds of features in the world—physical and phenomenal. For this view and reason, it is referred to as 'dualism' (Latin word for two). In addition to the above classifications of consciousness theories, there are also higher order theories of consciousness. They include Higher Order Perception Theory and Higher Order Thought Theory by actualists; Higher Order Thought Theory by dispositionalists; and Self-Representational Higher-Order Theories (See Stanford Encyclopaedia of Philosophy, Higher–Order Theories of Consciousness. Accessed from http://plato.standford.edu/entries/consciousness-higher, on 24 May, 2016).

All these theories (somewhat similar to cognitive theories in general), assume that the right level at which to seek an explanation of phenomenal consciousness is a cognitive one; providing an explanation in terms of some combination of causal role and intentional content. To elaborate, all such theories claim that phenomenal consciousness consists of certain kinds of intentional or representational content figuring in certain distinctive positions in the causal architecture of the mind (Stanford Encyclopaedia of Philosophy).

Some Western scholars such as Kirk (1994), Dretske (1995) and Tye (1995) claim that for a state to be conscious in the relevant sense is for it to be poised to have an impact on the organism's decision-making processes, perhaps also with the additional requirement that those processes should be distinctively rational (Block, 1995). Interestingly, the main motivation of higher order theories of consciousness derives from the belief that all (or at least most) mental-state types are both of conscious and unconscious varieties.

While attempts are made to highlight some consciousness theories, this paper will continue to delve into the concept of consciousness from multidisciplinary approaches (including social science, humanity, psychology and religion) before finally discussing its paradigms from Islamic dimensions.

2.1 Consciousness from Multidisciplinary Perspectives

Discussing consciousness from societal perspectives within rational thoughts, Berger and Luckmann (1966) in their *Social Construction of Reality* logically argue that the objective social reality, although produced in social action, appears to the individual as separate and independent from him or her. The subjective side consists of the consciousness an actor has, shaped in pervasive processes of socialisation, and sustained and modified in daily interactions (Eberle, 1992). Society must be grasped in its duality as an 'objective' and a 'subjective' reality. Berger and Luckmann further explained that persons and groups interacting in a social system create, over time, concepts or mental representations of each other's actions, and that these concepts eventually become habituated into reciprocal roles played by the actors in relation to each other. Berger and Luckmann's sociology of knowledge works have also introduced the term 'social sciences'. Their studies were strongly influenced by the work of Alfred Scultz (1899-1959), an Austrian social scientist whose works bridged sociological and phenomenological traditions to form a social phenomenology.

Next, in the rubric of consciousness within social sciences, C. Wright Mills (1916-1962), in *The Social Imagination* (1999) deliberates on how social science should be pursued and advocated for a humanist sociology connecting the social, personal and historical dimensions of human lives (Elwell, 2013). Mills sees links between the apparently private problems of the individuals and important social issues within one's consciousness. In his vision, Mills has in fact proposed the proper relationship in sociological scholarship between biography and history within his own self-consciousness to enhance socialisation processes in the 'classy' and differentiated American society during his time.

Interestingly, another Western scholar, Teun A. Van Dijk (1998), in his book, *Ideology - a multidisciplinary approach*, while discussing on the social dimensions of ideologies, linked consciousness with the cognitive context, very much related to the history of the study of ideologies since Marx and Engels. However, he recognised that this form of consciousness is mostly 'false consciousness.

False consciousness here means group ideologies that do not reflect the 'objective' socio-economic interests of a group (Van Dijk, 1998). "*Thus, workers or poor people may develop an ideology that is rather in the interests of, respectively, the ruling class, the elites, the company they work for, or the owners or managers of that company*" (Van Dijk, 1998 pp. 96).

Van Dijk explicitly advocated 'consciousness' as 'awareness'. However, he views that although the notion of 'consciousness' may be clear in its everyday sense, it is theoretically hardly explicit. Being conscious of, being aware or realizing something, first of all, is a 'state of mind' (Van Dijk, 1998 pp. 98). Hence, 'consciousness' is again defined with concepts like 'awareness', which is present in man through metacognition genes. Van Dijk further theorised 'consciousness' into many types or grades i.e. between totally explicit awareness and knowledge on one hand, and largely implicit knowledge and 'mere' use as 'lived experience' on the other hand. "*The distinction is socially often associated with 'ideologues' and the 'masses', a distinction we need to analyse as part of the social analysis of ideologies*" (Van Dijk, 1998 pp. 99). The chapters that follow Van Dijk's Consciousness chapter (Chapter 9) further examine on inter-related issues such as common sense, knowledge and truth, identity and social cognition.

Following Van Dijk's line of cognitive thoughts and grasping on earlier important indicators such as using the five traditionally recognised human senses (i.e. sight, hearing, taste, smell and touch), the concepts of 'will', awareness and being aware of things happening around a human being and having a sense of purpose for doing something, we may deduce that a conscious person will use his/her senses to conduct life as civilised beings.

While Van Dijk (1998) views consciousness as theoretically hardly explicit, another Western scholar, Jerry Foss (2000, pp. 4) openly admits that consciousness is a riddle. His works attempted to link the objective concept of science with the subjectivity of consciousness which he later referred to as the science of consciousness. He advocated that there are two main ways of understanding the world: *the manifest model* and the *scientific model*. The *manifest model* is the original representation of the world generated for each human being by the natural senses inherited from human evolutionary history. The *scientific model*, on the other hand, is a more sophisticated representation of the world generated for people in general, employing instruments and inference; the product of acquired systems that humans inherit from cultural history. The *manifest model* is privileged as it is the ultimate interface with reality; both as a human being and for scientific models. Conversely, there is power in the scientific model since it has increased the scope and depth of human knowledge. When the relationship between the two is functioning well, it achieves complementarity. The difficult relationship between *manifest* and *scientific* models gives life to the riddle of consciousness (Foss, 2000, pp. 1). Needless to say, his works may be useful for scientists working on cognitive science, if not for philosophers to understand consciousness from theoretical and modelling constructs.

Nevertheless, as much as substantial work has been done on consciousness in the West, more seem to agree that consciousness is the object of psychology (Struthers, 2001). While psychology has focused on consciousness, again scholars have viewed that it is a slippery term and many have attempted to formulate a coherent description of what consciousness represents (Struthers, 2001). In fact, American psychologist, William James (1890) avoided explicitly in defining consciousness. He believed that humans are all similar with consciousness through introspection; it is a self-evident phenomenon requiring no operational defining (Struthers, 2001). To quote Struthers' words, "It is this introspective technique which was used to excess by many early psychologists that, in part, led to a reactionary movement by the behaviourists." Later in his life, however, James changed his position, arguing that consciousness was a non-entity and had no place as an object of empirical investigation (Jaynes, 1978).

Another prolific psychologist, Sigmund Freud (1912) thoroughly explored the human mind and views that, "the conscious mind may be compared to a fountain playing in the sun and falling back into the great subterranean pool of subconscious from which it rises." Freud relates consciousness with the human mind and thought, action and interaction. Freud's notion of unconscious mentality is arrived at by pressing the distinction of the mental states from consciousness and combining it with the topographical model where all the psychological locales are spoken of as existing independently from their members at any given moment (De Sousa, 2011; Freud, 1915; Freud, 1937).

Freud's notion may be interpreted in terms of gaps in self-explanation. Freud believed that when we explain our own behaviour to ourselves or others (conscious mental activity), we rarely provide a true account of our motivation. This behaviour no doubt does not relate to lying, but although humans are deceivers of others; they are even better at self-deception. One's rationalisations of his/her conduct are therefore disguising or concealing the reasons. Freud's life work was dominated by his attempts to find ways of penetrating this often indirect and elaborate façade that obscures the hidden structure and processes of personality. Freud founded psychoanalysis, (i.e. a method for treating mental illness and also a theory explaining human behaviour) which regarded the entire mental being in the first place as unconscious, and hence for them (especially psychotic cases), consciousness might be present or absent. However, it is important that one understands the nature of the unconsciousness in broad holistic terms rather than the fine details that Freud gave, and also one must follow the coherence or logic of such a concept to understand the present day understanding of consciousness (Freud, 1912; Ricoeur, 1970; De Sousa, 2011).

The elaborative discussions have so far outlined the concept of consciousness from multidisciplinary approaches (i.e. from medical sciences, philosophy, social science, humanity and psychology) based on several significant works of Western scholars. Much as the above studies have attempted to outline the concept of consciousness, it is well-said that consciousness is an abstract concept that is beyond human experiments (James, 1890; Struthers, 2001).

Accepting the fact that consciousness is an implicit, abstract and a slippery concept and recognising the contributions of Western scholars from rational judgements, one issue remains to be answered. Why is it that even from these definitions and elaborative explanations from rational thoughts, conscious people still commit unethical and false acts? Rational thoughts have its limitations. Therefore, in order to answer this challenging question, discussions are engaged within religious perspectives to understand deeply the meaning of consciousness.

The author shall briefly explain from the viewpoint of Christianity, followed by Buddhism and Hinduism before dwelling in detail on how Islam views consciousness to put in context with the research study i.e. the consciousness of the Malay community on ethics.

Recognising consciousness as an object of psychology (Struthers, 2001), the words of Struthers (2001) is quoted as follows: "Christianity places an emphasis to study consciousness as the locus for decision-making and establishing relationships is significant for Christians in psychology, cognitive science, and neuroscience because these two abilities are near the heart of the Christian experience. While scripture has little to say about the link between the mind and the brain, the nature of the 'stuff that it is made of', or the architecture of mental life, it does highlight our mental life as an integral part of our relationship with Christ" (Hunt, 1995). In short, Christianity views consciousness as a link to Jesus Christ while embarking on studies within the psychology discipline. On the other hand, Buddhism views consciousness (i.e. *vijnana*) as life force, mind or discernment. It relates to the six internal bases i.e. the eye, ear, nose, tongue, body and mind and six external sense bases i.e. visible forms, sound, odour, flavours, touch and mental objects. Hence, sense specific consciousness arises and is dependent on an internal and an external sense base (Thanissaro, 1998).

Interestingly, Hinduism relates consciousness as to know God distinctly by realising His Existence through an intuitive experience and to speak to Him intimately (Swami Nikhilananda, 1985). What do Muslim scholars have to say about consciousness?

2.2 Consciousness from the Islamic Perspective

An Islamic scholar, Nevzat Tarhan (2012), in his explanation on consciousness, relates humans to their nature and role on earth. "Man, is a being that adds meaning to everything and that is aware of his existence. Consciousness is one of the properties that distinguishes man from the other living beings and the most important and distinct property forming the difference that distinguishes him from the other species. Consciousness is a step between spirit and body."

Consciousness, hence, is more than cognition. It appears that in order to relate the concept of consciousness from Islamic thoughts, one has to understand the creation of man in itself. Consequently, this paper will explore on Islamic thoughts to understand the nature of man in order to link with consciousness within Malay-Muslim and Islamic contexts to relate to the full doctoral study.

Among all creations in the heaven and earth, only man has been bestowed with many distinctive qualities (Mohd Abbas, 2011). Allah has created man as the best creation and gave him cognitive and reflective abilities as well as intelligence (*'aql*) and heart (*qalb*) to distinguish between good and evil (Khalidah *et al.*, 2014, pp. 362). As the recipient of these distinctive qualities from Allah the Almighty, man is placed on the highest pedestal in the hierarchy of creations. Some of these distinctive features are: man has been created in the best of forms

(The Quran, At-Tin, Verse 1-4), provided with creative intelligence (The Quran, Al Baqarah Verse 31-34), and given the power of speech and freedom of choice (The Quran, Ar- Rahman, Verse 4; The Quran, Al-Insan, Verse 3) (Refer to Mohd. Abbas, 2011 and Azfalur Rahman, Muhammad, 1988).

Because of his special creation, man is chosen to be Allah's *khalifah (vicegerent)* on earth. Man has been further honoured in his stature and position when Allah made all things created in the heaven and earth to be of service to him (The Quran, At-Jathiyah, Verse 13; Al-Isra', Verse 70) (Refer to Mohd Abbas, 1997). Therefore, he is expected to be self-conscious and be an ethical human being, following the rules of the Quran and the *Sunnah* due to his noble position in the sight of Allah, the Creator.

"We have indeed created man in the best of moulds."

(The Quran, Surah At-Tin, Verse 4)

In Islam, man is born with *fitrah* (i.e. primordial nature). In Arabic language, though diverse in meanings, *alfitrah* also explains that man comes to this world without any stain of sin (Mohd Abbas, 2011). However, the concept of *alfitrah* is not in congruence with the principle of the 'Original Sin'as espoused in Christianity (Note: Original sin is an Augustine Christian doctrine which views that everyone is born sinful. This means that they are born with a built-in urge to do bad things and to disobey God. It is an important doctrine within the Roman Catholic Church. The concept of Original Sin was explained in depth by St Augustine and formalised as part of Roman Catholic doctrine by the Councils of Trent in the 16th Century. Original sin is not just this inherited spiritual disease or defect in human nature; it is also the 'condemnation' that goes with that fault. (Sourced from http://www.bbc.co.uk/religion/religions/christianity/beliefs/originalsin_1.shtml, accessed on 10 October, 2015).

In the Quran, *alfitrah* explains that man has been born as a believer in Allah the Almighty (The Quran, Surah Ar-Rum, Verse 30). Nevertheless, parents have a primary role to ensure that their children are shaped with *aqidah* (firm belief/*I'tiqaad*), *akhlaq* (good character/ethics), and *adab* (good manners/morals). William H. Shaw (2008, pp. 12) has rightly emphasised that upbringing is the first critical factor influencing one's ethical values which is manifested from internalisation and self-consciousness (Khalidah *et al.*, 2014, pp.4-5). The pertinent role of parents in the upbringing of their children have also been duly recognised in Islam based on a relevant Quranic verse,

"O ye who believe! Save yourselves and your families from a Fire whose fuel is Men and Stones, over which are (appointed) angels' stern (and) severe, who flinch not (from executing) the Commands they receive from Allah, but do (precisely) what they are commanded."

(The Quran, Surah At-Tahriim, Verse 6)

Further exploration of The Holy Quran reveals the dual nature of man: his body and his spirit (*ruh*). This fact well supports and confirms Nevzat Tarhan's writings on consciousness, earlier discussed in this paper. Relating to this too, The Quran further explains that after creating Adam, Allah breathed into him His spirit:

"Behold! Thy Lord said to the angels: "I am about to create man, from sounding clay form mud moulded into shape; "When I fashioned him (in due proportion) and breathed into him of My spirit, fall ye down in obeisance unto him."

(The Quran, Surah Al- Hijr, Verse 28-29)

These verses further justify that man's conscience is felt from both his body (through the senses) and his spirit; that is his soul. These facts well support and confirms Nevzat Tarhan's writings on consciousness, earlier discussed in this discourse. Hence, it is appropriate to regard consciousness as a step between man's body and his *ruh* or spirit (Nevzat Tarhan, 2012), an immortal feature that is indeed beyond human explanation. Aptly, it is the *ruh* that leaves the human body upon death (Mohd Abbas, 2011); hence, he/she is no more conscious! Within consciousness, man, being made the most honoured and best creation of Allah, is also entrusted with the *Amanah* (trust) from Allah.

"We did offer indeed the Trust to the Heavens and the Earth and the Mountains: but they refused to undertake it, being afraid thereof: but man undertook it—he was indeed unjust and foolish."

(The Quran, Surah Al-Ahzab, Verse 72)

Amanah can be referred to as reason, intellect or the faculty of volition (Muhammad Asad, 1980; Mohd Abbas, 2011). *Amanah* can also be related to power and authority granted to man through his role as a vicegerent. Holistically, *amanah* may mean the sum total of all the decrees, commands (*amr*) and prohibitions (*nahi*) present in the *Shariah* (Afzalur Rahman,1988). This relates to *amar maruf nahi mungkar i.e.* promoting goodness and forbidding evil acts by religiously following the rules in The Quran and the *Sunnah* (The traditions of Prophet Muhammad SAW). Indeed, The Quran has revealed that humans are trustees of Allah. Hence, they have no choice but to be conscious of their role and display virtues such as honesty, trustworthiness and integrity while executing their responsibilities because they will be accountable for all their actions in this world and the hereafter (Syed Othman and Aidit, 1994; Khalidah *et al*, 2014).

There is another Islamic concept related to the nature of man as a special being, i.e. *mithaq* or covenant. Man has been granted the covenant by His Lord. Aptly, there is a contract between man and His Creator. Through this contract/agreement, humans as a whole have accepted Allah to be their Lord and Cherisher while they were in the form of spirits in *Lauh Mahfodz*. This contract between man and his Lord has been made long before he was born into this world (Refer to the Quran, Surah Al- A'raf (8:172). The content of this covenant is retained in the sub-conscious mind of man as potentiality that can be brought to consciousness (Mohd. Abbas, 2011; Afzalur Rahman, 1988).

Next elaborating on *mithaq* (i.e. the 'covenant'), interestingly, a Muslim scholar, Afzalur Rahman (1988) views that factors such as type of education, social environment and other influences could never totally efface or tarnish the covenant from coming to the conscious part of the human being. In spite of the perversion/misinterpretation and prevention caused by wrong education and negative social influences, the covenant as a commitment/contract lays embedded in the sub-conscious mind of man and it will emerge to reality in response to positive external factors (Afzalur Rahman,1988; Mohd. Abbas, 2011) that have always invited humanity to the remembrance of the covenant. Some of these factors are the prophets, heavenly books, inviters to the truth from among the followers of the prophets, good social environment, proper and correct education and so forth (Afzalur Rahman, 1988; Muhammad Asad, 1980, pp. 9-10).

In Arabic, 'nur' or light means knowledge. Knowledge is sourced from many channels including education and media through cognition. Knowledge opens to one's awareness and understanding on a subject matter. From behavioural and rational viewpoints, knowledge will consequently lead to practice, application and internalization as manifested by the consciously positive behaviours of ethical beings. Knowledge certainly contributes directly to human civilisation and wellbeing with practice and internalization. The works of Nasr (1981) has explored humanity's quest for knowledge and quest for the Divine and how these quests relate to one another throughout history. The importance of knowledge has also been discussed by Mahathir Mohamad in his book, *The Challenge* (1986) while linking Islam with the Malaysian Malay society and evolving capitalistic and liberal global trends.

However, several questions remain to be answered. Despite the covenant and knowledge, why are some people not conscious of the contract? Why are humans committing unethical acts despite having knowledge? Are they not conscious that they are accountable for their actions, be it good or bad, because of the covenant; the contract? As much as there are no concrete answers to these questions, they may be partly answered by differentiating between a conscious person and a knowledgeable person from Islamic viewpoints. A conscious person will know his role from Allah's sight. In contrast, a knowledgeable person need not necessarily be conscious in a true sense if he does not apply and internalise his knowledge and act virtuously with conviction in life. Quoting and translating from the words of Pahrol Md. Juoi, a Malaysian Malay scholar, "a knowledgeable person need not necessarily be conscious. However, a conscious person knows who he or she is; a servant to the Creator. As a servant, he/she is obligated to follow Allah's rules religiously." The Malay version of the translated quote reads, *"Seseorang yang tahu, dia tak semestinya sedar, tetapi bagi seseorang yang sedar, dia tahu siapa dia; seorang hamba Allah yang akan ikut perintahNya secara menyeluruh"* (IKIM.fm in the segment, *Indah Itu Indah)*.

To summarise, as much as the West has noted that 'consciousness' is a phenomenal concept, it is also implicit from Islamic viewpoints. Consciousness relates to other Islamic concepts such as *khalifah, al-fitrah, ruh, amanah and mithaq* (i.e. covenant). In fact, consciousness in Islam relates to the heart and soul. It appears that in order to be virtuously conscious beings, one has to develop his spirit that begins from a clean and good heart. The term 'heart' has different and multi-level connotations in the Islamic tradition. It is not simply the organ of emotions, 'feelings' and desires, or decisions and opinions, but goes to encompass the soul, knowledge, bravery and more.

When purified, it is the seat of the intellect, i.e. the ray of the Divine within the creatures (http://www.livingislam.org/m/qalb_e.html accessed on 12 October, 2015). As professed by Prophet Muhammad (*saw*), "Truly in the body, there is a morsel of flesh, which, if it is sound, all the body is sound and which, if it be diseased, all of it is diseased. Truly it is the heart" (*Hadith* by Imam Nawawi). Imam Nawawi said, "This *hadith* was used as proof that the seat of the mind is in the heart (*al- 'aql fi al –qalb*) and not the head" (http://www.livingislam.org/m/qalb_e.html, accessed on 12 October 2015). A good heart therefore leads to a good soul with high consciousness to do good deeds as ethical beings.

3 Conclusion

This paper has attempted to conceptualise and deliberate on the paradigms of consciousness as a discourse. It has investigated consciousness from scientific and Western perspectives, multi-disciplines and religion with a focus within Islamic contexts. Discursive paradigms of consciousness recognise it as an abstract concept. Consciousness is aptly beyond cognition. While cognitive thoughts, theories and ideologies relate consciousness to the five traditionally recognised human senses, (i.e. sight, hearing, taste, smell and touch), the concepts of 'will', awareness (i.e. being aware of things happening around a human being) and having a sense of purpose or intent in undertaking actions are relevant in the study of consciousness. It may be deduced that a conscious person will use his/her senses to conduct life as civilised beings. However, by linking the body and the soul in explaining consciousness from Islamic perspectives, consciousness is the awareness of oneself driven by his/her commitment as a servant of God and total submission to the Creator. Consciousness from Islamic contexts indeed relates to *iman* (faith) and *taqwa* (God consciousness).

4 References

Al Quran, King Fahd Complex: Madinah Munawawarah, K.S.A. For the Printing of the Holy Quran; www.qurancomplex.org

Azfalur Rahman, Muhammad. (1988). *Encyclopaedia of Seerah* Vol. 6 London: Seerah Foundation, pp1.

Berger, P.L. and Luckmann, T. (1966). *The Construction of Reality*, Garden City, New York: Anchor Books, 1966, pp.51-55, 59-61. Accessed from www.cf.a.c.uk/socsi/underg on 2 November 2015.

Block, N. (1995). 'A confusion about a function of consciousness,' *Behavioral and Brain Sciences*, 18: 227–247.

C. Wright Mills. (1999). *The Sociological Imagination (1916-1962),* Oxford University Press. Accessed from https://global.oup.com/academic/product/the-sociological-imagination-9780195133738?cc=my&lang=en&

Chalmers, D. J. (2002), *Consciousness and its Place in Nature,* Published in (S. Stich and F. Warfield, eds), *Blackwell Guide to the Philosophy of Mind* (Blackwell, 2003), and in (D. Chalmers, ed) *Philosophy of Mind: Classical and Contemporary Readings* (Oxford, 2002).

De Sousa, A. (2011), Freudian Theory and Consciousness: A Conceptual Analysis. *In: Brain, Mind and Consciousness: An International, Interdisciplinary Perspective* (A.R. Singh and S.A. Singh eds.) *MSM,* 9 (1), pp. 210-217. Unpublished work

Dretske, F. (1995). *Naturalizing the Mind.* Cambridge, MA: MIT Press.

Eberle, Thomas S. (1992). *A new paradigm for the sociology of knowledge: "the social construction of reality" after 25 years.* In: Schweizerische Zeitschrift für Soziologie 18 (1992), 2, pp. 493-502. URN: http://nbn-resolving.de/urn:nbn:de:0168- ssoar-24182

Elwell, F.W. (2013). "*C. Wright Mills on the Sociological Imagination,*" Retrieved Feb 21, 2018. From http://www.faculty.rsu.edu/~felwell/Theorists/Essays/Mills3.htm

Foss, J. (2000). *Science and the Riddle of Consciousness – A Solution*, USA: Kluwer Academic Publishers, pp.1-4.

Freud, S. (1912). *In: Totem and Taboo*, James Strachey, translator. London: Hogarth Press, Standard Edition.

Freud, S. (1915). In: *The Unconscious.* James Strachey, translator. London: Hogarth Press; Standard Edition.

Freud, S. (1937). In: *An Analysis Terminable and Interminable*. James Strachey, translator, London: Hogarth Press; Standard Edition

Hunt, T. W. (1995), *The Mind of Christ*, Nashville, TN: Broadman & Holman Publishers.

James, W. (1890). *Classics in the History of Psychology. Classics in the History of Psychology.* An internet resource developed by. Christopher D. Green. York University, Toronto, Ontario. (Return to index). Accessed from http://psychclassics.yorku.ca/James/Principles/prin4.htm.

Jaynes, J. (1978). *The Origin of Consciousness and Breakdown of the Bicameral Mind,* Cheektowaga, ON: University of Toronto Press, pp.1.

Khalidah Khalid Ali, Z. Ramly and T.C. Lau. (2014). *Business Ethics,* Shah Alam, Selangor, Malaysia: Oxford Fajar Publishing.

Kirk, R. (1994). *Raw Feeling*. Oxford: Oxford University Press.

Kriegal, U. (2006a). *'The same-order monitoring theory of consciousness*,' in U. Kriegel and K. Williford (eds.) 2006.

Kriegal, U. (2006b). Consciousness, Theories of. *Philosophy Compass* 1/1 (2006): 58–64, 10.1111/j.1747-9991. 2006.00008.x

Mahathir Mohamad. (1986). *The Challenge,* Ipoh, Malaysia: Pelanduk Publications (M) Sdn Bhd. pp. 1- 4. 10, 17.

Merriam Webster Online Dictionary, 2017.

Mohd. Abbas Abdul Razak. (2011), Human Nature: An Islamic Perspective, *Journal of Islam in Asia,* Special Issue. No.1 March 2011 pp 251-274.

Mohd Abbas Abdul Razak. (1997), *Human Nature: A Comprehensive Study Between Western and Islamic Psychology*, Kuala Lumpur: IIUM, 1997. pp. 171-173.

Muhammad Asad (1980). *The Message of the AlQuran,* Gibralta: Dar al-Andalus.

Nezvat Tarhan. (2012, Dec. 12). *Consciousness* [Online]. Available: http://www. questions onislam.com/article/consciousness#baslik2.

Oxford Dictionaries online, 2016

Ricoeur, P. (1970). *Freud and Philosophy: An Essay on Interpretation, New Haven.* Conn: Yale University Press.

Syed Othman Alhabshi and Aidit haji Ghazali. Islamic Values and Management. Kuala Lumpur, Malaysia: Inst. Islamic Understanding Malaysia (IKIM), 1994, pp.10.

Seyyed Hossein Nasr. (1981). *Knowledge and the Sacred,* Albany State Univ. of New York.

Swami Nikhilanada. (1985), *The Gospel of Sri Ramakrishna*, Ramakrishna- Vivekananda Centre, pp. 523 &1225.

Tye, M. (1995). *Ten Problems of Consciousness*. Cambridge, MA: MIT Press.

Thanissaro Bhikkhu. (Trans.) (1998). *Chachakka Sutta: The Sixth Sextets* (MN148). Retrieved 2007-06-17 from "Access to Insight."

The Holy Al Qur'an: Text, translation and commentary, Maryland: Amana Corporation, Abdullah Yusuf Ali, 1989.

Van Dijk, T. A. (1998). *Ideology: A Multidisciplinary Approach*. London: Sage Publications Ltd, pp. 96-97.

Struthers, W.M. (2001), *Defining Consciousness: Christian and Psychological Perspectives,* Perspectives on Science and Christian Faith 53, June 2001, pp. 102-106.

http://plato.stanford.edu/entries/qualia/, retrieved online on 6 September 2015.

http://www.brainyquote.com/quotes/quotes/s/sigmundfre403586.html#LOV747K1c6TjQgRp.99, retrieved on 18 December 2015).

http://www.questionsonislam.com/content/nevzat-tarhan-profdr, on 31 September 2015.

http://www.bbc.co.uk/religion/religions/christianity/beliefs/originalsin_1.shtml, accessed on 10 October 2015.

http://www.livingislam.org/m/qalb_e.html on 12 October, 2015.

SHS Web of Conferences **53**, 04004 (2018) https://doi.org/10.1051/shsconf/20185304004
ICHSS 2018

The Distribution of Zakat: A Comparison between the Rate of Niṣāb and the Rate of Poverty Line Income in Malaysia

Muhammad Pisol Mat Isa[1], Azhan Rashid Senawi[2], Azhar Harun[3]

[1, 2, 3] Management and Humanities, Universiti Teknologi PETRONAS Bandar Seri Iskandar, 32610 Tronoh, Perak, Malaysia

Abstract – Zakat plays a significant role in Muslim community as an agent in alleviating the poverty. The calculation of zakat payment depends of the calculation of nisab, which is the least amount of zakatable items of a Muslim wealth in one circle year. The zakat is an obligatory duty to a Muslim once the amount of wealth reaches the least level of nisab. This study aims to analyze the relationship between the niṣāb amount changes and the poor indication in Malaysia. This is based on the discussion of the contemporary scholars, who claimed that the price fluctuation in the world market impacts the amount of niṣāb of zakāt. Relatively, it could affect the poor once the price dropped. Therefore, the comparative analyses of the annually niṣāb and poverty line income data from 1993 to 2015 identifies that there is a significant correlation between the facts. Accordingly, it is an alert to the zakāt institution to take a precautionary step in preventing this situation.

Keywords- Niṣāb of zakāt, gold price, poverty, Malaysia

1 Introduction

Niṣab is an Arabic word which is linguistically derived from nasaba which means reference or standard, al-Dasuqiy said the word Nisab taken from al-nusub which means al-'Alamah or signage, where it shows the limit or border of wealth in paying a zakat (Al Dasuqiy, 2013). Technically it means a measurement of zakāt, where one is responsible to pay a certain amount once the wealth is equivalent and exceed the minimum level in that particular year. Niṣāb is also a tool of differentiating the poor and the affluent (Al Qaradhawi, 1999; Majah, 1952). As mentioned in Hadith as narrated by Muaz bin Jabal that Prophet Muhammad pbuh advised him to take the zakat from the wealthy and surrender to the poor in the region, and the amount of nisab (Imam Bukhari, 2015), Imam al-Bararati mentioned that, anyone who hold the amount of nisab and able to furnish their basic need, so they cannot receive zakat (al-Babarati,2002), the statement shows that the those who have amount of wealth above than nisab is considered as wealthy. In the issue of gold and silver price, it need to take an in depth revision, where the determination of silver or gold as a niṣāb, is mainly depended to the demand and supply in the world market. The fluctuation of its price, indeed varies the amount of niṣāb for that particular year. For the country that uses silver as the determination of niṣāb, then the zakāt of money tends to involve the poor as the contributor of zakāt due to the low of the silver price, as statistically reported during the recent period (Khan, 2000). Based on the facts, as long as the items are classified as the commodity items, then, it could not be averted from any rigorous market activity (Browne, Abbriano, & Lockwood, 2015). This situation proves that, as happened to the price of silver, it is also likely to happen to the price

[1]Corresponding author: mpisolmatisa@utp.edu.my

of gold. Thus, this could impact the household, who are categorized as poor and destitute (Abu-Saud, 1988; M Kahf, 1989; Sadeq, 1994). In order to justify the claimed situation by the contemporary scholars regarding the niṣāb of zakāt that associated with the poor and the destitute, this study will undergo some data comparison in particular to the amount of niṣāb changes and the poverty line income data. From this, it will conclude whether the argument meets the facts requirement or otherwise. As to facilitate the discussion, this paper has been structured as follows. The next section 2 and 3 will explore the literature reviews and the methodology of this study. Then, section 4 will present the results and discussion, and finally, section 5 summarizes the findings and the overall conclusion.

2 Literature Review

Since a century ago, paper money has been used predominantly all over the world. Gold and silver have no longer been a medium of the transaction since the collapse of Ottoman reign in Turkey. Apparently, the niṣāb of money which is 20 dinar of gold and 200 dirham of silver will not be the same as previous due to the factor of demand and supply in the world market (Al Qaradhawi, 1999; see also Kahf, 1989). Thus, it might cause a confusion on how to measure the niṣāb either base on the silver or the gold. In this matter, due to the fact of sunnah and ijmāᶜ, the tendency to use silver as the benchmark of niṣāb is definitely unarguable. It is explicitly determined in the holy book, al Quran and hadith and for this reason it is been used in the country of Egypt, Saudi Arabia, the gulf countries, India and Pakistan (Al Qaradhawi, 1999). Some scholars incline to use gold as the benchmark of niṣāb due to the aspect of stability compared to the silver. Those who favor to use gold as the niṣāb are Abu Zahrah, Khallaf and Hasan (Al Qaradhawi, 1999; Sadeq, 1994). This is due to the current situation where the price of gold is more suitable to be compared to another item such as camel, sheep, and crops than the silver. For instance, niṣāb of camel which is 5, or sheep which is 40, are very close to the niṣāb of gold, which is 85 grams of gold. Meanwhile, niṣāb of silver which is 595 grams not even equal to one camel price (Kahf, 1989).

There is a hadith that mentions five uqiyyah of the silver niṣāb was enough for one's household for their living in one year at very reasonable standard of living and at that specific time (Abu-Saud, 1988; see also Al-Dahlawi, n.d.). However, as the time passes, this measurement could not fit the living standard of today due to the fact of inflation and market instability. It would not be sufficient even for a week. Therefore, the benchmark of niṣāb is much more suited to gold, which is at 85 grams of gold. Otherwise, it would inflict the unjustness to the zakāt contributors (Abu-Saud, 1988; Mahmud & Shah Haneef, 2008; Sadeq, 1994). Furthermore, the instability value of money gives the bad impact to the silver niṣāb where the current price shows the low price of silver as compared to gold and other items of wealth. This low price of silver would also decrease the niṣāb of zakāt and consequently, it tends to give an effect to the recipients of zakāt especially to those in the group of poor (Sadeq, 1994). That is why gold is more preferable than silver due to the fact of time changing.

Nevertheless, there would also a matter of market price fluctuation, where the price of gold could not be avoided from affecting its value. Certainly, people will ask the matter of the availability from the aspect of shariah in discovering the other item besides the silver and the gold. In this condition, Al Qaradhawi (1999) views that the determination of zakat niṣāb should either be equivalent to one-half of fives camel or forty sheep in that particular country. This is proposed in order to moderate the standard and the purchasing power in the world market due to the tremendous fluctuation of gold prices that could impact the niṣāb. Meanwhile, the Muslim council in Pakistan, Majlīs Taḥqīq Masāil Ḥaḍirah has strongly opposed the intention in reviewing the niṣāb due to the reason that the determination of niṣāb is not derived from the ijtihad but it is from the holy book; al Quran, hadith and ijmak or Muslim Jurist's consensus (Mahmud & Shah Haneef, 2008; Sadeq, 1994). However, they admit that should the value of niṣāb increase, it results in a decrease of zakāt contributor and lead to a high of zakāt recipient. Thus, it will create an excess in demand for zakāt and tendency to have a begging culture in the society (Mahmud & Shah Haneef, 2008; Sadeq, 1994).

On the other side, Raqibuzzaman (1980) insists that the niṣāb should be allowed to be changed according to the situation of the country. It is due to the condition that varies from country to another country. In addition, the niṣāb is a portion that value the adequate standard of living on that particular time. Thus, the burden should not be borne to the poor and destitute as a result of the low niṣāb. This practice also uses the analogy of what has been done by Caliphate Umar, who measures the horse as a zakāt rate (Sadeq, 1994). Narrated in a hadith that reports "Zakāt will not due if it less than 5 camels, while, for grain, it will not due less than 5 awsuq (1300 kilogram) then, for money, it will not due less than 200 dirhams" (Majah, 1952). However, Abu-Saud (1988) affirms that these benchmark was only appropriate for that time being rather than its use today. As for an example, during the time of the Prophet, 5 awsuq was sufficed for a family of three for a year, however, 5 awsuq for a planter in Malaysia nowadays cannot even satisfy their basic needs in a year at all. Due to changing times, it is impractical to use an akin standard of benchmark in order to fulfill the needs of the poor and the destitute (Mahmud & Shah Haneef, 2008; see also Abu-Saud, 1988). Thus, it needs to be revised in accordance to the principle of Islamic Jurisprudent. Furthermore, Abu-Saud (1988) posits that contribution for zakāt should be based on some criteria, in which, it should take into account the mandatory household expenses (ḥawāij al aṣliyyah), time value of money, inflation and deflation as well as income tax deduction for the zakāt payer. The rationale to review this issue is due to the time changes that affect the value of the niṣāb amount, in which it is not parallel to the rise in cost of living and significantly, it changes the basket of essential needs especially to those who are categorized as poor and destitute (Kahf, 1989). Thus, there is a need to tackle this issue in signifying the claim and to clarify the matters that could affect the lives of the poor and the destitute.

Meanwhile, in many countries including Malaysia, the mechanism in identifying the poor is through the Poverty Line Income (PLI). In the case of Malaysia, the PLI is issued by Economic Planning Unit under the Department of Prime Minister. According to the PLI data, there is two categories of poor, known as the absolute poverty and relative poverty (Jamil & Che Mat, 2014; Mohamed Saladin Abdul Rasool, Mohd Fauzi Mohd Harun, Ariffin Mohd Salleh, 2011; Ragayah, 2010). Under the circumstances of absolute poverty, a person whose income is equivalent or lower than the poverty line is considered as poor. While, the relative poverty quantifies the average of the living standard within the community in that particular year (Jamil & Che Mat, 2014; Ragayah, 2010). Thus, this study will run a comparison analysis of both data in order to have the significance of the claim by the contemporary scholars. It may consider a general and concrete analysis of the niṣāb and the poverty line income in Malaysia in justifying the matter that arises.

3 Methodology

This study uses the data from the Economic Planning Unit, Prime Minister's Department for the poverty line income and Pusat Pungutan Zakāt (Center for Zakat Collection), Wilayah Persekutuan for the niṣāb of zakāt in comparing the relevancy of the both data. It will cover the year of 1993 to 2015 of the analysis period. It is a qualitative approach that used a content analysis technique in analyzing the data sets. By comparing the observed figures and applying the logical inference, it will determine the relevance and the implication that might appear in this analysis. From here, the study concludes the result for the purpose of a future research.

4 Result and Discussion

For a discussion, the table of niṣāb and the poverty line income are placed in order to facilitate the analysis. The table 1 below shows the details of the both data.

Table 1: The Amount Gap between the PLI and Niṣāb of Zakāt from 1993 to 2015 (RM/annually)

Year	PLI (RM)2	Niṣāb	PLI - Niṣāb
1993	5160	2900	2260
1994	5160	2900	2260
1995	7320	2700	4620
1996	7320	2700	4620
1997	7320	2800	4520
1998	7320	2900	4420
1999	7800	2900	4900
2000	7800	3200	4600
2001	7800	2900	4900
2002	7800	3000	4800
2003	7800	3000	4800
2004	8280	3700	4580
2005	8280	4200	4080
2006	8280	4600	3680
2007	9000	6200	2800
2008	9000	6400	2600
2009	9600	8000	1600
2010	9600	9300	300
2011	9600	10300	-700
2012	10320	13113	-2793
2013	10320	14066	-3746
2014	11400	12223	-823
2015	11400	11308	92

Source: Pusat Pungutan Zakāt (PPZ) and Economic Planning Unit, Prime Minister's Department

The above table is generalized from the original of both the niṣāb of zakāt and PLI data. The data have shown the amount trend of the niṣāb and the PLI from 1993 to 2015. Apparently, from this comparison, it is proved that the PLI has been at the level of more than the niṣāb of zakāt. It could be detected from the amount of PLI from 1993 to 2010, where these years, the amount of PLI is greater than the niṣāb of zakāt. It is not just a tens of RM difference, but it reaches thousands RM. This could clearly be seen from the fourth column. From table 2, it has explained well that the low niṣāb is indeed exploiting the poor that based on the PLI data. Accordingly, Kahf(1989), Sadeq (2004) and Mat Isa (2013) argument's on the issue of niṣāb, that claim the niṣāb changes could ruin the standard of living of the poor is proven. As denoted from the above table, the highest amount gap between the PLI and niṣāb is RM4900,

[2] The PLI data have been transformed to annually in order to draw the best fit comparison with the niṣāb of zakāt. It is not a yearly assessed data. It is, in fact, a data of once in-four-year. Therefore, the above data produced is to suit the yearly data of niṣāb of zakāt in order to have a better comparison between both data. For a quick search, the PLI could be excessed at www.epu.gov.my/.

which is at 1999 and 2001. While, the lowest amount gap is at 2015, where the amount is only RM92. However, on average, most of the accounted year, the PLI has a higher amount than the niṣāb. Except for the year 2011 to 2014, where the PLI is amounted low by RM700, RM2793, RM3746 and RM823 respectively. This situation will be more clearly seen from the line graph of figure 1 below.

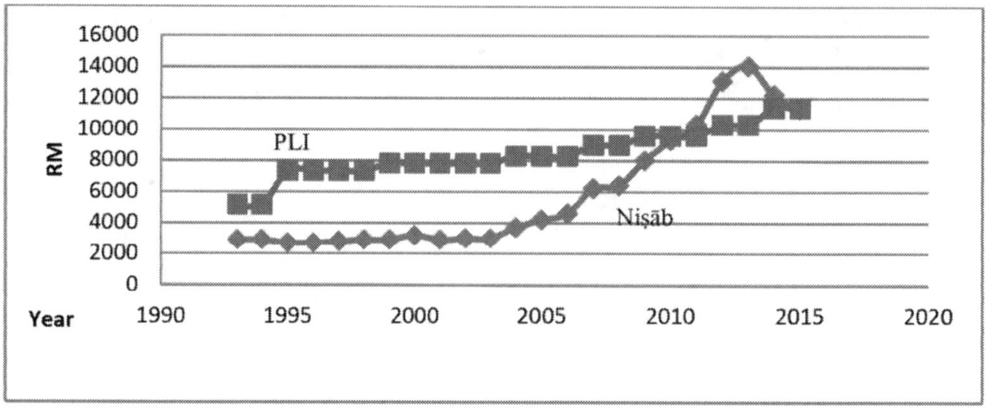

Figure 1: Comparison between the Nisab and the PLI from 1993 to 2015

From the above figure, it clearly shows that the gap between the PLI and the niṣāb are significantly explained that the poor earnings exceeds the amount of niṣāb as shown by the square dotted line from 1993 until 2010. In other words, this excess signifies how the role of the gold price and the macroeconomic determinants affect the well-being or the standard of living of the poor in terms of the eligibility to pay the zakāt even though they are not eligible (Kahf, 1989; Sadeq, 2004; Mat Isa, 2013). Whereas, in 2011 to 2014, the diamond dotted line of niṣāb is seen higher than the PLI due to a high price of gold in the world market. As the Majlīs Taḥqīq Masāil Ḥadirah, the Muslim Council in Pakistan claim, this will reduce the collection of zakāt and simultaneously, inhibit the distribution of zakāt to the poor (Mahmud & Shah Haneef, 2008; Sadeq, 2004). Therefore, from this comparison, it can be summarized that there is a linkage between the niṣāb and the poor group. Significantly, the arguments by Kahf (1989), Sadeq (2004) and Mat Isa (2013) as well as Mahmud and Shah Haneef (2008), are reasonable and justify the relationship between the niṣāb and the poor, in which, it relates the impact assumption of the niṣāb amount changes and the poor group.

Nonetheless, the study opines that these two items (the PLI and the niṣāb) need to be scrutinized further. This is due to the progressive and the fluctuation pattern shown by the PLI and the niṣāb item respectively. If observed the PLI from 1993 to 2015, it shows a moderate-increase trend from RM430 in 1993 to RM950 in 2015. Undeniably, it could be increased more due to the socioeconomic situation such as the cost of living and the status of the developed country in 2020. Furthermore, the current proposal of the minimum wage in Malaysia, which is at RM1000 to RM1200 by the Malaysian Trades Union Council (Solomon, 2016; Ling, Yusof, Hasnaa, Mahmood & Soon, 2014) could be also the pull factor of the PLI indication. These could be one of the huge factors that might elevate the PLI in the future because the current PLI is, in fact, portrays the level of the minimum wage for the support civil servants, which is around RM600 to RM900 monthly (Suruhanjaya Perkhidmatan Awam, 2014).

The concern now is the niṣāb, where the amount very much depends on the world market price. It could be at a high level and it could be low. As portrayed in the niṣāb data from 1993 to 2015, it begins with RM2900 and gradually increases to RM11308 in 2015. But, during the years of 2012 to 2015, the niṣāb is floating from RM13113, RM14066, RM12223 and lastly at RM11308 respectively. This can be clearly viewed in the figure 1, where the graph line plotted the fluctuation of the niṣāb higher than the PLI. It raises an issue that, what would happen if the PLI keeps accelerating, but the niṣāb continues to down trend as a result of the gold prices volatile in the world market. It seems to bring an indication that the niṣāb could probably ruin the virtue of the maqāṣid of sharīʿah of the zakāt. Therefore, it could be concluded that the claim by the Kahf (1989), Mat Isa (2013) and Sadeq (2004) have its own ground to be proved. In fact, it is illustrated well to point out the basis of the argument, when it is assessed in figure 1 of the graph line as well as the tables of the niṣāb and PLI above.

5 Conclusion

Alms are for the poor and the needy, and those employed to administer the (funds); for those whose hearts have been (recently) reconciled (to Truth); for those in bondage and in debt; in the cause of Allah. and for the wayfarer: (thus is it) ordained by Allah, and Allah is full of knowledge and wisdom' (Quran, 2011). Thus, it is a responsibility of all Muslims to remain alert to any possibility that could affect the poor caused by the economic condition. This is to avoid any possible implication to the poor and destitute by endowing as a rich group, unintentionally due to the changes of niṣāb. From the above discussion, it shows that a low niṣāb from 1993 to 2010 and 2015 caused the poor to be among the contributor cluster, although, they are among the group, who deserve to be the zakāt recipient. All of this is the consequence of a series, where the PLI that represent the poor is higher than the niṣāb. Hence, the claim that the niṣāb has a tendency to affect the poor group has its own basis. Therefore, this study provides an opportunity to further a research in signifying the possible patterns of niṣāb of zakāt on the poverty line income. An in-depth study could be a novelty contribution to the Muslim society as well as to the zakāt institutions in justifying the issues highlighted.

6 References

Abu-Saud, M. (1988). Contemporary Zakāt. Zakāt and Research Foundation, 113.

Al Qaradhawi, Y. (1999). Fiqh al Zakah. Beirut: Dar Al Taqwa, pg. 14.

Al-Dahlawi, S. (n.d.). Hujjatullah al-Balighah. Maktabah al-Kautsar, vol 2, pg. 506.

AL-Dasuqi, Ibn Irfah. (2013), Hasyiah al-Dasuqiy ala Shar al-Kabir, vol. 4.pg 211.

Browne, R., Abbriano, S., & Lockwood, R. (2015). Gold & Silver Marketwatch.

Jamil, N., & Che Mat, S. H. (2014). Realiti Kemiskinan: Satu Kajian Teoritikal. Jurnal Ekonomi Malaysia, 48(1), 167–177.

Kahf, M. (1989). Zakāt: Unresolved issues in the contemporary fiqh. IIUM Journal of Economics and Management.

Kahf, M. (1999). The principle of socio-economic justice in the contemporary Fiqh of Zakah. IQTISAD Journal of Islamic Economics.

Khan, M. (2000). Some Accounting Issues Relating to Zakāh. Islamic Studies, 104–105.

Ling, T. P., Yusof, M. F., Hasnaa, N., Mahmood, N., & Soon, T. H. (2014). The Implementation of Minimum Wages in Malaysia. Sains Humanika, 2, 43–46.

M. Isa, M. P. (2012). The Implementation of Gold and Silver as the Value of Niṣāb for Zakāt Calculation: A Comparative Study between the Present Practices and those during the Period of Prophet Muhammad Pbuh. In Kepentingan Dirham Perak dalam Zakāt (pp. 1–7).

Mahmud, M. W., & Shah Haneef, S. S. (2008). Debatable issues in Fiqh al-Zakāt: A Jurisprudential Appraisal. Jurnal Fiqh, (5), 118–141.

Majah, I. (1952). Sunan Ibn Majah. Darul Kutub Al Ilmiyyah, Beirut.

Quran, A. (2011). At Tawbah (verse: 60).

Raqibuzzaman, M. (1980). Some Aspects of the Economics of Zakah. Indiana: American Trust Publication

Sadeq. (1994). A Survey of the Institution of Zakah: Issues, Theories, and Administration (Research Paper).

Al-Bukhari, Imam Abu ʻAbd ʻAllah Muhammad b. Ismaʻil (1989), Sahih al-Bukhari, Tahqiq Mustafa Bugha, . (Hadith no. 1308),, c. 2, j. 1, Beirut: Dar Ibn Kathir/

SHS Web of Conferences **53**, 04005 (2018) https://doi.org/10.1051/shsconf/20185304005
ICHSS 2018

Usability of QiraahBot for extensive Arabic reading activities

Al-Muslim Mustapa[1,*], *Zulkifli* Nawawi[1], *Salamiah* Ab. Ghani[1], *Maryam* Abdul Rahman[1], *Zuraida* Shaadon[1], *and Nur Syazwina* Mustapa[1]

[1]Arabic Language Unit, Faculty of Islamic Studies, The National University of Malaysia, Malaysia

Abstract. Today's changing learning landscape urges the integration of mobile phone technology and Arabic learning as a foreign language. Learning through mobile phones is seen as a potential teaching aids which demonstrate effectiveness in teaching and learning activities. A QiraahBot prototype was built to aid the implementation of extensive Arabic reading activities. The user's perspective on QiraahBot's usability is essential for the effectiveness of learning through mobile phone applications. This study aims to identify the level of usability of the QiraahBot application prototype and to explore the deeper experience of users using the application. Therefore, a group of 30 student users was involved in the study. They were assigned tasks to be completed within the specified periods via the application, and then given an open-ended questionnaire adapted from the System Usability Scale. The findings show that the QiraahBot usability level is moderate. The study also revealed some of the advantages of QiraahBot as easy to use and the ability to help the learning process. In addition, there are weaknesses in applications, especially aspects of application usability, which needs to be improved in the next version. The researcher has provided some suggestions for further studies that are more comprehensive in the future.

1 Introduction

Reading skills are an important skill for second language learners including Arabic. Having these skills help students to master Arabic better. Students who master the reading skills will acquire rich language information and also assist in the acquisition of the language (Krashen, 2004). Additionally, enhanced reading skills also help to improve other language skills such as listening, speaking and writing (Cummins, 2003). This situation leads to the mastery of the language as a whole. Extensive reading (ER) is one of the ways to improve reading efficiency. The concept of ER is someone reading a lot of preferred text (Jeon & Day, 2016; Krashen, 2004). ER aims to develop the ability to read and to acquire as much knowledge as possible about the language and subjects being read. Studies have shown that ER has helped students, as in vocabulary, motivation, textual comprehension, reading fluency and reading efficiency. The meta-analysis of ER's effectiveness on reading efficiency by Jeon and Day (2016) shows that ER affects reading efficiency starts with

* Corresponding author: almuslim@ukm.edu.my

small to medium size. According to both researchers, the results of the study are consistent with other meta-analysis. Hence, previous studies suggest that ER should be included in the curriculum besides the Intensive Reading tradition (Jeon & Day, 2016; Lee & Mallinder, 2017).

Today, times have changed. Students do not have to read text on paper anymore. Now, students can read a text on the Internet that can be accessed through the mobile phone screen, respectively. In Malaysia, internet usage is widespread with the percentage of consumers aged 15 years and above at 71.1 per cent in 2015. Interestingly, 97.5 per cent of users use mobile phones to access the Internet (Department of Statistics Malaysia, 2016). Therefore, mobile phones have changed the landscape of learning in a much different fashion because of the extensive access to the cheap and sophisticated devices (Nezarat & Mosavi Miangah, 2012). In Malaysia, the use of mobile phones in all sectors of education is increasingly acceptable and enters the era of mobile learning research extension (Nordin, Embi, Norman, & Panah, 2017). This situation requires changes in terms of ER's implementation in Arabic language teaching as a second language in line with the current technology development of students. Teaching and learning that still use traditional methods in reading classes have negative effects on students such as lack of motivation to learn, relying entirely on teachers, coming to class without preparation and weak competition (Hazaea & Alzubi, 2016).

ER activities in Arabic language can be carried out through the Mobile assisted language-learning concept (MALL). MALL refers to integrating mobile phones and their applications to help and improve language learning inside or outside the classroom. MALL is considered as a tool in the field of language acquisition (Chinnery, 2006; Kukulska-Hulme & Shield, 2008). Researchers believe that mobile phone integration or applications can improve the learning effectiveness and ability of students. Chen and Hsu (2008) claim that mobile phone applications can help students to improve their vocabulary information as well as help them to improve reading comprehension. Chang and Hsu (2011) also point out that the second language student considers translation applications, PDA-based annotations or the web useful and easy to use for individual reading activities. Supporting the use of MALL in reading activities as in the study by Huang and Lin (2011) which emphasizes that reading on mobile phones is more popular than the traditional reading activity.

2 Literature Review

The development of mobile phone technology and its applications in the education world cannot be denied. Mobile phone applications are getting the attention of educational researchers, including instant messaging and bot applications.

With IM's popularity among users, instant messaging applications in 2014 introduced a third-party service support known as a bot. Today, popular IM applications such as Telegram, We Chat, Kik, Facebook Messenger and Skype support the bot function, which is a communication application system that interacts with users where it works in the messaging application itself. More interestingly, this bot is located in a very common place for the user, namely in the communication thread. Bots are placed together in a list of personal conversations with friends and relatives. Most users actually use IM several times a day. They already have good experience and understanding of the interface and how the IM application works (Klopfenstein et al., 2017). Hence, installation and use of bots in messaging applications is not a new thing.

According to Cuno Klopfenstein and Bogliolo (2017) and Klopfenstein et al. (2017), this bot works as an automated agent that communicates with users via simple text communication. At this time, the bot application has mastered most of IM. It is increasingly used as a platform to access certain information or to use the services offered in that way. Due to the widespread popularity of IM through the mastery of the use of smart phones, bots offer attractive platforms, direct access to a large number of users and low value access (Cuno Klopfenstein & Bogliolo, 2017; Klopfenstein et al., 2017). In other words, bot offers the same advantages as IM to developers and users. In addition, bot users are reportedly inclined to use bot due to productivity factors. Bot efficiently makes it easier for them to get help or timely information (Brandtzaeg, 2017)

Bot can be considered as an autonomous application that ride on IM. Bot can be built with the Application Programming Interface that is available on common applications such as messaging, payments, bot directories, authentication and others. Bot also accepts UI such as buttons, locations, images and others. This ability provides space for developers to provide innovative services through conventional consumer experience (Klopfenstein et al., 2017). In addition, the Bot application does not need to be downloaded and installed. It is automatically installed when called through a unique and functional code once the conversation is started in the messaging application. Device storage capacity is almost unnecessary. Installation is fast and easy. Accordingly, complex configuration steps are not required. This makes bot as a fast and light application when compared to traditional mobile applications.

As discussed earlier, many studies confirm the importance of bringing today's technology, particularly in relation to IM in teaching and learning. The use of IM can stimulate better learning through communication and collaboration between student groups or between students and teachers. With the support of IM to bot, there is a new learning and teaching space to be captured through this development. A study by Cuno Klopfenstein and Bogliolo (2017) has introduced a bot system that contains educational quizzes in the field of online programming. This quiz bot works in IM Telegram. The study explains the detailed system of bots developed including functional programs, application strengths and weaknesses. The findings suggest the effectiveness of coding quiz operated by the bot and its implementation program. The study also noted the ability of bot to be used more effectively in many educational activities.

A study by Fryer and Carpenter (2006) was conducted to identify the potential of bot in foreign language learning. 211 student respondents were asked to use two bots for foreign language learning in the classroom and their feedback was recorded in writing. The results showed that most students like using the bot. They are also generally found to feel more comfortable communicating with bot than friends or teachers. This shows that there is a tendency for students to use bot for foreign language learning and some of them prefer to learn the language or practice it through bot than their own teachers or friends.

Supported by literature review, this study builds a bot-based application as a tool for ER activity. Teachers in the process of teaching Arabic reading skills as a second language use this bot. The bot named QiraahBot works on the IM Telegram platform. The main objective of the study is to determine the extent of usability of these applications in terms of acceptance of students who use it.

3 Construction of QiraahBot

QiraahBot application is a bot-based automatic system that can interact with users individually or in groups. An interaction occurs via an interface based on the given option button. See figure 1. This means that user interaction and QiraahBot application are guided and strict. Based on the nature of extensive reading beyond the classroom, the QiraahBot system works based on the 'send and monitor' concept. Thus, there are two main functions of the application. 1) Supplying extensive texts, and 2) monitoring the progress of the student reading process. The 'send' function built into QiraahBot is supplied with the following features:

a. Existing reading materials uploaded into the application. This material can be accessed through the buttons set on the interface by category. Users can browse this material using persistent buttons. Material selection based on Narrow Reading approach recommended by Krashen (2004).
b. Automatically post from selected service accounts to QiraahBot users. Some outside services such as Facebook, Twitter and web pages have been linked to the application. Based on the Narrow Reading approach, several Facebook accounts, Twitter and certain web pages have been selected. Any post in the selected account will be automatically displayed as a message in the application. This message is additional texts besides existing materials. As a message, it is assisted by the mobile notification system. New message alert will appear and disappear after reading like regular message in IM.
c. Builder or QiraahBot admin has a button to send the reading material. This material works as a Message. The message will be sent to the application user immediately or at the time specified by the admin in the application.

The 'monitor' function on QiraahBot was built to facilitate the teacher as an admin. Monitoring through instant messaging conversations is in text form. Conversations occur according to the strict "questions and answers" format. Admin assigns a question through a special button where the user initiates the interaction by touching the button. The user typed the response and sent it. Then, the system will collect the local response based on the question. Admin can click the entire response at any time.

Fig 1. QiraahBot Interface

This application is implemented as a bot using IM Telegram platform. The action does not develop its own application based on the current trends of mobile phone users. Research shows that most smart phone owners barely install new apps on their devices every month

(Lella & Lipsman, 2014). Therefore, using existing platforms is a good choice. In addition to the above factors, bot is easily installed and functioning virtually without the use of telephone storage, which is the driver of bot system selection as the application medium.

4 Methodology

QiraahBot's usability study uses qualitative and quantitative methodologies. A set of questionnaires containing the System Usability Scale (SUS) instrument and the open-ended question was used. 30 student respondents were given a QiraahBot boat application for a week and a brief description of the installation and use. An Extensive Reading assignment was also given for a week. ER monitoring was also conducted through QiraahBot once. The QiraahBot application provided for the study is prototype version. After 7 days of use, respondents were given an online questionnaire built on the Microsoft Form platform.

SUS Instruments developed by Brooke (1996) contains 10 items using 5 scales. SUS is designed to get an instant overview of the applicability of an application. According to Lin et al. (2011), SUS is a questionnaire used to assess the user's views and the level of satisfaction with particular systems or products. SUS is usually given after the user uses a system or product that is evaluated and before they are involved with any orientation or discussion (Suominen, 2013). Item statements on odd numbers are framed positively, and items on even numbers are negative. For research purposes, the word 'system' of the original SUS instrument was changed to QiraahBot. At the end of the questionnaire, there are open-ended questions. It aims to get respondents qualitative inputs about the strengths, weaknesses and scope of improvements that can be made.

The data obtained were analyzed using SPSS statistical software. Descriptive analysis and frequency applied to the SUS instrument response. Qualitative data results from open-ended questions were also discussed at the end of the findings based on quantitative data results.

5 Findings

5.1 Total SUS score

SUS is built to get usability scores quickly and easily. Researchers can know the usability of a system or product through the overall SUS score ranging from 0 to 100. SUS maker, Brooke (1996) has provided a standard scoring formula for calculating the overall score of SUS. Each SUS score obtained from the respondents was assessed using the mean to obtain the overall score of SUS.

The overall SUS score for QiraahBot application is 65.5. The descriptive analysis for the SUS score in this study is shown in table 1. The SUS mean score for QiraahBot usability is 65.5, the minimum score is 55 and the maximum score is 85. Bangor, Kortum and Miller (2009) suggests that good score is between 70 and 80. For scores value of 90 and above, it indicates a high usability. If the overall score of SUS is between 50 and 70, the system or product analyzed is considered marginally acceptable. However, the score below 50 is considered unacceptable. Based on this guide, SUS score of 65.5 indicates that the QiraahBot application assessed by respondents can generally be marginal.

Although the QiraahBot application is deemed to be used according to Bangor et al. (2009), a study by Sauro (2011) on 500 previous usability studies shows a more stringent read score. According to him, the average SUS score is 68. The score above 68 is considered above average, and the score below 68 is considered below average. This score reading illustrates the usability of the QiraahBot application that is slightly below the score rated as average usability by respondent's experience as a user. It also reveals that the application still has a room that can be improved before it is officially adopted.

Table 1. SUS Descriptive Statistics

	N	Minimum	Maximum	Mean	STD
Statistics	30	55.0	85.0	65.50	5.66

5.2 Descriptive analysis of each SUS item

Descriptive statistics on each item in the SUS give a clearer picture of the respondent's assessment. Based on Table 2 below, item number 5 reaches the highest mean value, namely 4.17 (mean). The mean value of item 5 indicates that most respondents agree that the diversity of functions in QiraahBot is well integrated. Item number 1, 3 and 7 also got a good score above 4.00. Item number 1 (mean = 4.10) represents the desire of most users of QiraahBot to continue using it. Possibly because most respondents consider that QiraahBot is easy to use as interpretation of score 3 (mean = 40). Not only QiraahBot's respondents are able to use it, most of them also think that others are also learning how to use this application quickly as shown by item number 7 (mean = 4.00). In addition, descriptive analysis of all positive SUS items showed mean values between 3.63 to 4.17. In general, respondents gave a simple agreement on the various aspects of the application of QiraahBot.

Table 2. Descriptive Statistics of Each SUS Items

		N	Minimum	Maximum	Mean	STD
1	I think that I would like to use this QiraahBot app frequently.	30	2	5	4.10	.712
2	I found the QiraahBot app unnecessarily complex.	30	1	5	2.87	.937
3	I thought the QiraahBot app was easy to use.	30	2	5	4.00	.871
4	I think that I would need the support of a technical person to be able to use this QiraahBot app.	30	1	4	2.13	.860
5	I found the various functions in the QiraahBot app were well integrated.	30	3	5	4.17	.592
6	I thought there was too much inconsistency in this QiraahBot app.	30	1	5	3.23	.858
7	I would imagine that most people would learn to use this QiraahBot app very quickly.	30	2	5	4.00	.743
8	I found the QiraahBot app very complex to use.	30	1	5	3.57	1.006
9	I felt very confident using the QiraahBot app.	30	2	5	3.63	.765
10	I needed to learn a lot of things before I could get going with this QiraahBot app.	30	1	5	1.90	.923

Items negative statement, namely item number 2, 4, and 10 indicate a lower score. The score for this item ranges from 1.90 to 2.87. Low values on negative statements suggest that most respondents disagree with the negative aspects of the QiraahBot application. Almost all respondents disagree that they need more learning to continue using this application (mean = 1.90). Most respondents also did not consider it difficult to use (mean = 2.87) and they did not need technical assistance for it (mean = 2.13). However, most respondents agreed that the QiraahBot application was still unstable for use (mean = 3.32). Hence, some respondents are uncomfortable using this application (mean = 3.57).

In conclusion, the analysis of each item indicates that the QiraahBot application is generally acceptable. However, there are significant weaknesses that cause users to feel uncomfortable.

5.3 Qualitative results

This study instrument besides containing SUS also includes open-ended questions to respondents. This question asks respondents to reflect on the experience of using QiraahBot and suggestions for improvement. According to Albert and Tullis (2013), the use of open-ended questions in usability studies can help identify areas for improving a product or system.

Generally, most respondents provide a good reflection throughout the experience using the QiraahBot application. According to respondents, QiraahBot is a good application (P06, P09, P11, P13, P15, P24, P29 & P43). The qualitative data also recorded some positive features such as easy-to-use the application of QiraahBot (P03, P14, P23, P26, P28 & P38). In the context of the use of QiraahBot as a learning tool for Arabic Language Extensive Reading, some respondents agreed that this application helped them in the learning process.

Behind the positive elements of QiraahBot, there are comments and suggestions by respondents who make room for improvement of this application system. The bot application, which is still at the prototype level and in the beginning of the test, certainly requires constructive comments and comments before formally and widely used.

The most significant weakness in this application is technically viewed. The respondents reported that they were easily caught (P05) and slightly complex (P04 & P01). Quantitative data reported that respondents did not agree that they needed to learn more before using them. However, qualitative comments suggest that learning is still needed. This application system is found to have a slightly complex system or setting. This situation requires users to seek help or try several times before using it. Some respondents also say something about IM Telegram platform that supports bot applications. They '... *feel a little confused* ...' (P44) and '... *lack of understanding* ...' (P45) QiraahBot application because '... *less skilled* ...' (P45) and '... *rarely use* ... ' (P44) IM Telegram platform.

The respondents' comments throughout QiraahBot's experience provide useful inputs and suggestions, especially in subsequent versions of improvements. Improvements are necessary to make QiraahBot more user-friendly and effective in the user's perspective.

6 Discussion

According to descriptive analysis, QiraahBot is considered to be used at a moderate level by respondents (mean = 65.5). If the value 68 is placed as usability measure as in the study by Sauro (2011), QiraahBot is slightly below average. However, according to Bangor et al. (2009), QiraahBot's reading position indicates that it is still OK with some aspects of the application requiring improvements before it can be fully used.

A fairly low usability score may be due to respondents' immature usage experience. One task within a week can be considered brief so students are still unable to familiarize themselves with the app. Sauro (2011a) and Orfanou, Tselios and Katsanos (2015) emphasis that there are significant differences between first-time users and experienced users support this assumption. The study also revealed that the SUS score would be high if the students were more frequently using technology or learning-related systems. Hence, there is a higher probability of SUS QiraahBot being valued if more experience and time is given. From observation by researchers, students have been using the last-minute times to finish the work from a weeklong period.

However, for newly built applications and still considered prototypes, the level of "OKAY" usability based on the given period may be considered a good start (Ganapathy, Shuib, & Azizan, 2016). Tanja (2011) in a study of usability personal video recording that earned a score of 64 states that it should be considered a candidate for continued improvement. Study by Ganapathy et al. (2016) has scored 64.17 for the MyGraTe prototype. For them, the modest usability value is acceptable for applications that are still at the prototype level. In addition, Nieminen, Runonen, Tyllinen and Nieminen (2014) also agreed that a prototype system with some technical problems and lack of features, but obtaining a marginal SUS value is acceptable. This is acceptable because the preliminary test of a prototype is the first step for further action. Therefore, the lack of early application construction is normal.

Some respondents expressed support for using this application as it is easy to use and does not require technical assistance from other parties. However, details on quantitative and qualitative descriptive analysis reveal a lack of usability when using QiraahBot. This can be considered normal because usability is typically considered from the point of view of issues or problems encountered by users (Kukulska-Hulme, 2007). Bangor et al. (2009) when commenting on the score at OKAY level (52 to 72) asserted that the system is okay, but still clearly deficient in terms of perceived usability.

According to Mostakhdemin-Hosseini (2009), the usability of mobile learning applications should consider several aspects, including device usability, application usability and content usability. This study did not find any deficiencies in terms of device usability. This is likely due to the lightweight, easy-to-install feature of the bot application, which does not require complex storage and configuration capacity and is supported by IM applications (Brandtzaeg, 2017; Klopfenstein et al., 2017). Similarly, in the context of content usability. The study did not receive any negative comments on the content. In fact, the respondents were praised for the content of the application that was considered productive (P15 & P18). The lack of QiraahBot reported is on application usability. Respondents noted the issue of time lagging and the need to facilitate system travel. Although a SUS descriptively detailed score implies the QiraahBot app is generally rated as easy to use, there is still a complaint about inconsistency and complexity when the application works. This condition is also normal when there are other studies that are almost identical to the findings of study by Al-Khalifa (2010) and Ganapathy et al. (2016). Their study also reported a survey system or product that was seen to be user friendly through the SUS scores but faced with application usability issues such as inconsistency.

Additionally, there are respondents who still need time and guidance to use QiraahBot skillfully. Student needs for help and clarification of usage are supported by a study conducted by Chen and Denoyelles (2013). They pointed out that students still need technical assistance to use mobile phone technology for learning needs. Technical training and skills development are an important factor in the advancement of learning technology. This is because students have considered both of them more important than the technology itself. Therefore, Ganapathy et al. (2016) recommends that developers for mobile-based learning applications take into account the need to provide support when developing an effective mobile learning environment.

7 Conclusions

The education world is now witnessing the rapid adoption of mobile phone application technology in learning sessions, including Arabic language learning as a foreign language. This technology changes the learning environment to be more dynamic and open. The QiraahBot application was developed to embellish these changes and bring Arabic students to the latest in the education world. It serves as a tool to facilitate the implementation of extensive Arabic reading activities. However, it is important for us to realize that learning technology through mobile phones depends on the acceptance of the students involved. Any new technology including QiraahBot needs to be 'adopted' by the student as a user so that it remains and is effective as expected. Studies on the usability of the QiraahBot prototype reported moderate acceptance by respondents. The findings also reveal some aspects of usability that need to be improved for the next step.

This study was conducted with certain limitations. The study involved only 30 respondents using the QiraahBot prototype at a relatively short time, as this study is an initial study of newly built applications. It may yield more in-depth and rich insights when involving more respondents, longer usage times and covering multiple tasks. This study involves only one task that needs to be resolved by the respondents. The study also focuses on quantitative on SUS instruments. In the future, it is proposed that the usability study using the mix method; qualitative and quantitative as proposed by Albert and Tullis (2013). Qualitative data such as focus group interviews, observations and usage analysis can provide in-depth information for developers or researchers for further action. Therefore, studies on the application of QiraahBot in the future should involve more extensive and deep usability ratings as well as beyond the limitations of previous studies.

8 References

Al-Khalifa, H. S. (2010). A first step in evaluating the usability of Jusur learning management system. In *3rd Annual Forum on e-Learning Excellence in the Middle East.*

Albert, W., & Tullis, T. (2013). *Measuring the user experience: collecting, analyzing, and presenting usability metrics. Interactive Technologies Series* (2nd ed.). Elsevier Inc.

Bangor, A., Kortum, P., & Miller, J. (2009). Determining what individual SUS scores mean: Adding an adjective rating scale. *Journal of Usability Studies*, *4*(3), 114–123.

Bouhnik, D., & Deshen, M. (2014). WhatsApp Goes to School : Mobile Instant Messaging between Teachers and Students. *Journal of Information Technology Education: Research*, *13*(1), 217–231.

Brandtzaeg, P. B. (2017). Why People Use Chatbots Petter. *Insci 2017*, *9934*(November), 377–392. https://doi.org/10.1007/978-3-319-70284-1_30

Brooke, J. (1996). SUS-A quick and dirty usability scale. *Usability Evaluation in Industry*, *189*(194), 4–7.

Calvo, R., Arbiol, A., & Iglesias, A. (2014). Are all chats suitable for learning purposes? A study of the required characteristics. *Procedia Computer Science*, *27*, 251–260.

Chang, C.-K., & Hsu, C.-K. (2011). A mobile-assisted synchronously collaborative translation–annotation system for English as a foreign language (EFL) reading comprehension. *Computer Assisted Language Learning*, *24*(2), 155–180.

Chen, B., & Denoyelles, A. (2013). Exploring students' mobile learning practices in higher education. *Educause Review*, *7*.

Chen, C. M., & Hsu, S. H. (2008). Personalized intelligent mobile learning system for supporting effective english learning. *Educational Technology and Society*, *11*(3), 153–180. https://doi.org/10.1109/ICSMC.2006.385081

Chinnery, G. M. (2006). Emerging technologies. Going to the mall: mobile assisted language learning. *Language Learning & Technology*, *10*(1), 9–16.

Church, K., & De Oliveira, R. (2013). What's up with whatsapp?: comparing mobile instant messaging behaviors with traditional SMS. In *Proceedings of the 15th international conference on Human-computer interaction with mobile devices and services* (pp. 352–361). ACM.

Cummins, J. (2003). Reading and the bilingual student: Fact and friction. In *English Learners: Reaching the Highest Level of English Literacy* (pp. 2–33). International Reading Association.

Cuno Klopfenstein, L., & Bogliolo, A. (2017). the Quiz - Master Bot : a Persistent Augmented Quiz Delivered Through Online Messaging. *11th International Technology, Education and Development Conference*, (March). https://doi.org/10.21125/inted.2017.2328

Department of Statistics Malaysia. (2016). ICT Use and Access By Individuals and Households Survey Report.

DeVellis, R. F. (2016). *Scale development: Theory and applications* (Vol. 26). Sage publications.

Fryer, L., & Carpenter, R. (2006). Bots as language Learning Tools. *Language Learning & Technology*, *10*(3), 8–14.

Ganapathy, M., Shuib, M., & Azizan, S. N. (2016). Malaysian ESL students' perceptions on the usability of a mobile application for grammar test: A case study of ESL undergraduates in universiti sains Malaysia. *3L: Language, Linguistics, Literature*, *22*(1), 127–140. https://doi.org/10.17576/3L-2016-2201-10

Grinter, R. E., & Palen, L. (2002). Instant messaging in teen life. In *Proceedings of the 2002 ACM conference on Computer supported cooperative work* (pp. 21–30). ACM.

Hazaea, A. N., & Alzubi, A. A. (2016). The Effectiveness of Using Mobile on EFL Learners' Reading Practices in Najran University. *English Language Teaching*, *9*(5), 8. https://doi.org/10.5539/elt.v9n5p8

Huang, L., & Lin, C. (2011). EFL learners' reading on mobile phones. *The JALT CALL Journal*, *7*(1), 61–78.

Jeon, E.-Y., & Day, R. R. (2016). The effectiveness of ER on reading proficiency: A meta-analysis. *Reading in a Foreign Language*, *28*(2), 246–265. https://doi.org/10.1017/SOI42716411000932

Kim, H., Lee, M., & Kim, M. (2014). Effects of mobile instant messaging on collaborative learning processes and outcomes: The case of South Korea. *Journal of Educational Technology & Society*, *17*(2), 31.

Klopfenstein, L. C., Delpriori, S., Malatini, S., & Bogliolo, A. (2017). The Rise of Bots. In *Proceedings of the 2017 Conference on Designing Interactive Systems - DIS '17* (pp. 555–565). New York, New York, USA: ACM Press. https://doi.org/10.1145/3064663.3064672

Krashen, S. D. (2004). The case for narrow reading. *Language Magazine*, *3*(5), 17–19. https://doi.org/10.1016/0346-251X(95)00054-N

Kukulska-Hulme, A. (2007). Mobile usability in educational contexts: what have we learnt? *The International Review of Research in Open and Distributed Learning*, *8*(2).

Kukulska-Hulme, A., & Shield, L. (2008). An overview of mobile assisted language learning: From content delivery to supported collaboration and interaction. *ReCALL*, *20*(3), 271–289.

Lee, H., & Mallinder, M. (2017). Role of Extensive Reading in EFL Vocabulary Development : Review and Recommendation. *The English Teacher*, *XL*, 145–163.

Lella, A., & Lipsman, A. (2014). *The US mobile app report. Tech. Rep.*

Lenhart, A., Rainie, L., & Lewis, O. (2001). *Teenage life online: The rise of the instant-message generation and the Internet's impact on friendships and family relationships*. Pew Internet & American Life Project Washington, DC.

Lin, H.-C. K., Hsieh, M.-C., Wang, C.-H., Sie, Z.-Y., & Chang, S.-H. (2011). Establishment and Usability Evaluation of an Interactive AR Learning System on Conservation of Fish. *Turkish Online Journal of Educational Technology-TOJET*, *10*(4), 181–187.

Marçal, E., Andrade, R., Melo, R., Viana, W., & Junqueira, E. (2016). Using mobile message to improve student participation in blended courses: a Brazilian case study. *ACM SIGAPP Applied Computing Review*, *16*(3), 15–25.

McKitterick, W. (2015). *The messaging app report: How instant messaging can be monetized*. Technical report, BusinessInsider.

Mobile Ecosystem Forum. (2016). *Global Insights into Chat Apps and SMS Usage*.

Mostakhdemin-Hosseini, A. (2009). Usability considerations of mobile learning applications. *International Journal of Interactive Mobile Technologies (iJIM)*, *3*, 29–31.

Nezarat, A., & Mosavi Miangah, T. (2012). Mobile-Assister language. *International Journal of Distributed and Parallel Systems (IJDPS)*, *1*(1), 309–319. https://doi.org/10.1002/9781405198431.wbeal0768

Nieminen, M. P., Runonen, M., Tyllinen, M., & Nieminen, M. (2014). Collaboration space for creative knowledge work–Analysis of industrial pilots. In *International Conference of Design, User Experience, and Usability* (pp. 653–662). Springer.

Nordin, N. M., Embi, M. A., Norman, H., & Panah, E. (2017). A Historical Review of Mobile Learning Research in Malaysia and Its Implications for Malaysia and the Asia-Pacific Region BT - Mobile Learning in Higher Education in the Asia-Pacific Region: Harnessing Trends and Challenging Orthodoxies. In A. Murphy, H. Farley, L. E. Dyson, & H. Jones (Eds.) (pp. 137–150). Singapore: Springer Singapore. https://doi.org/10.1007/978-981-10-4944-6_7

Orfanou, K., Tselios, N., & Katsanos, C. (2015). Perceived usability evaluation of learning management systems: Empirical evaluation of the System Usability Scale. *The International Review of Research in Open and Distributed Learning*, *16*(2).

Pallant, J. (2011). *SPSS survival manual, 3rd*. Allen & Unwin.

Sauro, J. (2011a). *A practical guide to the System Usability Scale (SUS): Background. Measuring Usability LLC*. Denver, Colorado.

Sauro, J. (2011b). Measuring Usability With The System Usability Scale (SUS). *Measuring Usability*, 1–5.

Suominen, M. (2013). *Evaluating usability in video conferencing service in Metso* (unpublished master's thesis). JAMK University of Applied Science. Finland.

Tanja, R. (2011). *A usability evaluation of a personal video recorder: Navigation and channel list management from a user perspective*. Bachelor thesis of Cognitive Science. Linköping University, Sweden.

Towards ensuring inter-religious harmony in a multi-religious society of Perak

Mohd Ikhwan Izzat Zulkefli[1,*], *Mohd* Nuri Al-Amin Endut[1] , *Muhammad* Ridhuan Tony Lim Abdullah[1] , and *Azizan* Baharuddin[2]

[1]Department of Management and Humanities, Universiti Teknologi PETRONAS, 32610 Bandar Seri Iskandar, Perak, Malaysia
[2]Institute of Islamic Understanding Malaysia, 2, Langgak Tunku Off Jalan Tuanku Abdul Halim, 50480 Kuala Lumpur, Malaysia

Abstract. Inter-religious harmony is crucial in maintaining political and national stability. Living in a peaceful and harmonious environment is important when the society consists of diverse races and religious adherents. The existence of minor tension in the society to some extent has jeopardised inter-religious relations. The failure to preserve religious harmony could expose the society to conflict and chaos. Thus, this paper aims to discuss and elaborate the pertinent elements in sustainable harmony among multi-religious society in Perak. The paper elaborates on the fundamental factors contributing to socio-religious harmony among multi-religious society. Library research method is used in addressing the situation of religious harmony in Perak and the pertinent element should be fostered in order to preserve religious harmony. This paper also emphasises that elements such as acceptance, understanding, co-operation as well as right and justice are significant elements in sustaining religious harmony. Hence, the discussion is aimed to build a sustainable concept of multi-religious harmony in Perak. The findings could also encourage multi-religious society to maintain and build a strong relationship among them.

Keywords: Religious harmony, the element of harmony, conflict, challenge, Perak

1 Introduction

Malaysia is one of the countries that has a pluralistic society in the region of Southeast Asia. It is a developing country that has 32.3 million population in 2017 (Malaysia, 2018) with various ethnicities and religious backgrounds. The Malaysian nation comprises of multiracial, multi-religious, multicultural, and multilingual society. Historically, the formation of Malaysia as a pluralistic country can be traced back to the Malay Sultanate of Malacca in the early fifteenth century (Ahmad, 2007). During that time, Malacca is one of

* Corresponding author: ikhwanmuslimin92@gmail.com

the most strategic and famous centres of trade in the Malay Archipelago (Ahmad, 2007). Besides, this country has also been occupied by several western countries, namely, Portuguese, Dutch and British. On 31 August 1957, this country achieved independence from the British. Even though this country has reached her independence for over 6 decades, this country has been experiencing major and minor conflict of inter-religious adherents since its independence.

Religious harmony is crucial in ensuring unity and solidarity among multi-religious adherents. In 2017, Malaysia is placed as the twenty-ninth most peaceful country in the world and the fifth most peaceful country in the region of Asia Pacific (Peace, 2017). However, it is not immune from inter-religious conflict whereby it could interrupt social relation and national stability (Abdullah et al., 2016). The conflict between religious adherents is an inevitable issue which could become a great challenge for the government, religious leaders as well as society to preserve social harmony among the Malaysian communities. As emphasised by Sintang, Baharuddin, and Khambali (2012), they stated that the government faced a challenge to create religious harmony and peaceful environment among multi-religious adherents. It is very important for the government to manage every inter-religious conflict wisely in order to avoid it from becoming worse (Samad, 2016). Hence, strengthening unity and solidarity among multi-religious adherents is considered as a major instrument to ensure social harmony and peaceful co-existence in the society.

In maintaining inter-religious harmony, the government plays an important role by enforcing various efforts and policies. There are a number of programmes, activities and policies that have been invented by the government to ensure all Malaysian society with different ethnic and religious backgrounds to mingle with each other (David, McLellan, Meng, Li, & Tien, 2010). Preserving social harmony has become a national agenda for the government since the past decades. It is important to note that religious harmony is vital in preserving inter-religious relations. Among government efforts to preserve a harmonious society are the implementation of Rukun Negara (National Principles), 1 Malaysia Concept, the formation of the Department of National Unity of Integration (JPNIN), Institute for Ethnic Studies (KITA) (Abdullah, Al-Amin, Ur-Rahman, Baharuddin, & Ahmad, 2017), and the celebration of religious festivals (Bakar, 2013), to name a few. Therefore, this paper is aimed to discuss and elaborate the pertinent elements in sustainable harmony among multi-religious society in Perak.

2 The background of multi-religious society in Perak

The formation of Perak as a multi-racial and multi-religious society began when the Chinese and Indian residents were granted full citizenship. Perak is one of the thirteen states of Malaysia and the fourth largest state in the country. This state consists of three main ethnic groups namely, Malay, Chinese, Indian and followed by other ethnic groups. Religious diversity is a main characteristic of Perak because its population comprises of heterogeneous religions and ethnicities.

Furthermore, every ethnic group in Perak has their own identities and backgrounds in terms of languages, cultures as well as customs and beliefs system. With the diversity of religion, race, language, and culture, Perak has shaped its own cultural identity as a multi-cultural state. Almost all major world religions exist in this state namely Islam, Buddhism, Hinduism, Christianity, Confucianism, and Taoism. Today, Perak's population is estimated to be at about 2,496.4 million people with 56.4% Bumiputera, 27.9% Chinese, 11.2% Indian and 0.4% others. Meanwhile, the total population of non-Malaysian citizens is 4.1% (Malaysia, 2018). In terms of religious adherents, Islam recorded the highest percentage of religious adherents at 55.3%, followed by Buddhism at 25.4%, Hinduism at 10.9%,

Christianity at 4.3%, Confucianism, Taoism and Tribal/folk/other traditional Chinese religion took 1.7% and other religions being at 0.8% (Malaysia, 2010).

Historically, the formation of Perak as a multi-ethnic and multi-religious society was closely associated to the large-scale migration of workers and labourers from mainland China and India to the Malay Archipelago in the nineteenth century (Mun, 2013). According to Jamil (2010), the Malay Archipelago was mainly mono-ethnic inhabited by the Malays in early nineteenth century. However, between 1874-1957, there was a great social landscape change of Malaysian society from homogenous into heterogeneous society during British colonisation (Jamil, 2010). The major factor has led the British to bring Chinese and Indian labourers to the Malay Archipelago due to the need of workers to work in tin mines, plantations, and estates (Mun, 2013). Thus, the migration of Chinese and Indians from their countries is a main factor that formed Perak as a multi-religious society in nowadays.

It is important to emphasise that the existence of diverse religious adherents in Perak is one of the challenges for the government and society in order to retain peace and harmony. Although Perak seems like a peaceful state, the emergence of conflicts in Perak has a great potential to interrupt national and political stability. Sintang et al. (2012) highlighted that inter-religious issues and incidents became more complicated when it deals with the issues of living together and social interaction rather than dealing with the theological issues. Nowadays, there are numerous religious conflict cases especially those related to the place of worship, the right of child custody (Ahmad, Talib, Isa, Gill, Jawan & Mohad, 2014) and the issue of kalimah Allah which can be used by non-Muslim (Nor, 2011). Besides that, other sensitive issues related to religious matters include the use of Quranic and Arabic terminologies, Islamic propagation, funeral process, and the use of halal certificate by non-Muslim businessman (Sintang et al., 2012). Thus, it is clearly showed that religious issues nowadays are no longer dealing with the theological aspect of religion.

3 Religious harmony in Perak

As a multi-religious state, Perak is a peaceful and harmonious state in Malaysia. In order to create a harmonious society, a lot of efforts should be implemented to maintain peaceful coexistence. The Sultan of Perak, Sultan Nazrin Shah asserts that "Malaysians were from different races and practised different religions and cultures, hence building harmony in the country is a collective responsibility" ("Sultan Nazrin: respect religious boundaries to maintain peace and harmony," June 12, 2014). This shows that preserving and creating peaceful coexistence is a collective responsibility among government, religious leaders, and society. There are numerous efforts enforced by the government and non-governmental organisation (NGOs) in order to build inter-religious harmony. Among the initiatives are inter-religious dialogue, 1 Malaysia concept and the formation of the Department of National Unity of Integration (JPNIN).

However, religious harmony in Perak has also been disturbed by minor tension or conflict. There are several incidents that could interrupt inter-religious relations in Perak. There was an incident at Jalan Hospital Ipoh where a man went to the Sree Muneeswaran Amman Kovil Temple and broke the statues (Zahari, 2016). The statues in the Hindu temple are considered as sacred among the Hindus and such incident were regarded as an insult to their religion. This incident has created a tense situation and anger among Hindus. Besides, there was also the issue of Taiping MP who wished Hari Raya celebration featuring a picture of US flag (Fai, 2016) and a caricature of giving out cash in the form of "duit Raya" to children and guest during Hari Raya as a bribe (Ghazali, 2016). These issues have been regarded as insulting the noble practice of Muslim community in Malaysia. The implications of the incidents pertaining to religious tension could lead to hostility and enmity between religious adherents.

Furthermore, the case of conversion of three children to Islam by their father Mohd Ridzuan Abdullah could also jeopardise inter-religious relations. The main issue in this case is the father who is a Muslim attempted to convert his children to Islam without the knowledge of their mother who is a Hindu. However, the mother named Indira Gandhi applied for judicial review in the High Court of Ipoh because the certificates of conversion of her three children to Islam is inconsistent with section 106 (b) of the Islamic Administration Religious Enactment of Perak 2004 and section 5 and 11 of the Guardianship of Infant Act 1961 (Kusrin, Hamjah, & Sham, 2015). In 2018, the Federal Court declared that the conversion of these three children to Islam is null and void (Karim, January 29, 2018). Based on these issues and incidents, it shows that religious issue is a sensitive matter that should be resolved wisely. Even the smallest of incident could spread very fast and become a trigger to the immense incident. Therefore, to those living in a multi-religious society, respecting other religious adherents is significant to preserve inter-religious relationship.

4 Fundamental elements of sustaining religious harmony

4.1 Acceptance

A multi-religious society is a community with diverse religions such as Islam, Buddhism, Hinduism, Christianity and others. In order to live in a peaceful and harmonious society, acceptance of other religious adherents in a particular community is very important. In Islamic perspective, Islam is a religion that accepts the diversity of religions in this world but rejects that all religions are of truth. Religious plurality can be defined as the diversity of religion in this world. However, in the concept of religious pluralism, it can be defined as a belief that all religions in this world is absolute truth (Sintang, 2014). Pluralist viewed that all religion in this world is true and should be respected. For John Hick, he accepts the understanding that all truth claims are valid and then should be given equal respect and it only means that the Christian has to accept other truth claims such as Muslims, Jews, Hindus and others (Sintang, 2014). However, in Islamic perspective, Islam accepts the diversity of religions but it does not recognise all religions as truth.

Moreover, acceptance of different religious background among multi-religious society is a key to ensuring harmony and solidarity in the country (Talib, Gill, & Ramli, 2014). Appreciation of the religious diversity in a multi-religious society is an essential factor in maintaining social harmony. This is in line with Abdullah et al. (2016), where appreciating kindness and common values of other religions is among the important elements in maintaining harmony. For example, Muslim people appreciating the existence of other religious groups such as Buddhist, Hindus, and Christians can strengthen inter-religious relationship and nurture kindness towards them. Therefore, accepting and appreciating the existence of religious diversity should be regarded as a binding factor of the particular community.

4.2 Understanding

Every individual who lives in a multi-religious society should understand and accept religious diversity. Understanding of the faith and teachings of other religious groups could encourage an individual to accept and respect the beliefs of others religions. According to Rahman and Khambali @ Hambali (2013), the celebration of all religions in Malaysia is a sign of respecting the existence of multi-religious adherents in Malaysia as stated in the concept of 1 Malaysia. However, when the community in Malaysia do not understand the faiths of other religions it could contribute to the misunderstanding of concept of religious tolerance (Rahman & Khambali @ Hambali, 2013).

The concept of mutual understanding seems vital as people should not interrupt the practices of other religious adherents (Kasmo, Usman, Taha, Salleh, & Alias, 2015). Every individual or religious adherents should respect the doctrines of other religions. Mutual respect among inter-religious adherents is also considered as an important principle to preserve a harmonious society in Malaysia (Abdullah et al., 2016). Respecting the existence of religious diversity can contribute to political and national stability. Attitudes of respect could also improve inter-religious relations by acknowledging the existence of multi-religious groups. The value of mutual respect, cooperation, tolerance, understanding, and goodwill are the significant elements to be practised while living in a multi-racial and religious society (Seok et al., 2013).

4.3 Co-operation

The element of co-operation is also important to build good relationship among multi-religious adherents. With the existence of this element in their life, they will respect and tolerate other religious adherents. Co-operation can be seen through their action by respecting the celebrations, rituals as well as beliefs of other religious adherents in their community. Hence, every individual living in a multi-faith society should tolerate and co-operate with each other. The Theory of *'Asabiyah* by Ibn Khaldun emphasised that *'Asabiyah* (spirit of kinship within the family or tribe) is important in order to strengthen the relationship among individual in the community. Ibn Khaldun viewed *'Asabiyah* from positive side. Moreover, Ibn Khaldūn emphasised that the spirit *of 'Asabiyyah* basically exists due to the primitive life possessed by certain groups or nations when they face difficulties (Halim, Nor, Ibrahim, & Hamid, 2012). When these groups or communities face difficulties, this will force them to stand together and protect their people from any harm from outside the group (Halim et al., 2012). The feeling of *'Asabiyyah* also can lead people to protect and improve their group or society towards a better nation. The spirit of *'Asabiyyah* can also emerge as a solidarity within large group anonymous (Halim et al., 2012).

4.4 Right and Justice

In order to preserve a harmonious society, the right and freedom in practising religions should be given to all multi-religious adherents. The right and freedom of every religious adherents should not be abandoned (Abdullah et al., 2016). The right and freedom of practising religion is a basic human right for each individual and it should not be neglected. However, in Malaysia, defining the rights of Muslim and non-Muslim is still not clear (Rahman & Khambali @ Hambali, 2013). The vagueness of the civil and Shariah court jurisdiction has contributed to the problems of conversion of religions. The overlapping of the civil and Shariah court jurisdiction contributes to the difficulties to solve the case.(Rahman & Khambali @ Hambali, 2013).

In Malaysian context, the right of practising a religion is mentioned in the Federal Constitution of Malaysia. This is based on Article 3 (1), "Islam is the religion of the Federation; but other religions may be practised in peace and harmony in any part of the Federation" (*Federal Constitution of Malaysia* 2010). Based on this provision, it can be understood that Islam is declared as an official religion in Malaysia and specifically in Perak. However, at the same time, other religious adherents were also given freedom to practise their beliefs in this country. In addition, Article 11 (1) also declared that "every person has the right to profess and practise his religion and, subject to Clause (4), to propagate it" (*Federal Constitution of Malaysia* 2010). Based on this article, it is also mentioned that all religious adherents can practise their understanding but they cannot propagate their beliefs to Muslim. This showed that the right and freedom should be given in sustaining peaceful and harmonious society.

In addition, Islam always empahsised on the concept of justice (*al-'adl*). The term 'justice' may refer to "administration of law according to prescribed and accepted principles, conformity to the law, legal validity the quality or fact of being just" (Ismail, 2015). In Islam, it is important to emphasise the concept of justice. Islam emphasises on justice towards inter-personal relationships within the family, within the community, in the interaction between communities and nations, in the interface between the human being and nature (Qur'an, 4: 58, 65,105,135). There is justice to kith and kin, to the orphan, to the destitute, to the slave, to the wayfarer, to the needy (Qur'an, 2: 177). Justice is to bond and hold society together and transforms it into one brotherhood (Ismail, 2015). In order to have justice in harmony, the right to practise religion by an individual or religious group should be permitted.

5 The challenges of religious harmony

It is important to note that living in a harmonious society without conflict and tension is very important. No single person in this world would prefer to live in tension, conflict, and war. Thus, strengthening unity and solidarity among religious adherents is essential to create a harmonious society. Inter-religious harmony should be built based on trust and mutual respect among multi-religious adherents. The existence of harmony could prevent any conflict which contributes to fragility of inter-religious relations and political instability.

However, there are various challenges to improve and strengthen the relationship between multi-religious adherents. The disagreement between religious adherents happened due to the lack of mutual respect. Mutual respect among inter-religious adherents are crucial nowadays to strengthen the relationship between different religious adherents. Based on the study conducted by Abdullah et al. (2016), the study highlighted that mutual respect is considered as an important element to build a good relationship between religious adherents. It is also very significant to be practiced, especially in daily interaction. Meanwhile, Rahman and Khambali @ Hambali (2013) asserts that people should not only respect the doctrines of religions but also should respect the adherents of the respective religions. Social interaction and communication between different religions are essential to guarantee a harmonious society. Besides, a study conducted by Talib et al. (2014) also emphasized that all people should respect the doctrine and teaching of other religious adherents. Mutual respect in this context is that all religious adherents such as Muslim, Buddhist, Hindu, and Christian should respect each other, specifically in the aspect of belief or doctrine, teaching, and festival.

Furthermore, the study shows that inter-religious conflict also happens due to the issue of right and freedom of religion. Freedom of religion is applied to all religious adherents regardless of Muslim and non-Muslim. Based on article 11 (1), the Constitution of Malaysia declared that "Every person has the right to profess and practise his religion" (*Federal Constitution of Malaysia* 2010). Through this provision, it is highlighted that all religious adherents have an equal right to practise their doctrines. However, other religious adherents such as Christian and Hindu is not allowed to disseminate their faith towards Muslim. This is based on Article 11 (4) "State law and in respect of the Federal Territories of Kuala Lumpur, Labuan, and Putrajaya, federal law may control or restrict the propagation of any religious doctrine or belief among persons professing the religion of Islam" (*Federal Constitution of Malaysia* 2010). The restriction is imposed on all other religious groups except Islam. Although other religious adherents are guaranteed their rights but there are several of restriction imposed on them. This restriction also contributed to the tension. The minority group starts to question the privilege granted to Islam and Muslim. Khambali @ Hambali and Haled (2008) mentioned that issues related to freedom of religion should be discussed openly because all religious adherents are not facing any discrimination in terms of ethnic group and religion.

Besides that, maintaining harmony in the society is also a challenge when there is no clear guidance of tolerance practise. In this issue, there is no clear guideline and rule which describes the right of Muslims and non-Muslims (Rahman & Khambali @ Hambali, 2013). This issue occurred when it is related to the conversion of religion from Islam to other religions. Rahman and Khambali @ Hambali (2013) stated that enactments of law is difficult to make due to the overlapping of the jurisdiction of court between civil and Shariah on certain issues. The ambiguity of laws would hence contribute to tension.

6 Conclusion

This paper presented that there are several minor issues which could jeopardise inter-religious relations in Perak. The situation of religious harmony in Perak has also been discussed in order to show that there is tension although it is not as serious as other states in Malaysia. In addition, creating and ensuring a harmonious co-existence is the main agenda for the government and society. The government has implemented various efforts and policies to maintain religious harmony.

The occurrence of religious conflict had begun since Malaysia achieved its independence in 1957. Among religious issues that has always been discussed are religious conversion, the use of kalimah Allah, and the construction of worship places. However, there are various challenges in ensuring religious harmony such as lack of mutual respect, the abuse of rights and freedom of minority group. As a conclusion, apart from discussing the current situation of religious harmony in Perak, this paper also discussed the fundamental factors contributing to religious harmony. Among the important elements are acceptance, understanding, co-operation, as well as right and justice. These elements are important to strengthen inter-religious relations as well as building a harmonious co-existence.

This paper was financially supported by the Institute of Islamic Understanding Malaysia (IKIM).

7 References

Abdullah, M. R. T. L., Al-Amin, M. N., Ur-Rahman, A., Baharuddin, A., & Ahmad, Z. (2017). Sustainable socio-religious harmony development in Malaysia: An interpretive structural modelling for multi-religious society *Journal of Al-Tamaddun, 12*(1), 53-64.

Abdullah, M. R. T. L., Al-Amin, M. N., Yusoff, A., Baharuddin, A., Khir, F. A., & Talib, A. T. (2016). Socio-religious harmony index instrument indicators for Malaysia *Journal of Al-Tamaddun, 11*(2), 29-44.

Ahmad, Z. (2007). Multiculturalism and religio-ethnic plurality: the Malaysian experience. *Culture and Religion, 8*(2), 139-153.

Ahmad, Z., Talib, A. T., Isa, N. A. M., S.Gill, S., Jawan, J., & Mohad, A. H. (2014). *Tahap kepentingan isu-isu antara agama di Malaysia*. Paper presented at the Seminar

Antarabangsa Dakwah & Etnik 2014: Da'wah & Ethnicity: Multidisciplinary Perspective, Pusat Kajian Dakwah Orang Asli dan Pribumi, UKM, Bangi, Selangor. .

Bakar, I. A. (2013). The religious tolerance in Malaysia: an exposition. *Advances in Natural and Applied Sciences, 7*(1), 90-97.

David, M. K., McLellan, J., Meng, N. Y., Li, L. M., & Tien, W. Y. M. (Eds.). (2010). *Ethnic relations and nation building: the way forward*. Petaling Jaya: Strategic Information and Research Development Centre (SIRD).

Federal Constitution of Malaysia (2010). The Commissioner of Law Revision, Malaysia.

Halim, A. A., Nor, M. R. M., Ibrahim, A. Z. B., & Hamid, F. A. F. A. (2012). Ibn Khaldun's Theory of 'Asabiyyah and Its Application in Modern Muslim Society. *Middle-East Journal of Scientific Research, 11*(9), 1232-1237.

Ismail, K. (2015). Islam and the Concept of Justice. *Jurnal Intelek, 5*(2).

Jamil, H. (2010). *Historical overview of Malaysia's experience in enhancing equity and quality of education: focusing on management and mediation of multiethnic issues*. Paper presented at the Africa-Asia Experience Sharing Seminar: Efforts towards Improving the Quality of Education, Accra: Ghana.

Karim, K. N. (January 29, 2018). Indira Ghandi wins 9-year legal battle to nullify children's conversion to Islam by ex-husband. *New Straits Times* Retrieved from https://www.nst.com.my/news/crime-courts/2018/01/329967/indira-ghandi-wins-9-year-legal-battle-nullify-childrens-conversion

Kasmo, M. A., Usman, A. H., Taha, M., Salleh, A. R., & Alias, J. (2015). Religious tolerance in Malaysia: A comparative study between the different religious groups. *Review of European Studies, 7*(3), 184-191.

Khambali @ Hambali, K. M., & Haled, M. H. M. (2008). Toleransi beragama dan amalannya di Malaysia: rujukan kepada artikel 11 perlembagaan persekutuan Malaysia. *Jurnal Usuluddin, 27*, 81-92.

Kusrin, Z. M., Hamjah, S. H., & Sham, F. M. (2015). Muslim father's rights in determining his minor child's conversion to Islam in Malaysia. *European Journal of Social Sciences Education and Research, 4*(1), 15-21.

Malaysia, D. o. S. (2010). Total population by ethnic group, religion, sex and state, Malaysia, 2010.

Malaysia, D. o. S. (2018). Demographic statistics fourth quarter (Q4) 2017, Malaysia. Retrieved from https://www.dosm.gov.my/v1

Malaysia, D. o. S. (2018). Population by age group, sex and ethnic group, Perak, 2018.

Mun, C. Y. (2013). Ethnic socialization: A case of Malaysian Malay and Chinese public universities students experiences. *International Journal of Social Science and Humanity, 3*(6), 582-585.

Nor, M. R. M. (2011). Religious tolerance in Malaysia: An overview. *Middle-East Journal of Scientific Research, 9*(1), 23-27.

Peace, I. f. E. a. (2017). *Global peace index 2017: measuring peace in a complex world* St Leonards Institute for Economics & Peace

Rahman, N. F. A., & Khambali @ Hambali, K. M. (2013). Religious tolerance in Malaysia: Problems and challenges. *International journal of Islamic thought, 3*, 81-91.

Samad, Y. (2016). *Peaceful co-existence in the light of Holy Qur'an: a case study of Christians living in Kuala Lumpur*. (Doctoral thesis), University of Malaya, Kuala Lumpur. Retrieved from http://studentsrepo.um.edu.my/6935/1/THESIS_20__-__2__-.pdf

Sintang, S. (2014). Peaceful co-existence in religious diversity in Sabah, Malaysia. *Global Journal of Human-Social Science Research, 14*(1).

Sintang, S., Baharuddin, A., & Khambali, K. M. (2012). Dialogue of life and its significance in inter-religious relation in Malaysia. *International journal of Islamic thought, 2*, 69-79.

Sultan Nazrin: respect religious boundaries to maintain peace and harmony. (June 12, 2014). *Malay Mail Online* Retrieved from http://www.themalaymailonline.com/malaysia/article/sultan-nazrin-respect-religious-boundaries-to-maintain-peace-and-harmony

Talib, A. T., Gill, S. S., & Ramli, M. R. (2014). *Pengaruh Agama Terhadap Identiti Politik: Kajian Kes Belia di Malaysia.* Paper presented at the Seminar on National Resilience: Diversity in Creating Unity Premiera Hotel Kuala Lumpur.

Talib, A. T., Isa, N. A. M., S.Gill, S., Ahmad, Z., Jawan, J., & Mohad, A. H. (2014). *Persepsi terhadap toleransi sosio agama di Sabah dan Sarawak.* Paper presented at the Seminar Antarabangsa Dakwah & Etnik 2014: Da'wah & Ethnicity: Multidisciplinary Perspective, Pusat Kajian Dakwah Orang Asli dan Pribumi, UKM, Bangi, Selangor.

Technological infrastructure and human culture: Appropriating innovative teaching methods to 21st century classrooms

Chong *Su Li*[1,*] and *Roselind* Wan[1]

[1]Department of Management and Humanities, Universiti Teknologi PETRONAS, Seri Iskandar, 32610 Perak Malaysia

Abstract. Often, it is assumed that the introduction of technological infrastructures into the 21st century classroom will provide the catalyst to appropriating innovative methods befitting modern educational frameworks. One of these appropriations is the introduction of technology like wireless connectivity, interactive smartboards and mobile devices in the shift from applying teacher-centredness to student-centredness in the classroom. However, this shift is often discussed in terms of technological infrastructure, to the exclusion of examining the cultural change that occurs as a result. This paper discusses findings from a study which set out to examine the effectiveness of electronic marginalia (e-marginalia) in a university classroom for Academic Writing, with e-marginalia being regarded as a technological infrastructure. Drawing from principles of Action Research, this study adopted a constructivist perspective which framed its research design. In the course of 14 weeks, 2 groups of 25 Engineering and Technology undergraduates respectively, underwent draft-writing through the use of Microsoft® track-change function and e-mail communication with two lecturers. This method of classroom and online engagement tested the extent to which infrastructure and human culture could align in a writing course. Data was also drawn from students and lecturers' feedback with regards to their 14-week experience of learning and teaching through new methods. Findings from this study show that when human culture and behaviour is considered, technological infrastructure can be capitalised through the cultural appropriation of human-relationship-sustenance and technological-content-identification. These findings answer the question of teacher-relevance in the Education 4.0 framework. Thus, not only are teachers still very relevant to current pedagogies, it is their sociocultural sensitivities that are tested and confronted. As such, in accommodating new ways of teaching and learning that must chime with Education 4.0, this paper advocates for new forms of openness in 21st century education.

* Corresponding author: chong_suli@utp.edu.my

1 Background of research

Much is being debated about how educational experience will alter as transformation is expected to take place in the educational landscape, worldwide. In the 21st century, digital technology is considered to be one of the main drivers undergirding this transformation. Thus, in answering the clarion call of Industrial Revolution 4.0, academics and educationists have chimed in through the Education 4.0 framework so as to think about and prepare for the shift in educational matters.

One of the main dilemmas in confronting educational transformations in the era of 4.0 is the question of whether teachers or lecturers may be replaced. This comes in light of new advances in technology which allows for the democratisation of teaching materials and methodology. Massive open online courses (MOOCS) for example has transformed the way we understand the physical classroom and face-to-face teaching. In allowing for learners to learn from anywhere, MOOCS' philosophy has reduced or completely done away with the face-to-face component. This raises the question of whether human communication may become indispensable in 21st century classrooms.

This paper raises the question of how educators can appropriate innovative teaching methods which account for both the advantage of technological infrastructure and the indispensability of human communication. This question is particularly pertinent within the field of engineering education where there appears to be a shift from teacher-centred teaching to student-centred learning. In this paper, this question is set within the context of how Academic Writing for undergraduate students in an Engineering and Technological university can be optimised in the current context that assumes that digital technology plays an integral part in the pedagogical process. This paper is a follow on from a previous study that also examined Academic Writing but through the pedagogy of Active Learning (Chong, 2017). As such, this current project sets out to gauge the effectiveness of the use of electronic marginalia (e-marginalia) for the draft writing process in Academic Writing. The significance of this research lay in documenting the response of the students and lecturers who engaged in the use of electronic marginalia across one semester. The research question for this project is: How do undergraduates and lecturers, engaged in the use of electronic marginalia, experience the draft writing process? This paper highlights the lecturers and students' emergent perspectives in order to understand how student-lecturer communication occurs within an educational setting that is supported by technological infrastructure.

2 Selected literature

2.1 Teacher-centred teaching and student-centred learning

The move from teacher-centred teaching to student-centred learning has been present in the field of Social Science for more than a century. Believed to have mainly begun with the work of educational philosopher John Dewey, this movement became popularised at the latter part of the 20th century. This has been largely due to the human rights movement that strove to locate the child at the heart of educational efforts thus providing the child with equal opportunities to not only learn but more importantly, to develop and thrive (Bromley, Meyer, & Ramirez, 2011).

Although it initially took root in the field of Social Science, this pedagogy of student-centredness gained ground in the field of Science education. Particularly, the academic field of Engineering pedagogy found this school of thought to be relevant in the latter decades of the 20th century. An initial drop in interest among potential engineering students and a major shift in the way science was understood within the Engineering profession meant that universities dedicated to training and shaping future engineers had to re-think the way they taught their courses. This had a positive impact on the field of Science, Technology, Engineering and Mathematics (STEM) because it encouraged educationists to think about how school and university students approached these subjects especially among learners in the early stages of their education experience (Clements & Sarama, 2016). One of the outcomes of this re-thinking is the understanding that STEM subjects is as much about hard science as it is about the softer side of learning, negotiating and communicating scientific knowledge (Dejarnette, 2012).

Another important outcome of this approach is the teaching fraternity's sensitivity towards an increasingly culturally-diverse student body across Engineering faculties, worldwide (Committee on the Engineer of 2020 National Academy of Engineering, 2005; Pretz, 2016). However, this shift has raised important questions about whether teacher-centred teaching and student-centred learning are necessarily dichotomous. The Engineering Education fraternity has not been able to agree on such differing paradigms which tend to pit the objectivist/post-positivist perspective against the subjectivist/constructivist perspective (Felder, 2011). The unexpected outcome of such thinking is an imbalanced view which imagines the teacher teaching in silo, or the student learning in silo. In arguing for middle ground, Mascolo (2009) suggests that the teaching and learning process is a constantly iterative and dynamic phenomenon. Even if student-centredness as a philosophy is applied to classroom teaching, the pedagogy is still based on the fundamental logic of what is called "guided participation". In other words, the role of the teacher is still critical to the success of the lesson. However, the efficacy of the lesson is based largely on "(T)he discursive use of language…between teachers and students (which) is a key aspect of the developmental process" (Mascolo, 2009, p. 15). In other words, how the teacher plans and carries out the lessons will shape how independently the students can develop. The title is set in bold 16-point Arial, justified. The first letter of the title should be capitalised with the rest in lower case. You should leave 22 mm of space above the title and 6 mm after the title.

2.2 Teaching of writing to Engineering and Technology undergraduates

In the present time, we see that student-centred learning has been applied to many Engineering and Technology-based courses that focus on technical knowledge. However, in the Engineering fraternity, there also exists courses that are not technically substantive but yet, revolve around technically substantive content. One example of this subject is Academic Writing. Because of its position as being outside the circle of technical courses, less focus tends to be given to writing (Chong, 2017). This then has brought about a gap in terms of writing-readiness amongst Engineering undergraduates. Jenkins, Jordan, and Weiland (1993) for example, investigated the writing-readiness of postgraduate engineering students in selected engineering faculties in America and found that they were far from ready. One of the main reasons was because their previous training at undergraduate level was insufficient when it came to Academic or Technical Writing. On top of this, language pedagogy lecturers also found that in some cases, the issue was not with writing, but "thinking" (Jenkins et al., 1993, p. 63). This meant that what was initially understood as 'writing challenges' was in fact, a complex form of language inadequacy and cognition building. Thus, classroom pedagogy in some contexts require re-aligning. Yet, when language pedagogy lecturers are given the responsibility to teach writing to Engineering and Technology students, there is a gap in terms of content matter because language lecturers are not Engineering content experts. As such, the question of how we as language practitioners can access Engineering and Technology students' thinking of substantive content as it relates to writing, is at the heart of this study.

2.3 Marginalia and the Electronic Shift

One way in which this idea of writing and thinking can come together is through the use of marginalia. Marginalia has existed from the time writing and printing were made possible. It is defined as notes written in the margins of texts and has been a main source of historical information for hundreds of years (Manguel, 1996). Marginalia is important for tracing both the reading and writing process of the person who apprehends text. In the 21st century, marginalia, traditionally hand-written, has evolved into electronic versions. Electronic books use the concept of marginalia but call them annotations (Wagstaff, 2012).

The issue with the use of marginalia and its electronic shift is with how instructors and even students may not be immediately open to the idea. McKenna (2016) points out that in keeping with technological advancement, the teaching of reading and writing must make use of innovative ways of making sense of print. If marginalia and other reading tools were found in print form, they can also be transferred to electronic forms. McKenna (2016) also points out that the field of STEM has been more accepting of these changes. Thus, the application of electronic marginalia can be usefully applied for STEM students.
Microsoft Word has track-change function that allows the user to insert comments into texts. This is the most basic definition of electronic marginalia. Current literature on electronic marginalia shows that fairly little has been discussed, save for Edwards, Goodwin, O'Connor, and Phoenix (2017) who explain that marginalia and technology can be difficult to implement because "calling up electronic note functions and inserting them into online text is impersonal" (Edwards et al., 2017, p. 162). However, in this current study, the research participants will be communicating their thoughts with their respective lecturers. This current study sets out to explore the potential of using electronic marginalia.

3 Research context

As part of the Scholarship of Teaching Learning (SoTL) initiative, this project was underpinned by principles of action research. Action research assumes that researchers who are themselves teachers/lecturers in the class will carry out an intervention within a research cycle, usually understood as one semester or academic year.

For this study, the research cycle was designed to occur within a 14-week academic semester of first year undergraduates from the Engineering, Technology, Geoscience and Applied Science faculties of a private university in Malaysia. The subject involved in this study was Academic Writing. Two groups of 25 first year undergraduates and two lecturers respectively were involved in this study.

Although Academic Writing has been taught in this university for very many semesters, both lecturers only began to systematically experiment with using electronic marginalia about two semesters prior to the semester described in this study. In this semester, we systematically crafted our respective classes to be conducted through the following format. In the first eight weeks of the semester, face-to-face lectures were carried out. These lectures addressed the fundamental building blocks of Academic Writing which are summarising and paraphrasing as well as citing and referencing. In the remaining six weeks, the lectures were replaced with multiple draft-writing which was done by the individual student. At this point, the students no longer met as a class, but met with the lecturers on a consultative, one-to-one basis. Emails were used as the means by which the softcopy drafts were sent. During these consultation sessions, students receive feedback, continue drafting, receive more feedback and finalise their essay. It is important to note that drafting process in previous semesters involved the use of hard-copy drafts which were submitted drafted using Microsoft Word and then printed and submitted by students. However, lecturers' feedback was handwritten on those printed copies. Often, those feedbacks were brief so as only to flag up the categorical nature of the error. For example, an orthographical error will be represented by "sp" to indicate a spelling error. Little to no further explanations will be given. It was assumed that further explanations will be given during the one-on-one consultation. In this study, however, feedbacks given in the electronic marginalia were done in greater detail. Particularly, there was greater propensity for the lecturers to provide error feedback not just in the categorical sense but through a communicative manner.

4 Findings

Findings of the study are presented through two sub-sections; one sub-section is derived from the lecturers' perspective and while the other is from the students' perspective.

4.1 Lecturers' perspectives

Data from lecturers' perspective is drawn from their reflections, observations and on-going discussions. Although the use of technology in the process of drafting was relatively new in its application, both lecturers found that the use of e-marginalia allowed for corrective feedback to be given to the students in a more systematic way. There were three points that illustrated why the lecturers considered the exercise to be significant. The key points are convenience, tracking ability and relationship-building.

First, both lecturers noted in their reflections that providing feedback using Microsoft's track change infrastructure was a convenient means of guiding the students with their draft writing. This was because all drafts were kept in softcopy version and all feedbacks were located in these softcopy forms. What this meant was the lecturers were not bogged down by hardcopy prints. In an increasingly seamless environment where professional and personal lives co-exist, the lecturers could easily bring their work along with them through a USB port. As long as they had a word processor, they could work on providing timely feedbacks to their students even and especially when they were away from their institution.

Second, both lecturers found Microsoft's tracking ability to be crucial in building a coherent narrative of feedback. This was because Microsoft's track change function provides time stem in terms of the dates in which the comments were made. What made the process systematic was how the corrective feedback was built in a cumulative sense. This is because the corrective feedback process visually occurs on the 'same' sustained platform that captures the lecturer-student communication from the beginning until the end of the draft writing process. Figure 1 illustrates the dates and hierarchical fashion of the dialogue that takes place within the margins of word document. For purposes of confidentiality, the names of the lecturer and student have been removed. The dialogue here can be differentiated according to the colours of the author icon.

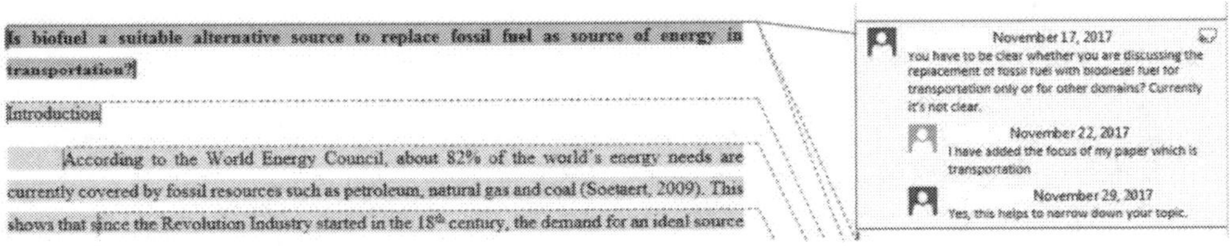

Fig. 1. Dates and hierarchical fashion of dialogue in e-marginalia

Third, the lecturers found that the infrastructure of e-marginalia became a bridge that facilitated in student-lecturer relationship-building. This was because the lecturers could provide communicative feedback through the language functions of questioning, clarifying, affirming and encouraging. The lecturers found that because they were comfortable and more importantly, skilled at using the keyboard to type their queries and comments, the technology afforded by Microsoft Word could optimally facilitate their process of communicating with the students via the electronic medium. Because the lecturers used these language functions in the marginalia, students also modelled after the lecturers and equally responded with language functions of questioning, confirming and informing.

4.2 Students' perspectives

All 50 students in this study organised their draft writing according to the lecturers' requirements. However, the students were differently meticulous with their drafting process. Some were more systematic than others. After the semester ended, the students were asked about their experience of draft writing using electronic marginalia. It is important to note that their response about the use of e-marginalia is with reference to and contrast with their other experience of using Microsoft word for other projects and assignments. All 50 students informed that they had up until then, not been asked to use the track change function for any other project or assignment. The students' responses were gauged through an open-ended questionnaire and semi-structured interviews. This section discusses their emergent response from both research methods. Specifically, their response revolved around three points – ease of technology, detailed feedback and association with social media.

First, all the students who responded to the open-ended questionnaire cited the ease of use and convenience of utilising e-marginalia for their drafting process. Because of their basic ability with using Microsoft Word, the students found the track change function to be relatively easy to handle. Once they could use the technology, they benefitted from its convenience. The sense of convenience stemmed from the elimination of paper printing, being able to type their questions into the softcopy draft and being able to email the drafts to the lecturers without having to walk across the campus to submit hardcopies. They found this to be time-saving.

Second, all the students' response in the open-ended questionnaire indicate that they found the combination of softcopy submissions and one-on-one consultations to be well organised in terms of how they could already view the lecturers' feedback even before they turned up for the face-to-face consultation. The communication and exchange that took place during the face-to-face sessions were important for how the students could clarify the error feedback with the lecturers. This meant that students received detailed and clarified feedback so that they could continue to work on the following draft. The students also felt that the face-to-face component of the draft writing process was crucial to the overall experience of writing. The detailed nature of the received feedback was shaped by the relationship built especially during the consultation sessions. Although written feedback was documented through e-marginalia and could be sent via email, the students overwhelmingly stated that they still required the face-to-face component of the drafting process. This points to the importance of human relationship formation in the pedagogical context.

Third, during a semi-structured interview, one student elaborated that he found the use of e-marginalia to be positive because it reminded him of the communications and exchanges he encounters in social platforms like Facebook® and WhatsApp®. Coming from a generation that is comfortable with dialoguing through speech bubbles, this student was pleasantly surprised by how the draft writing process through e-marginalia resembled his social exchanges with his friends through social media. He eased into this process very quickly and was able to carry out the language functions of questioning, clarifying and explaining in order to improve his draft. Figure 2 below is an example of this student's e-marginalia exchange with the lecturer. The layout of speech bubbles on the right side of the page, juxtaposed with the Academic Writing genre on the left side of the page illustrates how two seemingly incompatible worlds can come together to advance technical knowledge and improve writing skills.

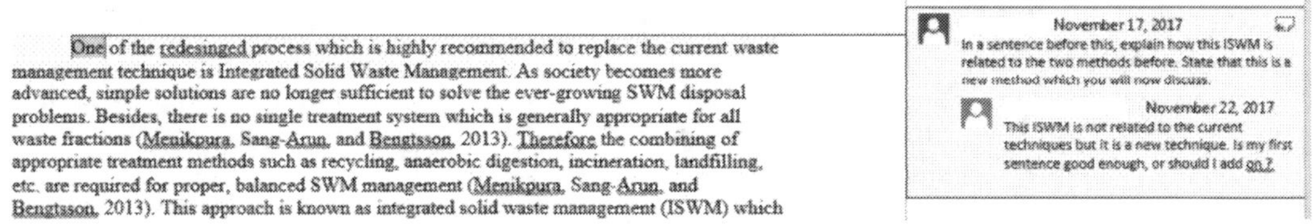
One of the redesinged process which is highly recommended to replace the current waste management technique is Integrated Solid Waste Management. As society becomes more advanced, simple solutions are no longer sufficient to solve the ever-growing SWM disposal problems. Besides, there is no single treatment system which is generally appropriate for all waste fractions (Menikpura, Sang-Arun, and Bengtsson, 2013). Therefore the combining of appropriate treatment methods such as recycling, anaerobic digestion, incineration, landfilling, etc. are required for proper, balanced SWM management (Menikpura, Sang-Arun, and Bengtsson, 2013). This approach is known as integrated solid waste management (ISWM) which

November 17, 2017
In a sentence before this, explain how this ISWM is related to the two methods before. State that this is a new method which you will now discuss.

November 22, 2017
This ISWM is not related to the current techniques but it is a new technique. Is my first sentence good enough, or should I add on?

Fig. 2. Juxtaposition of speech bubbles and Academic Writing genre

5 Analysis and implications of findings

Findings from this study were drawn from the students and lecturers involved in the use of e-marginalia within one semester of Academic Writing. The emergent themes are convenience, tracking ability, relationship-building, ease of technology, detailed feedback and social media. Emergent themes (as discussed in the section above) can be further analysed and understood in relation to overarching categories (Saldaña, 2013). Therefore, from six emergent themes, four emergent categories are proposed as overarching notions that can be abstracted from the data. The four emergent categories therefore, are lifestyle norms, organization norms, technological norms and social norms. Here, the use of 'norms' is significant for how it can account for the habituation process that takes place during the teaching and learning environment. This is because the habituation process forms the foundation of human culture and behaviour. Thus in understanding how the lecturers' and students' feedback is located within human culture and behaviour, we can better appropriate modern teaching and learning methods in our ever-changing classrooms. Table 1 presents emergent themes and corresponding categories.

Table 1. Emergent themes drawn from the participants of this study

Study participants	Emergent themes	Emergent categories
Lecturers	Convenience	Lifestyle norms Organization norms Technological norms Social norms
	Tracking ability	
	Relationship-building	
Students	Ease of technology	
	Detailed feedback	
	Social media	

Findings from this study suggests that even as technological infrastructure expands at remarkable speeds, the application of such infrastructures are experienced in relation to lifestyle and organization norms of those involved in the pedagogical process. Students for example will locate the technological infrastructure within the ambit of what they consider to be acceptable in their day-to-day lives. This is what Miller (2010) refers to as the complexities of human-material relationship where focus should be placed not on the 'stuff' of technology, but the appropriation of stuff through human enculturation. Thus, it is important to understand young people's cultural acceptance especially when engineering educators pledge to locate students at the centre of the pedagogical process.

Further, the question of how language practitioners can access students' thinking of substantive content as it relates to writing can be addressed through the appropriation of technology with technological norms. This refers to the intangible but powerful cultural practice of lecturers and students with regard to their levels of acceptance of technological change (McKenna, 2016). For example, lecturers who are skilled and up-to-date with technological change will find themselves more amenable to new norms of imparting knowledge and skill through the medium of technology (McKenna, Labbo, Kieffer, & Reinking, 2012). In line with McKenna's (2016) contention that the field of STEM has been more accepting of these technological changes, this study encourages lecturers and students to be open to cultural changes as deep-seated human behaviour intersperse with modern technological advancement especially within the field of STEM.

Finally, underpinning these findings lies the role of social norms which form the foundation of human relationship. Even as technology progresses and democratises education and knowledge production, it is the human-social connection that is key to meaningful and valuable exchange (Chong, 2018). Insofar as the technological device can aid the teaching and learning process, the innovativeness of modern teaching methods is mediated by the strength of trust that is built within the student-lecturer relationship. Without this trust, any technological advancement in the 21st century classroom will not be suitably appropriated.

Through the prism of human culture and behaviour, we understand that technological infrastructure can be better capitalised through the cultural appropriation of sustaining human relationships with technological content.

6 Concluding remarks

In confronting the modern educational framework through Education 4.0, educators and policy makers must be prepared for the tension resulting from the intersections that occur when technological infrastructure connects with human culture. Minutiae data from this study suggest that the endeavour of teaching and learning in a subject like Academic Writing is still in need of the communicative, face-to-face connection of human exchange. Particularly, technological advancement must be matched with sociocultural norms of those involved in the educational process. Perhaps to the ultimate question of whether technology will supersede human connection, it appears that human communication is still valued.

This study is funded by Universiti Teknologi PETRONAS' Scholarship of Teaching and Leaning (SoTL) grant (0152AA-A52). We are most grateful to UTP's Centre of Excellence for Teaching and Learning (CeTAL) for the support.

7 References

Bromley, P., Meyer, J. M., & Ramirez, F. O. (2011). Student-Centeredness in Social Science Textbooks, 1970-2008: A Cross-National Study. *Social Forces, 90*(2), 547-570.

Chong, S. L. (2017). *Student-centredness in Engineering Education: Where goes the teacher?* . Paper presented at the 7TH World Engineering Education Forum (WEEF) 2017 (14th - 16th November, Berjaya Hotel, Kuala Lumpur, Malaysia.

Chong, S. L. (2018, 22nd February). Of technology, time and the 21st century, Opinion, *New Straits Times*. Retrieved from https://www.nst.com.my/opinion/columnists/2018/02/337817/technology-time-and-21st-century

Clements, D. H., & Sarama, J. (2016). Math, science and technology in the early grades. *The future of children, 26*(2), 75-94.

Committee on the Engineer of 2020 National Academy of Engineering. (2005). *Educating the Engineer of 2020 : Adapting Engineering Education to the New Century.* . Washington: National Academies Press.

Dejarnette, N. K. (2012). America's children: Providing early exposure to STEM (Science, Technology, Engineering and Math) Initiatives. *Education, 133* (1), 77-84.

Edwards, R., Goodwin, J., O'Connor, H., & Phoenix, A. (Eds.). (2017). *Working with paradata, marginalia and fieldnotes*. Cheltenham, UK: Edward Elgar.

Felder, R. M. (2011). Engineering education: A tale of two paradigms. In B. McCabe, M. Pantazidou & D. Phillips (Eds.), *Shaking the foundations of Geo-engineering education* (pp. 9–14). Leiden: CRC Press.

Jenkins, S., Jordan, M. K., & Weiland, P. O. (1993). The Role of Writing in Graduate Engineering Education: A Survey ofFaculty Beliefs and Practices. *English for Specific Purposes, 12*, 51-67.

Manguel, A. (1996). *A history of reading*. London: HarperCollins.

Mascolo, M. F. (2009). Beyond Student-Centered and Teacher-Centered Pedagogy: Teaching and Learning as Guided Participation. *Pedagogy and the Human Sciences, 1*(1), 3-27.

McKenna, M. C. (2016). Literacy instruction in the brave new world of technology. *The Phi Delta Kappan, 96*(3), 8-13.

McKenna, M. C., Labbo, L. D., Kieffer, R. D., & Reinking, D. (Eds.). (2012). *International Handbook of Literacy and Technology* (Vol. 2). London: Routledge.

Miller, D. (2010). *Stuff.* Cambridge, UK: Polity Press.

Pretz, K. (2016). A Look at the State of Engineering Education Worldwide. *The Institute: The IEEE news source*.

Saldaña, J. (2013). *The coding manual for qualitative researchers* (2nd ed.). London: SAGE.

Wagstaff, L. K. (2012). The Evolution of Marginalia. Retrieved 5 January 2017, from http://www.wkiri.com/slis/wagstaff-libr200-marginalia-1col.pdf

Investigation the effects of comparison between bilingual instruction and the use of L2

Fatemeh Mirzaei*[1], *Muhammad Ridhuan* Tony Lim Bin Abdullah[2], *Zulqarnain* B Abu Bakar[3], *Waseem* Sattar[4]

[1]Management and Humanities Department, Universiti Teknologi PETRONAS, Perak, Malaysia

[2]Management and Humanities Department, Universiti Teknologi PETRONAS, Perak, Malaysia

[3]Management and Humanities Department, Universiti Teknologi PETRONAS, Perak, Malaysia

[4]Management and Humanities Department, Universiti Teknologi PETRONAS, Perak, Malaysia

Abstract. This paper is a comparison of the impact of using bilingual instruction and the use of L2 on learners' learning improvement. The mixed mode is used to determine the use of mother language (L1) as a resource helped the students in learning. The respondents were selected from faculty of Management in UTM, Malaysia. They used rarely English language outside of the school and had average level of English language proficiency. In this paper, it also distinguished resource, difficulties and challenges which the students faced in using bilingual instruction. The questionnaire was used to determine students' perceptions and data was analyzed by SPSS. In addition, the data was collected from interview which is transcribed and described. The results indicated that L1 in UTM can be used as a resource to teach specific subjects to students. According to the results, respondents had positive attitudes about using bilingual instruction in the class and it was effective on students' leaning improvement. Based on findings, it is hoped that positive feedback in this research provide a reason why L1 should be used in teaching and look it as a resource for language learning in the future.

1 Introduction

The use of first language (L1) is an important issue in learning language. There are different views on using L1 in teaching foreign/second language. A number of researchers argue that using L1 can be beneficial to learn foreign language and some disagree. Monolingual's advocates believe that learning foreign/second language should be used in the classroom (Jadallah and Hasan, 2009). Jones (2010) found out that beyond the L2 input essential role in learning second language, the use of the L1 can help the learners to learn L2 better. He mentioned that the teachers are struggling between theory and practicality. According to theory the teacher cannot use L1, in contrast they experienced that using L1 can be valuable tool in function of the L2 and help learners to feel secure.

There are many different viewpoints and arguments in using L1 and L2 coincidently in teaching and learning L2. This study will discuss about those arguments in following sections and will conclude that whether the use of L1 cab helpful in learning ESP beyond learning L2.

2 Literature Review

Most of the applied linguistics, linguistics and psychologists approve the influence of using L1 in second language acquisition. Despite that some agree to use L1 in learning L2 and others believes using L1 can be as hindrance to L2 learning.

*Corresponding author: fatemeh_16005850@utp.edu.my

These different views on the role of the L1 are considered in the concept of language transfer (Faerch and Kasper, 1987). They define it as the process when the students use L1 knowledge in improving or the use of interlanguage..They also note that that process can be positive or negative transfer.In accordance with Ellis (1984) many researchers support using English only in teaching. They believe that using L1is not necessary and also using of it too much cause to prevent appropriate input of L2. With regard to views are mentioned above, advocates of using L1 think that the use of L1 can provide students t to understand better and promote their skills. They believe that it can work as strategy to make students understand comprehensibly (Jadallah and Hasan ,2009)

The reason why some researchers oppose to the use of L1 is related to the Grammar Translation Method (GTM). This method focuses on structure, vocabulary and understanding literature. The rules are very important in GTM and the teachers teach deductively (Chastain, 1988). Harmer (2001) states Students' disability to communicate cause that many researchers concluded that learning target language is boring and without communicative point. Later on, Audiolingual Method and Direct Method avoid using L1 in teaching English language (Jadallah and Hasan, 2009).

2.1 Code Switching

The above two different views discussed on the role of the L1 has influenced teachers' views as to whether the students should use L1 in language classroom. The other phenomenon that is also draws attention due to its reliance on L1 knowing is code mixing and code switching. Code mixing can be defined as situation in which a word or an expression from is language is used in a group of words whose structure belongs to another language. Code switching is when learners use complete sentences from both languages.

Code mixing is a widespread phenomenon in bilingual communities where speakers use their native language (L1) and their second language (L2) in certain different domains (Celik, 2003). One strong view against the practice of code switching in the classroom is expressed by Wong-Filmore (1985). She has questioned the use of L1 in the language classroom. In her three years longitudinal study investigating language learning process, proficiency and acquisition. She concluded that teachers who switched to the L1 were a practice in unsuccessful classroom but never in successful ones. She states the use of the L1 did not aid students in their efforts to learn a new language.

On the other hand, Cook (2001, 2003) is today's leading and influential voice advocating ode switching. He argues that language teachers should break away from the long held tradition that marginalized the use of L1. Proponents of code switching believe that the L1 plays a cognitive role in helping students comprehend and achieve language tasks (Anton and Dilamilla, 1999 cited in Cook, 2001). They also view L1 as a mediating tool when learner works collaborating and cognitively effective strategy to develop language learning. Viakinna-Brinson (2006) study provided insights on the use of code switching as students liked being taught grammar in French/ English ... as it was easier to understand the smaller technical details students with insufficient vocabulary will benefit from such learning and progress foster in language acquisition. The L1 'role as resource and language learning facilitators is imminent besides being a strategy for learners to overcome their limitations in second language acquisition (Ellis, 1985).

In Malaysia English as a second language is used more than using in work places such as in education. Education ministry of Malaysia insists to teach English language in schools. Teachers have to use English language only in the classroom perfectly (Ahmad, H, 2009).

Based on Ahmad,H (2009) Malaysian students' low proficiency level in English language cause the teachers solve this problem. Teachers have concluded that code switching can help them as a tool to provide students to be self-confident in communicating in English. In addition, code switching provides students to understand better. It helps to time saving that it is important issue in teaching. Teachers can use it to clarify difficult points in English language instead of explaining that takes much time. Norrish (1997, cited in Ahmad,H, 2009) claims that teachers use code switching when the levels of the students are low and they tire out to adapt themselves to their levels.

2.2 Arguments on the Use of Bilingual Instruction

The role of the L1 either as a resource or as a hindrance have reviewed in previous sections. In this section, it aims to review the literature of bilingual instruction in second language learning classroom. The L2 have a large amount of knowledge when participate in L2 classroom and want to relate it to L2. Besides, they know the general knowledge and communicating strategy that help them to understand L2 input better (Ellis, 1997). Sapada and Lightbown (1999) considered the awareness of the differences between L1 and L2 can help learners to face difficulties in L2 learning. In addition, Huang (2007) claims that both linguistic students and teachers support the significant role of the L1 in L2 learning. The logical factor behind using English language only in ESL and EFL classroom is "neither conclusive nor pedagogically sound" (Auerbach, 1993). Maarof (2003) pointed out" teachers can reassure the judicious use of the L1 in ESL classroom is acceptable pedagogically and theoretically and can be efficiently utilized to assist students learning L2. Using bilingual instruction during"

In the following sections, the influence of bilingual instruction will be reviewed in two perspectives in second language learning. The role of the bilingual instruction in comprehension and visualization in learning language for specific purposes.

2.3 Bilingual and Language Awareness

Another argument for bilingual instruction comes from the development in the area of language awareness. There have been difficulty views on how grammar is acquire and how best to teach grammar. Chomsky (1979) theorizes that we do not "really learn the language rather the grammar grows in the mind". Then again, there is, however very little clear evidence in support of the claim that practice makes perfect or that structures can be taught (Ellis, 1988) Carder (1967), viewed that language teacher' primary function should be to facilitate the unfolding of what he recognized as the learner' powerful internal syllabus, not to try to impose an external one (cited in Fakharzadeh, 2009). As students become aware of the target language, they start noticing as noticing is related to the question of what is started in memory (Schmit,1993 cited in McGinnis, 2007). Learners may consider both target language and their own language from their own views (Swain, 1998).

Inability to produce the target language will reveal to the learners the 'hole' in one's interlanguage. Thus the learner will then push the limits of the interlanguage system to handle the output by comparing it with the internalized knowledge, the L1 rules. Noticing draws attention to the similarities and differences in form, for both the L1 and L2. Auerbach (1993) insists that the L1 should be used to complement the L2 learning process and to later 'generate attention to L2 production'. This claim is also supported by Jiang and Kuehen' s (2001) study findings on transfer in academic language development amongst low-intermediate ESL students. 62 percent of students in the older group said they thought about their L1 grammar when learning English grammar. In summation, since students refer back to their L1 knowledge to learn the same domains of knowledge in L2 to facilitate learning, using bilingual instructions in an ESL classroom will raise consciousness as students start noticing the L2 form and become aware of its language structures.

2.4 Bilingual and Comprehension

The use of L1 does not mean to use L1 to teach L2. Using L1 provides teachers to explain the difficult points of grammar and make clear the comparison between L1 and L2 equivalent (Jadalla and Hasan,2009). Atkinson states three reason why the use of L1 can be useful in EFL classroom: using L1 cause students to feel secure, different features of L1 and L2 prevent negative transfer and it can be useful in timesaving. He adds that the use of L1 help teachers to correct the comprehension of students and encourage them to cooperate in discussion in the classroom. Furthermore, develop and reinforce the presentation of the students. Harbord (1992 cited in Jadalla and Hassan, 2009) completes Atkinson' reasons. He states that L1 encourage students to participate in task of English language learning. Using L1 is also beneficial to teachers. He claims the teachers can chat with students, correct their errors, explain instruction and clarify the comprehension (Sharma, 2006).

Jiang and Keuhen (2001) advocate using bilingual instruction because it allows positive transfer of skills. They note that the teachers can teach students the ways that students use L2 effectively whether the teacher emphasizes on positive transfer. As a result, in a bilingual environment besides mentioned reasons for using L1, teachers and students can share their similar linguistic and cultural background. "It would be compatible with the general educational principle that learning involves developing new knowledge based on what is familiar" (Seidhofer, 2004. Cook, 2001).

3 Research design

Both quantitative and qualitative data are collected. This study will investigate how bilingual instruction affects students' learning of the specific subjects that are taught in English. It will also distinguish resources, difficulties and challenges the students faced in using bilingual instruction. Data are collected and analysed based on qualitative and quantitative method to explain the perception of using bilingual instruction and its resources and difficulties among management and human resource undergraduate students in UTM. The questionnaire (Maleki, 2005) includes two demography and perceptual sections are distributed among undergraduate students of Management and Human Resources Faculty at UTM. Then the data is analysed by SPSS. The interview data will be transcribed and described. For observing class, filled transcription and description will be used. Two sets of the specialist texts are provided to investigate the comprehension of students and then those texts are described. Data was collected through written survey questionnaire and semi-structured interviews with the participants. The use of these two data collection instruments help validate both the answers in the questionnaire and interviews.

4 Discussion on Research Question 1:

What are the perceptions of students towards using bilingual instruction?

To answer the first research question in this study, survey questionnaires were distributed among undergraduate students of Management and Human Resource Faculty. The aim of the questionnaire was to determine the students' perception toward using bilingual instruction and also to prove the first hypothesis that undergraduate students have positive attitudes toward using bilingual instruction in learning specific subjects.

According to the results of previous section, it is shown that most of the respondents agree with the use of bilingual instruction by lecturer to explain specific vocabularies (Item 2), explain concept of the subject (Item 3), and check the students' comprehension (Item 4). These findings are similar to the results of Ahmad's (2009) research when he discovered that university students had positive attitudes on using code switching in classroom which helps them to feel comfortable in the classroom.

But based on responses of interviewees who preferred the lecturers to use English language which are similar to the results reported by Ndamba (2008). In his research which examined the students and their parents' preferences of using mother tongue or English language in Zimbabwe, he found out that the respondents preferred English use in the classroom. His research respondents defined English language as 'Gateway' which is more close to answers of interviewees in this study. Interviewees in this research mentioned that they preferred to use English language in classroom because it is international language and proficiency in it helps them to find better job in the future.

Regarding to the results of presented study, it was found that the majority of respondents are neutral about using bilingual instruction to give instruction and on the other hand, Table 6 shows that the moderate number of respondents are neutral about presenting subject in bilingual instruction too. It might show that it is not important to them that lecturer' use of bilingual instruction to present subject or give instructions. It might indicate that they do not pay attention to language usage in the classroom. In addition, according to interview all of them said that they understand comprehensibly when the lecturer uses bilingual instruction in presenting and checking their comprehension. They emphasized that in answering the questions in examination, using bilingual instruction help them to understand the concept of test instructions and answer the questions confidently.

Totally, based on findings students agree that lecturers should use bilingual instruction in the classroom. On the other hand, it is needed to know the lecturers' perceptions on using bilingual instruction. A study done by Ford (2009) who carried out a research on teachers' perceptions in EFL context in Japan. He discovered that most of teachers tended to use English language but they mentioned that it depends on students' level and motivation. Teachers in her research stated that sometimes it is needed to use first language to provide relaxed atmosphere. It supported the results of this study which discovered that students preferred to use first language. Using Bahasa Malaysia help them to feel secure and comfortable which is more important point in learning improvement that will be discussed in next section.

According to the results students' use of bilingual instruction, it was found that the majority of respondents 'Agree' with using bilingual instruction. They agree that they should be allowed to use bilingual instruction. More than fifty percent 'Agree' that using bilingual instruction encourage students to be active in the classroom. On the other hand, it was found that the percentages of respondents are almost close together to answer the question 10 which referred to the use of bilingual instruction in class discussion. But it is clear that the percentage of agreement with the use of bilingual instruction is more than disagreement. It shows that the majority of the respondents 'Agree' to use bilingual instruction in class discussion because they can be active and comfortable._

It is supported by Ahmad's (2009) results; he found that there is significant relation between students' participation and using L1. He reported the students are more active and feel comfortable to participate in class discussion when they are permitted to use L1. In addition, Zabrodskaj (2007) investigated the Russian speaking learners' experiences during learning Estonian as foreign language in linguistic course. He discovered that code switching as mentioned in Literature Review of this study is learning facilitator beyond a strategy. He found out that using L1 helps learners to feel comfortable in their presenting in seminars which can be very useful in speaking skills.

Finally, by looking at the results of questionnaires and interview according the first question, the respondents from Faculty of Management and Human Resource's students surveyed lies in their perceptions and view on using bilingual instruction. Most of the respondents viewed English as a must in their field of study. It should be mentioned that most of the respondents from all groups (Management, Human Resource and Marketing) either 'strongly agree' and 'agree' that using Bahasa Malaysia is helpful but equally the respondents believe that there is a need in their classroom to take further attention to English language. The interviewees mentioned that English language as language of their textbooks must be used in the classroom to increase their proficiency in English language for getting job in the future.

4.1 Discussion on Research Question No 2:

How does the use of bilingual instruction promote students' learning of specific subject in English?

Most of the respondents of all groups agree the use of Bahasa Malaysia affect their learning and help them to comprehend the subject in the classroom. They 'Agree' that the English language should be accepted officially academically in UTM to conduct courses but on the other hand the lecturers use Bahasa Malaysia to teach comprehensibly. They were also in favor of having bilingual instruction that promote the use of English in UTM. It is supported by results of Fakharzadeh' (2009) research which proved that using first language or the use of translation in presenting lecture promotes the students' learning. And also, she stated "it can be used as a technique for raising consciousness at different linguistic and discourse.

Levels, including grammar, lexis, rhetoric and genre". As it was found in this study, students "Agree' that using bilingual instruction either in lecturer' presentation or in test instructions are clearer than using English only. Fakharzadeh' (2009) also noted that using first language in teaching motivate students and improve the students' autonomy. The students have self confidence in the classroom that it was found in presented research. The most of the respondents 'Agree' that using bilingual instruction give good feeling in the classroom and are self-confident to participate and be active in class discussion.

It should be mentioned that, there were close percentages between 'Agree' and 'Neutral' option among responses. It shows that the using of bilingual instruction for the moderate of respondents is not as a great matter in learning or comprehension of subject. It may show that using English language is not a challenge for these respondents because it is as a second language in Malaysia and they use it daily. But according to Ahmad (2009), low proficiency level of English language causes teachers to use Bahasa Malaysia though their teaching to explain the concept of subjects. Based on reported interview, two of three of participants agree that using Bahasa Malaysia can be helpful in concept understanding and answer in questions of examinations. But, it was found that it can be related to subjects or grade level of students. Because, based on responses of interviewees, except two or three general or compulsory subjects which are taught in Bahasa Malaysia another one are taught in English. They all agree that presenting in English is more useful than using Bahasa Malaysia because their books are in English. They mentioned that some 'English Terms' are not same in Bahasa Malaysia. In addition, according to interview discussion, the respondents agree that using English language help them to able to use English language and get self-confidence to get job. They believe that English language proficiency as a skill improve their communication and help them to find a good job in the future. And also, they added that using Bahasa Malaysia in brainstorming help more to understand the main concept of the subject.

One of the interviewees emphasized that English language as worldwide language must be used in the classroom especially for specific purposes. He believed that using English language help him to find a good job. He added that bilingual instruction also can help them to comprehend better but it should not be as first option to teach. He noted that using Bahasa Malaysia can be useful for weak students and impress them not to be passive in the classroom. On the other hand, one of them emphasized that because of the English language as original language in textbooks and specific words and terms, they cannot use Bahasa Malaysia most of the time. But using it to make clear the concept of the subject.

As Ndamba (2008) found out in his research most of the respondents (students, their parents and teachers) agree to use only English in learning foreign language. He mentioned that it is great concern for Education Ministry that mother tongue (L1) is ignored in language learning. It may concern in ESL context in UTM and it should be considered specifically.

5 Conclusion

The results of the study firstly indicate that the mother tongue (Bahasa Malaysia) can be used as a resource for teaching specific subject especially among ESP students. With the use of bilingual instruction students feel comfortable to comprehend the subjects and answer the questions of examinations confidently. They can focus on concept of subjects without being worry about level of English language proficiency. Although, most of them generally know English language because English is as second language in Malaysia and the people use English as daily language.

Secondly, the results of this study strongly confirmed the second hypothesis based upon using bilingual instruction influences students' learning significantly. According to responses of interview and questionnaires, it was discovered that with the use of bilingual instruction the students focus on concepts of subjects and reduce their stress during learning process. Interviewees agree that using bilingual instruction help weak students to improve in learning English beyond learning their subjects. Also, the use of L1 influences the interlanguage development and helps the students' language ability to improve in the target language learning much quicker. The use of L1 as a resource in ESL context may be a workable method especially for ESP learners which teachers should investigate on.

6 Implication of the Study

The results of this research showed that the majority of respondents surveyed indicated they had positive view on using bilingual instruction in the classroom. Most of the respondents agree that using bilingual instruction help and facilitate learning process. The faculties involved should consider coming up with syllabus based on the English for the Specific Purpose (ESP) approach to teaching the Marketing, Management and Human Resource. As the underlying aspects of this approach is based on the consideration of students' needs and a high motivational level to learn, the results of this study showed both elements were presented in the faculty of Management and Human Resource. It was mentioned in chapter two that the learner or student is center of ESP learning. It can be concluded that learners' needs should be considered as main part for teaching specific subjects that they prefer to use L1 and English language concurrently.

7 Recommendations

This research proceeded on students' perception which is as small scale in bilingual instruction. It is highly recommended that firstly using L1 should be taken to account to L1(Bahasa Malaysia) more in presenting lectures which improves students learning, based on findings of this study using L1 as resource help students to be comfortable and feel secure which is useful in learning. Next, future research can see the concept of subject clearly should be thinking aloud in L1. It shows that it. Thirdly, results in this research showed that how using Bahasa Malaysia influences learning improvement so, it can be recommended that technical Malay Standard Bahasa Malaysia should be taught in L1. Then, the number of subjects or sample of the study need to be increased to at least thirty percent of the entire intended population in order to get more findings. Finally, the relation between bilingual instruction with other factors such as awareness, imagination which can be more influenced in learning.

This work was funded by the YUTP grant cost centre no. 0153AA-E59. The authors would like to thank the reviewers for their valuable comments to improve this paper.

8 References

Ahmad, H. (2009). Teacher's Code Switching in classroom Instruction for Low English Proficient Learners. 2, (2).

Atkinson, D. (1987) The mother tongue in the classroom: a neglected resource?, *ELT Journal*, 41(4): 241-247.

Auerbach, E. R. (1993). Reexamining English only in the ELT classroom. *TESOL Quarterly*, 27(1), 29-32

Chastain, K. (1988). Developing Second Language Skills. Harcourt Brace Jovanovich. Inc.

Chomsky, N. (1979). Language and responsibility. New York: St Martin's Press.

Celik, M. (2003). Teaching Vocabulary Through Code- mixing *ELT Journal*, (57) 4: 361–369, October 2003. https://doi.org/10.1093/elt/57.4.361

Cook, V. J. (2001) Using the first language in the classroom. Carddian Modern Language Review, 57: 402-423.

Cook, V. J. (2003). Effects of the Second Language On the First. Tawnawanda, NY: Multilingual Matters Ltd.

Ellis, R. (1997). Second Language Acquisition. Oxford: oxford University Press.

Ellis, R. (1988). The role of practice in classroom language: learning Teanga 8: 1-28

Ellis, R. (1985). Understanding Second Language Acquisition. Oxford: Oxford University Press.

Ellis, R. (1984). Second Language Acquisition and Language Pedagogy Clevedon: Mulitilingual Matters.

Faerch, C. and Kasper, G. (1987). Introspection in Second Language Research Clevelind, OH: Multilingual Matters Ltd.

Fakharzadeh, M. (2009). Why's of Pro-First Language Use Arguments in ESP Context. PhD University of Isfahan, Iran.

Ford, K. (2009). Principles and Practices of L1/L2 Use in the Japanese University EFL Classroom. *JALT Journal*, 31 (1).

Harmer, J. (1991). The practice of English Language Teaching. London: Longman.

Huang, L. (2007). Strategies in ELL Literacy Development: effective Strategies for Teachers in english Language Learners' Literacy Development. Master' s Thesis. South Minnesota State University. Minnesota.

Jadallah, M. and Hasan, F. (2009). *A Review of Some New Trends in Using L1 in the EFL Classroom.* Al-Quds Open University.

Jiang, B. and Kuehen, P. (2001). Transfer in the academic language development of post-secondary ESL students. Bilingual Research Journal. 25(4).

Jones, H. (2010). *First Language Communication in the Second Language Classroom:* A Valuable or Damaging Resource? Education 6390 Memorial University of Newfoundland.

Lightbown, P. and Spada, N. (1999). *How Language are Learnt.* Oxford: oxford University Press.

Maarof, M. (2003). University the Native Language or L1 to comprehend L2 texts. In International Conference on Language and Nationhood. Putrajaya: *Intensive of Malaysian and International studies*, UKM.

Maleki, A. (2005). *ESP Teaching: A Matter of Controversy.* Preceedings of the First National ESP/EAP Conference (vol. 1). SAMT, Tehran, PP.169-178.

Ndamba, G. T. (2008). Mother Tongue Usage in Learning: An Examination of Language Preferences in Zimbabwe. *The Journal of Pan African Studies,* 2 (4).

Norrish, J. (1997). English or english? Attitudes local varieties and English language teaching. In TESL_ E (3) 1. Retrieved September 5. 2005, www. Writing. Berkerley. Edu/ TESL-EL/ ej09/ a2. Html.

Sharma, D. (2006). The Art of Communication. The Journal of English Language Teaching (India), XLIV/5, 41-43.

Seidlhofer, A. B. (2004), 'Research perspectives on teaching English as a lingua franca', Annual Review of Applied Linguistics, 24, 209–239.

Sapada. N and P. M. Lightbown, (1999). Instruction and the development of questions in L2 classrooms. Studies in Second Language Acquisition. 15, 2015-24.

Swain, M. (1998). *From on Form through Concouis Reflecyion.* In C. Doughty and J. Williams (Eds). Focus on Form in Classroom Second Language Acquisition. New York: Cambridge University Press.

Viakinnour-Brinson, L. (2006). To Teach or Not To Teach IN The Target Language Only? The effect of target language only and code-switching on the Grammatical Performance of Elementary Level French Students. *PhD thesis. Graduate School of Emory University.*

Wong-Filmore, L. (1985), *When Does Teacher Talk Was Input?* In S.M.Gass and C.G, madden (Eds), Input in Second Language Acquisition. Rowley. MA: Newbeny.

Zabrodskaja A. (2007). "Russian-Estonian code-switching in the university", SLA & Teaching, Vol. 14, pp. 123–139.

Continuous Quality Improvement for Calculus Using Mentor-Mentee Approach

Samsul Ariffin Abdul Karim[1, *], *Mohd Mughti* Hasni[2], and *Nooraini* Zainuddin[3]

[1,3]Fundamental Applied Sciences Department and Centre for Smart Grid Energy Research (CSMER), Institute of Autonomous System, Universiti Teknologi PETRONAS, Bandar Seri Iskandar, 32610 Seri Iskandar, Perak Darul Ridzuan, Malaysia.
[1]Center for Smart Grid Energy Research, Institute of Autonomous System, Universiti Teknologi PETRONAS, Bandar Seri Iskandar, 32610 Seri Iskandar, Perak Darul Ridzuan, Malaysia.
[2]Centre for Foundation Studies, Universiti Teknologi PETRONAS, Bandar Seri Iskandar, 32610 Seri Iskandar, Perak Darul Ridzuan, Malaysia.

Abstract. This paper proposes mentor-mentee event as a method to improve the performance of the students in Calculus subject for Foundation Third Semester in Universiti Teknologi PETRONAS (UTP). About 40 mentees are selected based on their course work (CW) marks and 20 mentors are chosen from UTP undergraduate students (UG). All the mentees must complete 12 hours within two-days event. In this event, active learning (AL) and cooperative learning (CL) are utilized as a learning method, this includes brainstorming, think pair share, think problem solving (TPS), pair work, turn to neighbor summaries, and etcetera. All the outcomes are analyzed through statistical measurements. From the results, it indicates that, Mentor-Mentee Event has improved significantly the performance of the mentees in the final exam.

Keywords: Continuous quality improvement (CQI), Mentor-Mentee, Active learning, Cooperative learning, Statistical analysis

1 Introduction

Calculus is a compulsory subject that must be taken by all Engineering and Science Foundation students in the university being studied. Based on the performance from previous semester, around 5% students have some difficulties in Calculus. It must be understood that students must have good basic in Calculus to excel in Differential Equation that will be offered in Year 1 Semester 1. A Pilot Project was tested in March 2016 Semester and July 2016 Semester where 40 Mentees and 20 Mentors were involved in both programs. Based on the Final Examination results for July 2016, 39 students passed the course except one pass after taking Supplementary Examination. The main outcome from the Mentor-Mentee Program indicates that this program can assist low achievement students to improve their results. In this paper, we discuss the development of a framework for Effective Mentor-Mentee Program to produce High Impact Mentor-Mentee Program.

As mentioned, it is compulsory for Engineering and Sciences as well as Technology Foundation students to pass Pre-Calculus in Semester I. Under this system, once they passed, Business Information System (BIS) and Information Communication Technology (ICT) students can register for the Calculus and Linear Programming while Engineering and Sciences students take Calculus in Semester II. Unfortunately, there are students who have some difficulties in some topics such as Limit and Integration in Calculus. How do we assist those students? Do we need to give some extra class to them? Do we need to innovate our teaching method? Maybe using some animation software can improve their basic understanding? There are many ways that we may assist the students. One of the ways is implementing Mentor-Mentee Program by selecting Mentees and suitable Mentor. Continuous Quality Improvement (CQI) is important for us to improve certain area not just in our teaching and learning but in the assessment of the courses. (Foster, 2011) indicated that youth development approaches i.e. mentoring, result in improvement behavior changes in interpersonal skills as well as the way they learn. Dittmar and McCracken (2012) proposed Mentoring, Engagement, Technology and Assessment Model or META Model to improve the e-learning for online teaching. Leidenfrost, Strassing, Schutz and Carbon (2014) studied the impact of peer mentoring style among all mentors. They found that any mentoring style can improve the performance of the mentees. Slack, Johnson, Dodor and Woods (2013) studied the mentoring for at risk middle school students. They conducted the survey involving 59 mentees and 43 mentors. Zales (1996) studied the active learning as a tool to improve student achievement in Mathematics. Adamu (2014) analyzed the relationship between Mathematics Anxiety and Calculus Achievement among 148 engineering students at Waziri Umaru Federal Polytechnic, Birnin, Kebbi, Nigeria.

* Corresponding author: samsul_ariffin@utp.edu.my

The main finding is the females' students has high anxiety than males' students. Meanwhile Hossain, Tarmizi and Ayub (2012) studied the cooperative and cooperative learning in Malaysian mathematics education. They found that by working in small groups, the students can improve their academic performance while promoting their interpersonal competencies through both types of learning. Tall (1993) discussed the students' difficulties in Calculus by given many types of the difficulties. For instance, in limits and infinite processes, among others.

Based on previous studies, active learning and mentoring can be used to improve the performance of the low achievement students. Thus, in this study, Mentor-Mentee Program will be used to improve the performance of the students incorporated with active learning as well as cooperative learning. Furthermore, the study does not focus on gender as discussed by Adamu (2014) but focus on students who have lower coursework (CW) marks in Calculus.

2 Methodology

This project is divided into several stages:

2.1 Phase 1:

The main objective of this stage is to answer research question number 1, which is, what topics in Calculus that the students (mentees) have most difficulties? Coursework (CW) breakdown will be carefully analyzed. In Calculus there are 5 quizzes; Quiz 1 and Quiz 2 – Differentiation, Quiz 3 and Quiz 4 – Integration and Quiz 5 – Series. Meanwhile Test 1– Differentiation and Test 2 – Integration. Any students who scores less than 27% out of 50% will be selected to be Mentee. There are also students also were selected as Mentees because their score is low for Test 1 or Test 2 even though their CW is more than 27%. This is to assist students that need help to improve their basic and concept in Calculus. Mentors are required to attend Training of Mentor (TOM) before they can be appointed as Mentor. For every semester, there are 40 mentees from Foundation and 20 mentors from undergraduates (UG). The ratio is 1 Mentor for 2 Mentee.

2.2 Phase 2:

The objective of Phase 2 is to determine whether the Mentor-Mentee Program to assist Mentees improve their main basic and concept in Calculus. The Mentor-Mentee Program was conducted on Week 12.

During Study Week, Mentees must undergo minimum 12 hours coaching sessions with their respective Mentor. Each Session is divided into several different topics. For instance, Session 1, they discussed Basic Differentiation and Its Applications. In 3 hours, Mentee will learn how to get better understanding in this topic. Active learning (AL) and cooperative learning (CL) were carried out throughout Mentoring and coaching session. The Mentees worked in pairs, where we utilized the think pair share, pair testing as well as reflection and at the end of the session, Mentees must be able to answer few questions to test whether they really understand or not the topics that have been discussed throughout each session. Mentees must submit their complete solution to their Mentor. Mentor marks the questions and gives appropriate feedbacks to Mentees within one day. Reflection will be given at the end of each session. They may meet again for further discussion on the matter. Based on our previous Mentor-Mentee event, it was noticed that Mentees were quite happy and can follow what the Mentors teach, as well as they discussed actively in a small group and as a pair. Think Pair Share, Pair Testing as well as Reflection provide the Mentees better understanding about difficult topics.

2.3 Phase 3:

This final phase looks at research question number 3. Can Mentor-Mentee Program improve mentees performance in Calculus final exam? At the end of the Program, mentees as well as mentor are required to give feedback of the Mentor-Mentee Program through questionnaire. Mentees final exam results and outcome from the questionnaire will be analyze in detailed through statistical and R software to determine the effectiveness of Mentor-Mentee Program.

2.4 Data Collection

The Mentor-Mentee Event are organized for two semesters, that is, March 2017 and July 2017 for Calculus subject offered in Semester 3 Foundation students. The deliberation of the mentor is made as follows: 1 Mentor for 2 Mentees. The (M2E) event is proceed as follows:

3 Data Analysis

Table 1. Students Grade (March 2017 & July 2017)

STUDENTS GRADE (COURSEWORK, FINAL AND TOTAL GRADE)						
GRADE	COURSEWORK (March 2017)	FINAL EXAM (March 2017)	CW+FE (March 2017)	COURSEWORK (July 2017)	FINAL EXAM (July 2017)	CW+FE (July 2017)
A	0	0	0	0	0	0
A-	0	0	0	0	0	0
B+	0	0	0	0	1	0
B	5	1	1	1	1	2
C+	1	6	7	8	7	5
C	6	1	3	2	2	4
D+	5	3	2	3	3	3
D	2	2	4	3	1	2
F	2	8	4	0	2	1
TOTAL	21	21	21	17	17	17

Table 1 above shows the student grade for March 2017. The total number of 21 students have been separated based on their grade. From CW grade, the highest grade that the students obtained was only B which is approximately only 23.81% from 21 students. This was followed by grade C+ which is 4.76%. Next, C where most of the students obtained C for their CW grade which contributed around 28.57%. For D+, D and F, they contributed 42.86% out of 21 students. For their FE, the highest grade that the students achieved was B which is the same as CW but with less students. Next would be C+ which is almost 28.57%. For the grade D+, D and F, it would be about 61.9%. For the grade D+, D and F, 61.9% is slightly higher as compared to CW which contributed only 42.86%. Now, if we look at the total FE grade, the total percentage of students got D+, D and F were about 47.62%. If we look at the median for CW in March 2017, the grade will be C. In their FE, the median would be D+. For the total final grade (CW + FE) in March 2017, the median would be C+. Notice that the median grade for the final grade which is total (CW+FE) is higher than the median grade for CW. By using the Mentor-Mentee approach, we managed to help the students be more prepared for their FE thus improving the total final (CW + FE) grade.

For July 2017 in Table 1, the results showed that the performance of students were getting slightly better. Using mentors to help mentee for the FE preparation helped the students to at least maintained their grade. Using grade median as our guideline for July 2017, we could see that the median grade for CW would be C+. The median grade for FE also managed to be C+. At the very least, the grade result has been maintained during CW and their FE.

Table 2. Students Coursework and Final Exam (July 2016, March 2017 & July 2017)

GRADE	COURSEWORK			FINAL EXAM			TOTAL (CW + FE)		
	Jul-16	Mar-17	Jul-17	Jul-16	Mar-17	Jul-17	Jul-16	Mar-17	Jul-17
A	0	0	0	4	0	0	0	0	0
A-	0	0	0	4	0	0	0	0	0
B+	1	0	0	4	0	1	0	0	0
B	1	5	1	5	1	1	8	1	2
C+	9	1	8	5	6	7	12	7	5
C	1	6	2	2	1	2	3	3	4
D+	7	5	3	1	3	3	1	2	3
D	5	2	3	0	2	1	2	4	2
F	6	2	0	5	8	2	4	4	1
TOTAL STUDENT	30	21	17	30	21	17	30	21	17

Table 2 shows the students CW, FE and the total (CW + FE) for July 2016, March 2017 and July 2017. For July 16, the median grade for CW, FE (CW+FE) was D+, B and C+ respectively. For July 16, the performance in the FE have increased significantly as compared to the median grade of CW. If we compared with March 17 and July 17 median grade, the performance was much better since during their 12 weeks' programs studying Calculus, the implementation of Mentor Mentee started earlier after Test 1 as compared to March 17 and July 17 which started only after Test 2. However, every batch still must complete a minimum of 12 hours Mentor-Mentee.

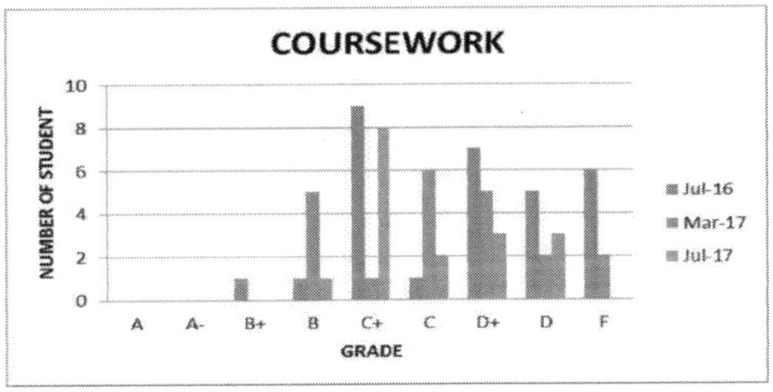

Fig. 1. Students Coursework (July 2016, March 2017 & July 2017)

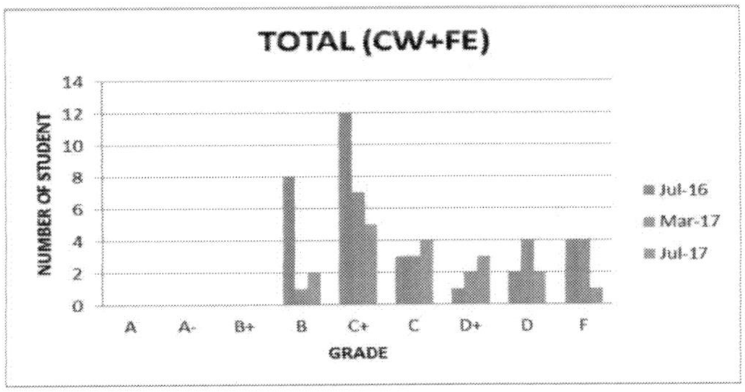

Fig. 2. Students Final Grade (July 2016, March 2017 & July 2017)

Fig. 1 shows the comparison graph of students CW grade of different batch which is July 16, March 17 and July 17 Different color would indicate different batch. For July 16 and July 17, most students would score C+ for their CW. For March 17, it would be C. From Fig 2, it gave us the comparison graph of students' total (CW+FE) grade of different batch which is July 16, March 17 and July 17. Most of the student's grade belong in C+. The comparison grade between their CW and total (CW + FE) shows that the result managed to be maintained and with some slight improvements for March 2017 where the final grade has improved from C to C+.

4 Conclusion

Thus far, the mentor-mentee approach has been implemented throughout three semesters namely July 16, March 17 and July 17. From the results as well as the descriptive statistics, it can be shown that the Mentor-Mentee Event has managed to help students to slightly increase their final grade by looking the median from the grades. We intend to complete the study by testing the hypothesis based on four semesters i.e. July 2016, March 2017, September 2017 and March 2018. This can be achieved by using ANOVA and chi-square tests.

Acknowledgement to Universiti Teknologi PETRONAS (UTP) for the financial support received in the form of research grant: Scholarship of Teaching & Learning (SoTL) **No. 0152AA-A67.**

5 References

Adamu, G.A. (2014). Mathematics Anxiety Among Engineering Students and Its Relationship with Achievement In Calculus. *International Journal of Psychology and Counselling*, *6*(1), pp. 10-13.

Dittmar, E. & McCracken, H. (2012). Promoting Continuously Quality Improvement in Online Teaching: The Meta Model. *Journal of Asynchronous Learning Networks*, *16*(2), pp. 163-175.

Foster, L. (2011). Effectiveness of Mentor Programs-Review of the Literature from 1995 to 2000. *California Research Bureau*. ISBN 1-58073-136-1.

Hossain, Md. A.H., Tarmizi, R.A. & Ayub, A.F.M. (2012). Collaborative and Cooperative Learning in Malaysian Mathematics Education. *IndoMS J.M.E.*, *3*(2), pp. 103-114.

Leidenfrost, B., Strassing, B., Schutz, M. & Carbon, C.-C. (2014). The Impact of Peer Mentoring on Mentee Academic Performance: Is Any Mentoring Style Better than no Mentoring at All? *International Journal of Teaching and Learning in Higher Education*, *26*(1), pp. 102-111.

Slack, R.S., Johnson, C.A., Dodor, B. & Woods, B. (2013). Mentoring "At Risk" Middle School Students: Strategies for Effective Practice. *North Carolina Middle School Association Journal*, *27*(1), pp. 11.

Tall, D. (1993). Students' Difficulties in Calculus. Proceedings of Working Group 3 on Students'Difficulties in Calculus, *ICME-7* 1992, Quebec, Canada, pp. 13-28, ISBN 2 920916-23-8.

Zales, C. R. (1996). Improving Student Achievement in Mathematics Through Active Learning. *In Electronic Proceedings of Conferences on the Teaching of Mathematics*, Bogacki, P. et al. (Eds.), pp. 1-7.

The Mediating Effects of Quality Learning on the Overall Learning Experience and Learning Outcomes of STEM Malaysian Students

S. Chee Choy , *Joanne Sau Ching* Yim*, and *Poh Leong* Tan

Centre for Educational Psychology and Teaching Strategies, Tunku Abdul Rahman University College, Perak Branch Campus, Kampar, Malaysia

Abstract This study examines the mediating effects of quality learning on the overall learning experience, and learning outcomes for a sample of 1180 Malaysian STEM students. Data were collected using the Quality Assurance Perception Questionnaire for Students (QAPQ-S), a self-reported questionnaire designed to measure students' perceptions of learning quality in higher education. The contribution of this study is the empirical examination of the main exogenous variables in greater depth using SEM-PLS. The research fills a gap in the literature through its analysis of the mediating role of quality learning. The results show that quality learning is a significant mediator of learning skills, teacher support and curriculum in the achievement of learning outcomes. Quality learning fully mediates learning skills and learning outcomes while partially mediating the other relationships. Limitation and implications for practice are also discussed.

Quality learning, learning experience, learning outcomes, mediating effects

1 Introduction

In the Malaysian Education Blueprint (2015-2025) for Higher Education (MEBHE) (Ministry of Education Malaysia, 2015), one of the aims for Malaysia is to raise the nation's higher education standards to one that is not only known for good value for money and quality of life but one that is also recognised, referred to, and respected internationally for its academic and research expertise. In order to achieve this aim, the MEBHE states six aspirations for students one which is for mastery and appreciation of Science, Technology, Engineering and Mathematics (STEM) subjects. Further to this, the recently launched MEBHE also articulated the need for holistic and balanced graduates with entrepreneurial minds suggesting that rethinking is necessary in making these shifts a reality and new ideas and models of learning are certainly needed so as to bring these aspirations into reality.

* Corresponding author: yimsc@taruc.edu.my

This aim is further articulated by the Malaysian Ministry of Higher Education setting new initiatives as part of its effort to cultivate holistic, entrepreneurial and balanced graduates to be globally competitive and meet the needs of Industry 4.0. Universities are focused on personalised learning outcome, defined as demonstrable and achieved learning that are aligned with a set of goals (Biggs & Tang, 2011) as well as pursuing technologies and innovations on campus that addresses students' learning needs, in short, quality learning (QL).

Quality in design and delivery of courses will determine the QL of an institution in terms of student performance and evaluation especially if these are based on the needs of students (Langstrand, Cronemyr & Poksinska, 2015). Although the idea is inherently workable, teachers may not be able to or know how to help students reflect on their learning situations because these teachers are more focused on attaining results that satisfy the standards set by the management and achieving their own key performance indexes (Anderson, 2006). QL is often viewed by universities as a concept of "standards" where the level of conditions must be met for the institution to be accredited by an accreditation agency, hence "benchmarking" established "standards" to other equivalent institutions where needed (Elassy, 2015). Hake (1998) and Handlesman (2004) noted that students using interactive-engagement methods designed to promote conceptual understanding through interactive engagement with peers and teachers resulted in the most improvement in performance of their STEM courses. Studies have also found that a supportive learning environment and perceived self-efficacy can significantly influence students' perceptions of course difficulty in STEM courses (Whalen & Shelley, 2010) and eventually their success in their chosen programmes. Instructor behaviour that are positive like confirmation, clarity and communicating to students that they are acknowledged as valuable, significantly influences student learning outcomes (Goodboy & Myers, 2008). Further to this, it was found that teachers' communication style and their availability for students encouraged more engagement when learning (Mottet et al., 2005) and development of positive attitudes (Christophel, 1990).

There have been many studies on STEM students in terms of their engagement and customer orientation, there are few on the mediating effects of QL on learning outcomes, hence this study will address the gap in the research. Studies have shown that the overall learning experience in a university comprising of the learning skills used by students to enhance their learning process (Hattie, Biggs & Purdie, 1996), delivery of lesson and the support of teachers (Goodboy & Bolkan, 2009, Samudra et al., 2016), the infrastructure and facilities of a university (Nicholson, 2011; Tagg, 2010), the teaching and assessment procedures of a university (Lizzio, Wilson & Simons, 2002, Entwistle & Peterson, 2004), the curriculum of the university (Biggs & Tang, 2011) and learning outcomes (Biggs & Tang, 2011) is mediated by the support structure provided by a university to help students with self-management and development, namely the QL in a university (Tagg, 2010, Biggs & Tang, 2011).

2 Literature review and research hypotheses

The conceptual framework for this study will consist of five exogenous variables (Learning Skills, Delivery and Support, Resources, Learning Environment and Curriculum), one endogenous variable (Learning Outcome) which are mediated by QL. A discussion of each variable with definition and relevant supporting literature follows.

2.1 Learning outcome and quality learning

All students have to use regulatory processes to help them achieve learning outcomes and will use specific strategies to achieve specific goals (Girash, 2014). Biggs and Tang (2011) suggested that achievement of learning outcomes are dependent on good teaching which is a function of an institutional-wide infrastructure and policies and procedures that ensures this must be put in place. Research has found that students who were less knowledgeable of course content benefitted from teacher initiated instructional prompts in the form of cognitive and metacognitive strategies like organisation skills, control of cognitive processes at the initial stages of learning (Lee & Kalyuga, 2014).This gives sufficient evidence to suggest that the type of guidance and support that teachers and the QL provided by a university must take into account students' cognitive characteristics to help them achieve their learning outcomes which are demonstrable and can be mapped and aligned with a set of goals (Biggs & Tang, 2011). The definition of learning outcome is the achieved learning by students that can be mapped and aligned with a set of goals.

Biggs and Tang (2011) also noted that the learning environment in the form of the infrastructure of an institution and the support that is provided to teachers and students will determine the overall QL. The focus for universities is to personalise learning outcomes as well as meet the needs of students using available technologies and innovations, in short, providing these students with QL (Kobayashi, 2005; Lee & Kalyuga, 2014). A supportive learning environment that is enhanced with QL, with appropriate teacher affirmations can significantly influence students' perceptions of course difficulty and eventually graduation rates from their programmes (Whalen & Shelley, 2010; Bolkan & Goodboy, 2011). In this study, QL is defined as the support structure for personal attribute development provided to students that will eventually allow them to develop the confidence to make decisions about their own learning that are consistent with the policies and procedures of a university and within an organised learning environment (Biggs & Tang, 2011)

2.2 Delivery and support

Teachers who behave inappropriately in terms of the way they manage the subject area taught, and exhibiting overall rudeness to students can significantly influence learning outcomes and the overall QL (Goodboy & Bolkan, 2009). According to Duschl and Hamilton (2010) the relationship between students and teachers is an interpersonal one that involves social influence and becomes more enhanced when they use appropriate communication behaviours (Goodboy & Bolkan, 2009). Kunter et al. (2013) noted that the QL for students increased when teachers were motivated to teach, provided more cognitively activating instructions and teaching support. Hence the following hypothesis:

H1. QL mediates the relationship between Delivery and Support and Learning Outcomes

Delivery and support is defined as the quality of teaching and the learning support provided by teachers to their students in the form of appropriate feedback and task learning strategies (Goodboy & Bolkan, 2009).

2.3 Learning skills

The results of many studies have shown that students use their learning skills when they are accompanied with appropriate and elaborate feedback which helps correct inappropriate task strategies and misconceptions (Harks et al, 2014). Hattie, Biggs and Purdie (1996) further notes that learning skills must be in the context of students' learning where there is a high degree of learner activity and metacognitive awareness focused on the self-management of learning. The development of learning skills in students involving feedback provided by the teacher on students' performance has been shown to be a powerful influence on student learning (Hattie & Timperley, 2007). Students with poor learning skills and beliefs are not able to achieve or do well in university settings (Hattie & Gan, 2011). Such learning skills and beliefs have been found to evolve over time and students given QL, can succeed (Ambrose & Lovett, 2014) in terms of achieving their learning outcomes. However, the mediating effect of QL on learning skills and learning outcome have received little investigation and hence the hypothesis:

H2. QL mediates the relationship between Learning Skills and Learning Outcomes

2.4 Resources

The most important ways of improving teaching is to recognise that teaching is a function of institutional resources (Biggs & Tang, 2011). Hence, policies and procedures that ensure and encourage good teaching and assessment across a university need to focus on students. Resources in a university must be put in place together with quality management to ensure that students can experience overall QL resulting in achievement of learning outcomes. In this study, resources are the infrastructure and facilities provided by a university in terms of quality of teaching staff, volumes available in the library and quality and type of equipment available to students (Nicholson, 2011) to enrich their overall QL.

H3. QL mediates the relationship between Resources and Learning Outcomes

2.5 Learning environment

Learning is interactional and underscore the dynamic quality of learning in which the characteristics of students, the learning environment, the timing and nature of the learning come together to influence what is learned (Duschl & Hamilton, 2010). In short when students are provided learning environments that support QL, they tend to achieve their learning outcomes. Ennis and Chen (2011) also note that when students learn in an environment that supports the complexities associated with learning, they are more likely to understand that process. Choy, Yim and Tan (2017) found that there was a significant relationship between learning environment of a university which they defined as the overall learning climate will support students' achievement of learning outcomes. The available research suggests the following hypothesis:

H4. QL mediates the relationship between Learning Environment and Learning Outcomes

2.6 Curriculum

According to Tyler (2013), a well-designed curriculum must contain educational goals to be achieved and learning experiences which are well organised as well as attainable learning outcomes. Learning experiences are the interactions of students with external conditions, that is QL, in the learning environment to which they can react. Doll (1993) notes that a curriculum that allows for student transformation, encourages the development of their natural capacity for complex organisation, hence the teachers' role is transformative and not causal, focusing on the process of learning. In this study, curriculum is defined as the arrangement and structure of courses that are for a specified duration to achieve stated learning outcomes and provides a 'map' to indicate what students can do after teaching (Biggs & Tang 2011). The available research suggested the following hypothesis:

H_5: QL mediates the relationship between Curriculum and Learning Outcomes

The conceptual framework of the mediating effects of QL on students' perspective of learning consisting of five exogenous variables and the endogenous variable, learning outcome is shown in Figure 1.

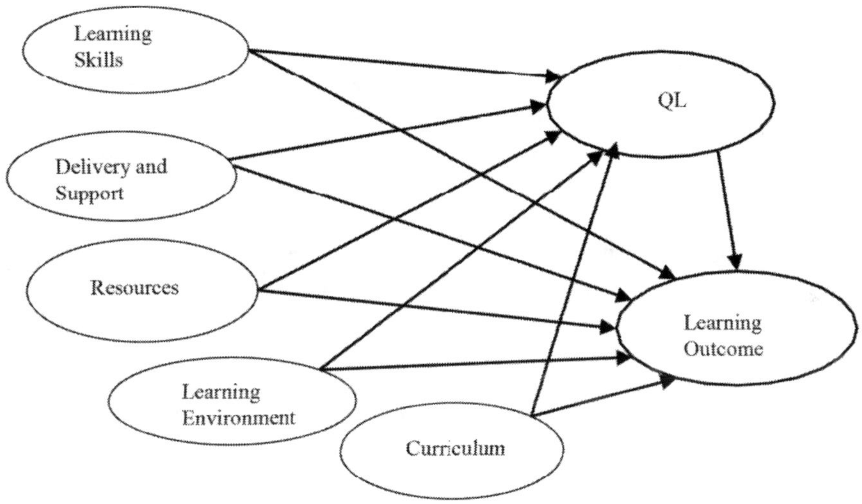

Fig. 1. Conceptual Framework of the Mediating Effects of QL on Student Perspective of Learning and Learning Outcomes

3 Method

3.1 Sampling and data collection

A total of 1043 students consisting of 739 males and 304 females with a mean age of 20 (SD = 1.361) enrolled in bachelor and diploma programmes at a university in Malaysia participated in this study. The samples were from one faculty in the University. The participants were assured their anonymity and confidentiality of information provided. They were also informed of the nature of the study and were told they could withdraw from the study anytime during or after the study.

3.1 Measurement scales

The study's main constructs were measured using items adapted from of the Quality Assurance Perception Questionnaire – Student (QAPQ-S) (Choy, Yim & Tan 2017), a questionnaire designed to measure the following areas: learning outcomes, curriculum design and delivery; student support, learning environment, academic staff, educational resources, and learning quality.

3.2 Limitation of the study

It must be noted that the data were collected from the social science faculty of only one university in Malaysia. Although the sample collected provided an acceptable quality in terms of outer loadings of 0.70 (Hair et al, 2017), it is still a self-reported instrument and hence the truthfulness of the respondents cannot be assured.

3.3 Data analysis

The present study employed the non-parametric Partial Least Squares (PLS) SEM approach to analyse data because the study is exploratory in nature. The model specified in this study has seven constructs with reflective measurements and the path model was estimated using SMART-PLS 3.0. Table 1 shows the internal consistency of constructs and variance inflated factor (VIF). The composite reliability for each construct was above 7.0 and showed satisfactory levels of internal consistency reliability. The AVE for all the constructs were between 0.567 and 0.782 satisfying the requirement for convergent reliability (Hair, et al., 2017). The VIF for the constructs are less than 3.3 (Diamantopoulos, & Siguaw, 2006) hence collinearity is not an issue.

Table 1. Internal Consistency of Constructs and Variance Inflator Factor (VIF)

Constructs	Items	Item loading	Composite Reliability	AVE	VIF	
					Learning Outcome	Quality Learning
Curriculum	CU1	0.814	0.815	0.596	1.629	1.552
	CU2	0.762				
	CU3	0.738				
Delivery & Support	DS1	0.874	0.878	0.782	1.59	1.461
	DS2	0.895				
Learning Environment	LE1	0.794	0.832	0.623	1.696	1.662
	LE2	0.833				
	LE3	0.738				
Learning Outcomes	LO1	0.804	0.869	0.624	N/A	N/A
	LO2	0.808				
	LO3	0.793				
	LO4	0.754				
Learning Skill	LS1	0.733	0.797	0.567	1.678	1.641
	LS2	0.735				
	LS3	0.789				
Quality Learning	QL1	0.824	0.806	0.675	1.714	N/A
	QL2	0.820				
Resources	R1	0.767	0.854	0.594	1.34	1.328
	R2	0.774				
	R3	0.729				
	R4	0.810				

In establishing discriminant validity, Table 2 shows the Fornell-Larcker criterion where the square-roots of AVE for all constructs were greater its correlations with other constructs (Fornell & Larcker, 1981). The square roots of the AVE are in the diagonal. In addition, Table 3 shows the heterotrait-monotrait (HTMT) criterion which imposes a more stringent assessment suggests that all constructs are distinctly different at $HTMT_{0.90}$ threshold (Hensler, Ringle & Sartstedt, 2015).

Table 2. Discriminant Validity – Fornell and Larcker criterion

	C	DS	LE	LO	LS	QL	R
C	(0.772)						
DS	0.411	(0.885)					
LE	0.513	0.47	(0.789)				
LO	0.520	0.462	0.606	(0.790)			
LS	0.483	0.475	0.487	0.461	(0.753)		
QL	0.497	0.522	0.482	0.529	0.482	(0.822)	
R	0.365	0.318	0.408	0.392	0.425	0.365	(0.771)

Note : C = Curriculum; DS = Delivery & Support; LS = Learning Skill; LE = Learning Environment; LO = Learning Outcomes; R = Resources; QL = Quality Learning

Table 3. Discriminant Validity – HTMT Criterion

	C	DS	LE	LO	LS	QL	R
C							
DS	0.593						
LE	0.751	0.659					
LO	0.717	0.608	0.807				
LS	0.723	0.629	0.679	0.604			
QL	0.849	0.851	0.798	0.82	0.764		
R	0.502	0.417	0.547	0.492	0.585	0.564	

Note : C = Curriculum; DS = Delivery & Support; LS = Learning Skill; LE = Learning Environment; LO = Learning Outcomes; R = Resources; QL = Quality Learning

Table 4. Mediation Analysis

Predictor –> Mediator -> Criterion		**Effects**		**Standard Deviation (STDEV)**	***t*-value**	***p*-values**	**Mediation type**
H1	DS -> QL -> LO	Direct	0.091	0.035	2.584**	0.010	Complementary partial mediation
		Indirect	0.05	0.013	3.860**	0.000	
H2	LS -> QL -> LO	Direct	0.054	0.034	1.605	0.109	Indirect only Full mediation
		Indirect	0.027	0.009	3.089**	0.002	
H3	R -> QL -> LO	Direct	0.077	0.031	2.492**	0.013	Complementary partial mediation
		Indirect	0.015	0.006	2.413**	0.016	
H4	LE -> QL -> LO	Direct	0.331	0.037	9.051**	0.000	Complementary partial mediation
		Indirect	0.026	0.009	2.957**	0.003	
H5	C -> QL -> LO	Direct	0.167	0.033	5.099**	0.000	Complementary partial mediation
		Indirect	0.039	0.010	3.837**	0.000	

Note: ** t-value > 1.96, (two-tailed)
C = Curriculum; DS = Delivery & Support; LS = Learning Skill; LE = Learning Environment; LO = Learning Outcomes; R = Resources; QL = Quality Learning

The structural model for this study involved seven constructs at two levels of relationships. The five predictor variables are Learning Skills, Delivery and Support, Resources, Learning Environment, and Curriculum. Learning Quality mediates the between the five predictor variable and the final target construct Learning Outcomes. The bootstrapping results (Table 4) for testing the mediation effects (Preacher & Hayes, 2004) of Learning Quality between Delivery and Support, Learning Skills, Resources, Learning Environment and Curriculum and Learning Outcomes at 95% confidence interval confirmed the mediator role of Learning Quality. Based on the interpretation of "complementary mediation" (Zhao et al., 2010), QL has a complementary mediating effect on the relationship between Learning Outcomes, and Delivery and Support (H1), Resources (H3), Learning Environment (H4), and Curriculum (H5), demonstrating partial mediation for these relationships. This is evident from the indirect and direct effects which are all significant and point in the same direction. It was also found that QL fully mediates the relationship between Learning Skills and Learning Outcomes (H2), implying that QL accounts for all of the observed relationship between the two variables. This is demonstrated by the insignificant direct effect and significant indirect effect obtained for this relationship.

3.4 Discussion

The study examined the role of QL in STEM students' achievement of their learning outcomes in a Malaysian university. In this study, Hypothesis 1 was supported, QL partially mediates between delivery and support and learning outcomes. Hypothesis 2 was supported, QL fully mediates the relationship between learning skills and learning outcomes. Hypothesis 3 was supported, QL partially mediates the relationship between resources and learning outcomes. Hypothesis 4 was supported, QL partially mediates the relationship between learning environment and learning outcomes. Hypothesis 5 was supported, QL partially mediates the relationship between curriculum and learning outcomes.

In this study, QL is defined as the structure and process that is provided to help students develop confidence and decision-making skills. QL is the support structure for personal attribute development that will allow them to make decision about their own learning, Evidence from this study showed that learning skills, defined as the strategies and tactics used by students to manage themselves (Hattie at al., 1996), was found to be fully influenced by QL, suggesting the structure provided to students which addresses personal development attributes will influence the overall achievement of learning outcomes. Hence, STEM students who are given challenging but achievable learning goals, experiences improved learning especially if they are allowed and trusted to make decisions and take responsibilities for their own learning in an environment that is organised and supportive of their learning experience (Biggs & Tang, 2011).

Another two factors partially influenced by QL is delivery and support, the support provided to students by their teachers and curriculum, the structured arrangement of courses. The finding in this study suggests that the structure and support provided by a university in the form of its curriculum can, together with the support given by teachers to their students in this case STEM students, influence the achievement of learning outcomes. Hence the design and delivery of courses is influenced by QL and care needs to be taken that it is tailored to the needs of students (Langstrad et al., 2015) and should not be viewed by universities as a means of achieving required standards (Elassy, 2015). A well organised and fully explained lesson helps students sustain their attention, leading to better understanding and subsequently more learning (Samudra et al., 2016).

Curriculum defined as the 'map' of what students can do after a stipulated period of learning, was also partially influenced by QL in the achievement of learning outcomes. It is possible that students perceive a supportive teaching environment with appropriate teacher affirmation can significantly mitigate course difficulty (Bolkan & Goodboy, 2011), and together with a well-structured curriculum can help them achieve their learning outcomes.

The results showed that QL also partially mediated the relationship between resources and learning environment, defined as the facilities and infrastructure of a university and the quality of the academic staff and procedures of a university respectively. These physical factors remain important even with the increased use of educational technologies such as web-based learning environments, augmented reality, digital games, and gamifications of lessons to improve STEM learning outcomes. Further research is needed to include educational technologies as part of the resources and learning environment, to examine their influence of STEM students' perception of QL.

QL fully mediated learning skills and partially mediated delivery and support and curriculum to influence the achievement of learning outcomes. The other two factors did not show significant mediation levels by QL. When applied to the Malaysian context, the findings imply the importance placed by STEM students on the support structure provided by a university in terms of the structure of their courses and how they are taught and delivered and the guidance given to help in the personal development of students.

4 Conclusion and implications for practice

The results of this study showed that QL fully mediates the relationship between learning skills and learning outcomes and partially mediated the relationship between delivery and support, curriculum, learning environment, resources and learning outcomes. Although universities aspire for students to develop mastery of their own disciplines as well as apply and connect all knowledge in order to possess marketable job skills to meet the needs of Industry 4.0, the results of this study indicate students are more focused on having institutions meet their individual needs in terms of self-development with their teachers providing the necessary guidance. This is supported by results indicating students' association of acquiring learning skills to the type of QL provided by the university. Although outcomes based learning has been the emphasis and has been implemented with relative success in many Malaysian universities, students may continue to be teacher centred in their views of learning. Hence students may tend to view QL more in terms of their teachers and the structure of their programme rather than the university environment and experience as a whole. These findings hint at the remnants of an examination oriented system where the teacher is notably seen as an authority of knowledge rather than as a facilitator of learning. Such perceptions will take time to change and the aims for Malaysian students to be critical thinkers continue to be a difficult but ultimately an attainable goal. These results have clear implications for university teachers and administrators alike. There must be more effort in place to help students experience holistic learning to mould and develop them to be graduates with the necessary skills that meet market demands for employability.

5 References

Ambrose, S., & Lovett, M. (2014). Prior knowledge is more than content: Skills and beliefs also impact learning. In V. Benassi, C. Overson, & C. Hakala, *Applying Science of learning in education: Infusing psychological science into the curriculum* (pp. 7-19). Retrieved from the Society for the Teaching of Psychology web site: http://teachpsych.org/ebooks/asle2014/index.php.

Anderson, G. (2006). Assuring Quality/ Resisting Quality Assurance: Academics' responses to'quality' in some Australian universities. *Quality in Higher Education,*, 12(2), 161-173. http://dx.doi.org/10.1080/13538320600916767.

Biggs, J., & Tang, C. (2011). *Teaching for quality learning at University: What the student does.* Berkshire: McGraw Hill.

Bolkan, S., & Goodboy, A. (2011). Behavioural indicators of teansformational leadership in the college classroom. *Qualitative Research Reports in Communication*, 12(1), 10–18. DOI: 10.1080/17459435.2011.601520.

Choy, S. C., Yim, J. S., & Tan, P. L. (2017). Students' and teachers' perceptions of learning quality and learning outcomes in a Malaysian university. *Journal of Institutional Research in South East Asia*, 15(1), 41-62.

Choy, S. C., Yim, S., & Tan, P. L. (2017). Students' perceptions of quality learning in a Malaysian university - A mixed method approach. *Quality Assurance in education*, https://doi.org/10.1108/QAE-12-2016-0088.

Christophel, D. (1990). The relationships among teacher immediacy behaviors, student motivation, and learning. *Communication Education*, 39:4, 323-340, DOI:10.1080/03634529009378813.

Diamantopoulos, A., & Siguaw, J. A. (2006). Formative versus reflective indicators in organizational measure development: A comparison and empirical illustration. *British Journal of Management*, 17, 263–282 ,DOI: 10.1111/j.1467-8551.2006.00500.x.

Doll, W. E. (1993). *A post-modern perspective on curriculum.* New York: Teachers College Press.

Duschl, R., & Hamilton, R. (2010). Learning science. In R. E. Mayer, & P. A. Alexander, *Handbook of research on learning and instruction* (pp. 78-107). London: Routlege.

Elassy, N. (2015). The concepts of quality, quality assurance and quality enhancement. *Quality Assurance in Education*, 23 (3), 250-261. http://dx.doi.org/10.1108/QAE-11-2012-0046.

Ennis, C., & Chen, A. (2011). Learning motor skill in physical education. In R. Mayer, & P. Alexander, *Handbook of research on learning and instruction* (pp. 148-165). London : Routledge.

Entwistle, N. J., & Peterson, E. R. (2004). Conceptions of learning and knowledge in higher education: Relationships with study behaviourand influences of learning environments. *International Journal of Education Research*, 41, 407-428.

Fornell, C., & Larcker, D. F. (1981). Evaluating structural equation models with unobservable variables and measurement error. *journal of marketing research*, 48, 39-50.

Girash, J. (2014). Metacognition and instruction. In V. A. Benassi, C. E. Overson, & C. M. Hakala, *Applying science of learning in education: Infusing psychological science into the curriculum.* (pp. 152-168). Retrieved from the Society for the Teaching of Psychology web site: http://teachpsych.org/ebooks/asle2014/index.php.

Goodboy, A., & Myers, S. (2008). The effect of teacher confirmation on student communication and learning outcomes. *Communication Education*, 57(2), 153-179. DOI: 10.1080/03634520701787777.

Goodboy, K., & Bolkan, S. (2009). College teacher misbehaviors: Direct and indirect effects on student communication behavior and traditional learning outcomes. *Journal of Communication*, 73:2, 204-219, DOI: 10.1080/10570310902856089.

Hair, J. F., Hult, G. T., Ringle, C. M., & Sarstedt, M. (2017). *A primer on partial least squares structural equation modelling (PLS-SEM).* Los Angeles: Sage.

Hake, R. (1998). Interactive-engagement versus traditional methods: A six-thousand-student survey of mechanics test data for introductory physics courses. *American Journal of Physics*, 66, 64-74. doi: http://dx.doi.org/10.1119/1.18809.

Handelsman, J., Ebert-May, D., Beichner, R., Bruns, P., Chang, A., DeHaan, R., et al. (2004). Scientific teaching. *Science*, 304, 521-522. DOI: 10.1126/science.1096022.

Harks, B., Rakoczy, K., Hattie, J., Besser, m., & Klieme, E. (2014). The effects of feedback on achievement, interest and self-evaluation: the role of feedback's perceived usefulness. *Educational Psychology*, 34:3, 269-290. DOI: 10.1080/01443410.2013.785384.

Hattie, J., & Gan, M. (2011). Instruction based on feedback. In R. E. Mayer, & P. A. Alexander, *Handbook of Research on Learning and Instruction* (pp. 249-271). London: Taylor & Francis.

Hattie, J., & Timperley, H. (2007). The power of feedback. *Review of Educational Research, Vol. 77, No. 1 (Mar., 2007), pp. 81-112*, 77(1), 81-112.

Hattie, J., Biggs, J., & Purdie, N. (1996). Effects of learning skills interventions on student learning: A meta-analysis. *Review of Educational Research*, 66(2), 99- 136. DOI: 10.3102/00346543066002099.

Hensler, J., Ringle, C., & Sarstedt, M. (2015). A new criterion for assessing discriminant validity in variance-based structural equation modelling. *Journal of the Academy of Marketing Science*, 43(1),115-135.

Kobayashi, K. (2005). What limits the encoding effect of note-taking?: A meta-analytic examination. *Contemporary Educational Psychology*, 30, 242–262. doi:10.1016/j.cedpsych.2004.10.001.

Kunter, M., Klusmann, U., Baumert, J., Richter, D., Voss, T., & Hachfeld, A. (2013). Professional competence of teachers: Effects on instructional quality and student development. *Journal of Educational Psychology*, 105 (3), 805–820. DOI: 10.1037/a0032583.

Langstrand, J., Cronemyr, P., & Poksinska, B. (2015). Practise what you preach: quality of education in education on quality. *Total Quality Management & Business Excellence*, 26 (11-12) , 1202-1212, DOI: 10.1080/14783363.2014.925290.

Lee, C. H., & Kalyuga, S. (2014). Expertise reversal effect and its instructional implications. In V. Benassi, C. Overson, & C. Hakala, *Applying science of learning in education: Infusing psychological science into the curriculum* (pp. 31-44). Retrieved from the Society for the Teaching of Psychology web site: http://teachpsych.org/ebooks/asle2014/index.php.

Lizzio, A., Wilson, K., & Simons, R. (2002). University students' perceptions of the learning environment and academic outcomes: Implications for theory and practice. *Studies in Higher Education* , 27(1), 27-52.

Ministry of Education Malaysia. (2015). *Malaysia Education Blueprint 2015-2025 (Higher Education).* Putrajaya: Ministry of Education Malaysia.

Mottet, T., Parker-Raley, J., Cunningham, C., & Beebe, S. (2005). The relationships between teacher nonverbal immediacy and student course workload and teacher availability expectations. *Communication Research Reports*, 22(4), 275-282. DOI: 10.1080/000368105000317482.

Nicholson, K. (2011). *Quality assurance in higher education: A review of the literature.* Ontario: Council of Ontario Universities Degree Level Expectations Project, McMaster University.

Preacher, K. J., & Hayes, A. F. (2004). SPSS and SAS procedures for estimating indirect effects in simple mediation models. *Behaviour Research methods*, 36(4) 717-731.

Samudra, P. G., Kai, I. M., Cortina, S., & Miller, K. F. (2016). No second chance to make a first impression: The "thin-slice" effect on instructor ratings and learning outcomes in higher education. *Journal of Educational Measurement*, 53(3), 313–331. DOI: 10.1111/jedm.12116.

Tagg, J. (2010). The learning-paradigm campus: From single- to double-loop learning . *New Directions for teaching and Learning* , 123, 51-61. DOI:10.1002/tl.409.
Tyler, R. (2013). *Basic prinicples of curriculum and instruction.* Chicago: The University of Chicago Press.

Whalen, D., & Shelley, M. (2010). Academic success for STEM and non-STEM majors. *Journal of STEM Education*, 11(1), 46-60.
Zhao, X., Lynch, J. G., & Chen, Q. (2010). Baron and Kenny: Myths and Truths about Mediation Analysis. *Journal of Consumer Research*, 37,(2), 197-206.

SHS Web of Conferences 53, 05005 (2018) https://doi.org/10.1051/shsconf/20185305005
ICHSS 2018

Influence of Student Isolation on Students' University Learning Experiences: Perspectives of Academic, Social and Psychological Development

Shing Yu Jolene Lim[1], and Vighnarajah[2*]

[1]English Language and Communications Department, Faculty of Social Sciences and Liberal Arts, UCSI University, Malaysia, jolenelsr@gmail.com

[2]English Language and Communications Department, Faculty of Social Sciences and Liberal Arts, UCSI University, Malaysia, vighnarajah@ucsiuniversity.edu.my, iolasus@gmail.com

Abstract. With the latitude of globalisation spreading across countries in the world, student transfer and studying abroad have been made easier and more economical for students who intend to further their studies abroad. This growing population of international students shows the need to start addressing critical underlying issues regarding accommodation of new environment of international freshmen, plausible barriers faced by international students at their host countries and measures to minimise negative impression of intercultural communication. The growing amalgamation of local and international cultures also evinces symptoms of isolation among the local students. In light of this aim, this descriptive-analytical study investigates the correlation between student isolation and their university learning experiences, with the latter variable indexed in terms of academic, social and psychological isolation. A total of 581 students across various disciplines participated in this study. Findings provide rich, descriptive understanding of the phenomenon of student isolation, and facilitate the awareness to all stakeholders to give more leverage in the engagement and management of such issue in the educational context. Conclusively, findings have shown the importance of prioritising students' needs from a holistic view as every aspect intertwines to enhance students' success rate during their learning journey in university, as well as employability rate after their graduation.

* Corresponding author: vighnarajah @ucsiuniversity.edu.my

1 Background to the Study

UCSI University has been welcoming students and educators from various parts of the world to disseminate and receive knowledge since 1986 (UCSI University, 2018). With a population of more than 10,000 students based in Kuala Lumpur, Kuala Terengganu and Kuching (UCSI University, 2018), UCSI University takes pride for its multicultural and multilingual learning environment.

The diversified education context has made students' university learning experiences extraordinary for students are able to immerse in different cultures in and out of their classrooms. In UCSI University, learning experiences are not only limited to students' academic activities, but also involve social interactions and psychological developments. A holistically nurtured student would pose higher chances in achieving success in all three aspects during his or her university learning experiences.

The variety of university learning experiences is a major contributing factor in developing all-rounded students who will be the nation's future human capital. However, studies have found that fruitless university learning experiences have been increasing among international students (Dickey, 2004; Andrade, 2006; Trice, 2007; Erichsen & Bolliger, 2011; Curtin, Stewart, & Ostrove, 2012). Student isolation has to be prioritised by all stakeholders involved in the educational context as it impacts students' overall achievement and future development on a long run.

2 Problem Statement

Review of prior studies have revealed the fact that very few research have been done to investigate the phenomenon of student isolation among students in traditional learning environment, since most studies were done in online and distance learning contexts (Dickey, 2004; Ali &Kohun, 2006; Vighnarajah, Wong, & Bakar, 2009; Croft, Dalton, & Grant, 2010; Vighnarajah, Hassan, Aziz, & Ooi, 2016).

Moreover, most studies employed a qualitative methodology in the investigation of student isolation in campus (Ali &Kohun, 2006; Gu, Schweisfurth& Day, 2010; Khawaja & Stallman, 2011; Ganapathy, Vighnarajah, & Kaur, 2015; Lim & Vighnarajah, 2018). In fact, very few studies have identified the impact of student isolation on university learning experiences (Zhao &Kuh, 2004; Kuh, Cruce, Shoup, Kinzie & Gonyea, 2008; Khawaja & Stallman, 2011; Hu, Ching & Chao, 2012; Umbach & Wawrzynski, 2014). Hence, this research attempts to fill the gap by employing quantitative methodology in deriving statistical significance to describe the severity of student isolation among tertiary students. Further, this study also investigates the statistical impact of student isolation on students' university learning experiences and the magnitude of such influence.

3 Research Objectives and Hypotheses of the Study

The research objective for this study is to investigate the influence of student isolation, and its magnitude, on students' university learning experiences in UCSI University. Based on prior studies and review of literature, the following hypotheses were formulated to guide the direction of this study:

$H_{1\theta}$: There is no statistically significant relationship between Academic Isolation (AI) and University Learning Experiences (ULE).

H_{1A}: There is statistically significant relationship between Academic Isolation (AI) and University Learning Experiences (ULE).

H_{2O}: There is no statistically significant relationship between Social Isolation (SI) and University Learning Experiences (ULE).

H_{2A}: There is statistically significant relationship between Social Isolation (SI) and University Learning Experiences (ULE).

H_{3O}: There is no statistically significant relationship between Psychological Isolation (PI) and University Learning Experiences (ULE).

H_{3A}: There is statistically significant relationship between Psychological Isolation (PI) and University Learning Experiences (ULE).

H_{4O}: There is no statistically significant relationship between Total Isolation (TI) and University Learning Experiences (ULE).

H_{4A}: There is statistically significant relationship between Total Isolation (TI) and University Learning Experiences (ULE).

4 Understanding Student Isolation

Engagement with diversity is no longer inevitable considering vast technological advancement in the growing landscape of education; in fact, "engagement with diversity is essential for 21st century learning as it provides opportunity for students and staff to interact across the limitations of culture, ethnicity, demography and many other aspects that typically define concerns of diversity." (Vighnarajah & Alias, 2018, p. 1).

However, one of the commonly underrated and overseen education issues in the context of diversity is the issue of student isolation. Past studies have shown the complexity of such issue in students' university learning experiences and the importance of addressing student isolation in campus (Bolliger & Martindale, 2004; Dickey, 2004; Andrade, 2006; Trice, 2007; Erichsen & Bolliger, 2011; Curtin, Stewart, & Ostrove, 2012). A student's progression from high school to tertiary education brings about many challenges, and even more so for high school students who intends to further their studies abroad in a foreign country (Peel, 2000).

Findings of a research conducted by Lim and Vighnarajah (2018) revealed that student isolation is not only apparent in Western educational context, but also in the Asian context where Malaysian students and International Chinese students were heavily affected by student isolation in their university learning experiences.

4.1 Academic Isolation

Experiences of anxiety and isolation are not uncommon for international students in their attempts to adjust to a new learning environment (Andrade, 2006). Isolation, in specific, is apparent among students who lack proficiency and competency in their medium of instruction as they failed to understand their peers and tutors, causing a breakdown in their communication channel (Turula, 2002). Sherry, Chui and Thomas (2009) highlighted problems international students experienced in their studies such as adapting to a new culture, lack of competency in the use of English language and lack of understanding from the university community. Clearly, these findings suggested that international students face major difficulties in their studies due to lack of English language proficiency.

In a qualitative study, Lim and Vighnarajah (2018) explored the effects of isolation experienced by international and host students - the results showed that students felt most isolated when they face difficulties in coping with the medium of instruction, as well as cultural terms used by local residents. Such isolation has caused deterioration in their academic performances as they were unable to communicate and effectively receive knowledge from their peers and instructors. International students have reported higher level of academic isolation as compared to local students for these international students were required to adapt to a distinctively different education system, pedagogy and medium of instruction.

The learning environment created by educators in a university has an immense effect on students' overall academic performances. Umbach and Wawrzynski (2014, p. 18) mentioned that "institutions where faculty engage students in and out of the classroom and place a high priority on enriching educational experiences had students who felt supported and were active participants in their learning". This only substantiates the fact that students who are isolated in and out of the classroom to (academically, socially and psychologically) withdraw from their studies. Attention and action to such issues call for measures to reconsider the role of instructors as they gradually change their teacher-centred pedagogy to a student-centred approach (Vighnarajah, Wong & Bakar, 2008). They further asserted that instructors shoulder critical roles in facilitating students into their learning environment as they take on roles extending the duty of teaching such as facilitator, guide, mentor and thinking catalysts.

4.2 Social Isolation

Students who just transitioned from high schools to tertiary education tend to establish closer rapport with their peers as they aim to be accepted as a part of the community (Robertson, Line, Jonas, & Thomas, 2000). However, new students, specifically international students, often voluntarily or involuntarily isolate themselves from their peers due to miscommunication and misunderstanding of one's culture (Robertson et al., 2000). Trice (2007) showed that international students face social isolation as they lacked the time pursue and positively establish closer relationship with their host (family).

International students need to put in more effort in their academic lives due to language difficulties and the need to negotiate with a foreign academic culture. Such unfortunate situations would force-oblige these international students to attach to culturally relevant groups of friends (Trice, 2007). Students who fail to fulfil their social needs are prone to experience negative emotions, such as loneliness or even depression, as they do not have the channel to reach for support and assistance (University of Nebraska-Lincoln, 2016).

4.3 Psychological Isolation

Students in universities experience isolation beyond their classrooms and social circles where they are forced to face conflicts within themselves, as they struggle to adjust their mindset and psychological state throughout the learning process (Imperial College London, 2016). Poyrazli's (2015) study revealed that international students experienced certain degree of psychological problems in relation to the stress they felt in their studies – and this was primarily due to lack of English language competency leading to difficulties to function in a new cultural setting.

Feelings of isolation eventually lead up to depression among affected students and bring a rise in student attrition rate. Goncalves and Trunk (2014) found that students who felt isolated and neglected in campus faced grave obstacles in pursuing academic success, which may cause them to eventually choose to withdraw from their studies. In other words, psychological isolation poses as much risk as other forms of isolation in a student's university learning experience.

5 Methodology

This study adopted a quantitative approach where numerical statistics and descriptive interpretations complement each other to generate findings that are substantiated with width and depth (Peel, 2000; Mark & Anthony, 2007; Croft *et al.*, 2010; Erichsen & Bolliger, 2011). A quantitative approach would allow the researchers to generate data and findings with width and apply the results in general context as the findings are supported with statistics of large scale (Andrade, 2006).

Specifically, this study adopted the correlational research design, and this particular research design was deliberately selected as it allows for proper investigation of relationship between the variables to be measured (Creswell, 2012; Cohen, Manion, & Morrison, 2011). In the context of this study, the correlational research design guides for rigorous planning, implementation and analysis of variables between student isolation in students' academic, social and psychological developments and students' university learning experiences.

5.1 Description of population and sample

The chosen population for this study is a cohort of undergraduate students with approximately 6,000-7,000 students housed in (one of the) nine faculties in UCSI University Kuala Lumpur campus (UCSI University, 2018). The nine faculties include Faculty of Medicine and Health Sciences (FMHS), Faculty of Pharmaceutical Sciences (FPS), Faculty of Applied Sciences (FAS), Faculty of Hospitality and Tourism Management (FHTM), Institute of Music (IoM), Faculty of Engineering, Technology and Built Environment (FETBE), Faculty of Business and Information Science (FOBIS), Faculty of Social Sciences and Liberal Arts (FoSSLA), and De Institute of Creative and Design (ICAD).

According to Krejcie & Morgan (1970), the ideal sample size for a population size of 6,000 would be 361. However, this study gathered a sample size of 581 through stratification of the chosen population for the purpose of achieving 95% of confidence level with 5% of confidence interval. This measure was imperative to establish the relationship and its impact between the variables of the study.

5.2 Sampling method

Stratified sampling is deemed to be the most appropriate sampling method as it includes representative samples from the target population with fixed criteria (Hardon, Hodgkin & Fresle, 2004). Also, this method of probability sampling ensures that the researchers gather data from samples who share similar characteristics that represent the entire population (Creswell, 2012). Due to the nature of the population where strata are heterogeneous between each other, stratified sampling fits the population characteristic where division of the strata is based on shared attributes (Sharma, 2017). There were two specific advantages of employing stratified sampling in this study. First, the sampling method ensures that the population is properly represented by the sample. Second, human biasness is reduced in this sampling process as this method takes into account the similarities in characteristics shared by the sample.

The stratification was done by dividing the population based on the nine faculties housed in UCSI University Kuala Lumpur Campus. The selection criteria for the participants are:

1. The participant must be an active and current student in either one of the nine faculties;
2. Participants must be enrolled in either undergraduate degree year one, year two or year three or year four;

3. Selection of participants must be random and equal in sample size; 70 sets of survey questionnaires were distributed to each of the faculty.

5.3 Instrumentation

The instrumentation in this study focuses on the use of survey questionnaire. This measure was deemed as the appropriate form of data collection as the chosen research design for this research is of quantitative nature. Moreover, use of survey questionnaires would allow for ease of access to the target population for a large scale study.

The survey questionnaire used in this study was based on two separate survey instruments, as well as literature governing the study of student isolation and university learning experiences. The combination of two questionnaires into a new set of instrument is done for the fitting of this study. The reason for combining the items is that Russell, Peplau and Ferguson (1978) addressed the aspect of psychological isolation whereas Erichsen and Bolliger (2011) addressed the aspects of academic and social isolation. A combination was needed as this study aims to measure the impact of student isolation on the aspects of academic, social and psychological isolation. Prior to the pilot study and distribution for data collection, the survey questionnaire was contextualized in terms of language to better suit the context of this study.

A pilot study was conducted to test the validity and reliability of the instrument. The pilot study comprised 40 participants who did not participate in the actual study. The selection of 40 sample size for the pilot study was based on recommendation given by Johanson and Brooks (2010) as they find the sample size of 36 suffices the necessary confidence levels. The finalised questionnaire was validated again by the identified experts before the instrument was administered for the actual data collection process.

The reliability results of the constructs in the questionnaire indicated that each construct has high internal consistency reliability as the alpha values range from 0.70 to 0.85. The individual academic, social and psychological isolation constructs were combined to form a new construct labelled Total Isolation (TI). The obtained Cronbach's coefficient alpha value for TI was 0.88. The final procedure is to measure the overall reliability of the entire instrument by combining AI, SI, PI and ULE. The obtained Cronbach's coefficient alpha value was 0.72, indicating that the overall instrument has high reliability and is able to yield data that are reliable for further analysis and interpretation.

Table 1. Reliability for constructs in questionnaire.

Construct	**Cronbach's coefficient alpha**
Academic isolation (AI)	0.85
Social isolation (SI)	0.83
Psychological isolation (PI)	0.70
Total isolation (TI)	0.88
University learning experiences (ULE)	0.81
Full instrument	0.72

6 Findings and discussion

The analysis of 581 questionnaires generated results for all four categorical variables, namely gender, year of study, faculty and nationality. A total of 293 males (50.43%) and 288 females (49.57%) participated in the survey, indicating that the data pose similar distribution of gender. This measure was necessary in reducing the risk of obtaining biased data which would interfere with the analysis and interpretation of data (Field, 2013). There were higher number of students who are currently in their second year of studies (n=222, 38.21%) as opposed to year one (n=165, 28.40%) and year three (n=157, 27.02%). Year four students (n=37, 6.37%) showed significantly fewer participants. Since most courses are based on a 3-year programme structure and students generally complete their courses within the 3-year frame. Hence, these year four students could potentially be those who subscribe to a four-year programme structure or typically extending the three-year programme structure duration.

The nationality of participants showed that the majority of the sample comprised of Malaysian participants (n=424, 72.98%), followed by participants from Mainland China (n=58, 9.98%) and Indonesian participants (n=54, 9.29%). Countries with comparatively smaller numbers of participants were Korea (n=6,1.03%), Nigeria(n=2, 0.34%), Pakistan (n=6, 1.03%), Bahrain (n=9,1.55%), Iraq (n=6, 1.03%), Morocco(n=4,0.7%),Kyrgyzstan(n=5, 0.86%) and Uzbekistan(n=7, 1.21%).

Pearson product moment correlation was used to investigate the relationship between the independent sub-variables academic isolation (AI), social isolation (SI), psychological isolation (PI) and the dependent variable university learning experiences (ULE). The analysis was then conducted between the main independent variable total isolation (TI) and the dependent university learning experiences (ULE). Data cleaning and preliminary analyses were conducted to eliminate any possibilities of anomalies and violation of assumptions.

The results in Table 2 shows that academic isolation (r = -.62, n = 581, p <0.001), social isolation (r = -.50, n = 581, p <0.001) and psychological isolation (r = -.44, n = 581, p <0.001) have moderate to strong and negative relationships with university learning experiences. Total Isolation, which is a combination of academic, social and psychological isolation, has the strongest, negative relationship with university learning experiences (r = -.72, n = 581, p <0.001). All relationships have reached statistical significance at the level of p<0.001.

Table 2. Correlation matrix.

	1	2	3	4	5
1. University learning experiences	-	-.622*	-.498*	-.442*	-.715*
2. Academic isolation	-.622*	-	.466*	.183*	.805*
3. Social isolation	-.498*	.466*	-	.212*	.816*
4. Psychological isolation	-.442*	.183*	.212*	-	.532*
5. Total isolation	-.715*	.805*	.816*	.532*	-

*p<0.001 (2-tailed)

Academic isolation indicates strong and negative correlation (r = -.62) with university learning experiences. That is, increase in academic isolation would have substantial decrease in university learning experience. This finding is supported with past studies (Turula, 2002; Andrade, 2006; Sherry, Chui & Thomas, 2009; Erichsen & Bolliger, 2011; Umbach &

Wawrzynski, 2014; Lim & Vighnarajah, 2018) that found isolation in students' academic activities lead to a decrease in their academic performances and satisfaction towards their university learning experiences. It is crucial for educators and policy makers to recognize the importance of mitigating academic isolation in attempt to decrease student attrition rate and maximise academic success.

The second variable, social isolation, indicates strong and negative correlation (r = -.50) with university learning experience. The result shows that an increase in social isolation leads to substantial decrease in students' university learning experiences. The result is similar to findings of past studies (Robertson, Line, Jonas & Thomas, 2000; Reynolds & Constantine, 2007; Erichsen & Bolliger, 2011) as students who suffer from social isolation in campus generally feel dissatisfied with their university learning experiences. These findings suggest that the university learning environment should be designed to promote active and positive engagement between students in and out of the classroom. This outcome could be achieved by conducting activities that enhance students' interpersonal and communication skills, and preferably with reference in the interests of diversity and culturally responsive pedagogy.

Psychological isolation, on the other hand, shows only moderate and negative relationship (r = -.44) with university learning experience. The result contrasts with prior studies (Kuh, Cruce, Shoup, Kinzie & Gonyea, 2011; Goncalves & Trunk, 2014; Poyrazli, 2015) as past findings indicated that psychological isolation is strongly associated with poor learning experiences. Upon further deliberation, one very possible reason for this is the early detection of student isolation and action taken by relevant departments in UCSI University in attempt to ensure the students experience positive university learning experience.

The overall isolation poses robust, negative relationship with university learning experiences (r = -.72). The result suggests that students' university learning experiences are influenced by student isolation as a whole to a great extent, proposing all stakeholders in the education field to pitch into student isolation as an entity consisting three areas of concern and not as separate aspects.

7 Implications for Future Studies

The findings of this study surfaced several crucial implications that require urgency and importance from various stakeholders. First, students would be more aware of the student isolation phenomenon i.e. it is very probable for students themselves not to be aware that they feel isolated by mistaking such feelings with mere stress. This is a common situation where students would miss their family and friends back in their home country – not realizing that this stress is due to their inability to familiarize with the learning environment, instead of missing the company of their loved ones.

Secondly, instructors should be aware of signs of student isolation stemming from students' academic performances, social lives and psychological development. Being aware of such signs would give instructors the proper guidance to mollify such unfortunate issues, and be more acclimatized to measures taken by the University in recognizing and mitigating the student isolation phenomenon.

8 Conclusion

This study highlighted the influence of student isolation on students' university learning experiences and the strength of impact towards different aspects in student's university learning experience. Findings showed that academic isolation and social isolation were strongly associated with poorer university learning experiences as compared to psychological isolation. Student isolation as a whole proffers extensive negative relationship with university learning experience. Stakeholders in the educational context ought to prioritise the impacts of student isolation among tertiary students as higher education institutions are the main producer of future human capital for a nation to continue developing its economy.

Our deepest gratitude to the participants for their cooperation in sharing their personal experiences of student isolation and university learning experiences. We would like to highlight that measures are being taken to mitigate effects of student isolation in the flourishing student diversity in the university.

9 References

Ali, A., & Kohun, F. (2006). Dealing with isolation feelings in IS doctoral programs. *International Journal of Doctoral Studies*, *1*(1), 21-33.

Andrade, M. S. (2006). International students in English-speaking universities: Adjustment factors. *Journal of Research in International Education*, 5(2), 131-154.

Bolliger, D. U., & Martindale, T. (2004). Key factors for determining student satisfaction in online courses. *International Journal on E-learning*, *3*(1), 61-68.

Creswell, J. W., & Creswell, J. D. (2012). *Research design: Qualitative, quantitative, and mixed methods approaches*. Sage publications.

Cohen, L., Manion, L., & Morrison, K. (2011). *Research methods in education*. Routledge.

Croft, N., Dalton, A., & Grant, M. (2010). Overcoming isolation in distance learning: Building a learning community through time and space. *Journal for Education in the Built Environment*, *5*(1), 27-64.

Curtin, N., Stewart, A. J., & Ostrove, J. M. (2012). Fostering academic self-concept: Advisor support and sense of belonging among international and domestic graduate students. *American Educational Research Journal*, *50*(1), 108-137.

Dickey, M. (2004). The impact of web-logs (blogs) on student perceptions of isolation and alienation in a web-based distance-learning environment. *Open learning*, *19*(3), 279-291.

Erichsen, E. A., &Bolliger, D. U. (2011). Towards understanding international graduate student isolation in traditional and online environments. *Educational Technology Research and Development*, *59*(3), 309-326.

Kuh, G. D., Cruce, T. M., Shoup, R., Kinzie, J., &Gonyea, R. M. (2008). Unmasking the effects of student engagement on first-year college grades and persistence. *The journal of higher education*, *79*(5), 540-563.

Goncalves, S. A., & Trunk, D. (2014). Obstacles to Success for the Non-traditional Student in Higher Education. *Psi Chi Journal of Psychological Research*, *19*(4), 22-34.

Gu, Q., Schweisfurth, M., & Day, C. (2010). Learning and growing in a 'foreign' context: Intercultural experiences of international students. *Compare*, *40*(1), 7-23.

Hu, Y. L., Ching, G. S., & Chao, P. C. (2012). Taiwan student engagement model: Conceptual framework and overview of psychometric properties. *International Journal of Research Studies in Education*, *1*(1), 69-90.

Imperial College London. (2016). *Pause*. Retrieved from http://www.imperial.ac.uk/student-space/pause/

Johanson, G. A., & Brooks, G. P. (2010). Initial scale development: sample size for pilot studies. *Educational and Psychological Measurement*, *70*(3), 394-400.

Khawaja, N. G., & Stallman, H. M. (2011). Understanding the coping strategies of international students: A qualitative approach. *Journal of Psychologists and Counsellors in Schools*, *21*(2), 203-224.

Lin, M. (2012). Students of different minds: Bridging the gaps of international students studying in the US. *US-China Education Review, A* (3), 333-344.

Mark, J. W., & Anthony, C. H. A. N. (2007). Reducing the effects of isolation and promoting inclusivity for distance learners through podcasting. *Turkish Online Journal of Distance Education*, *8*(1), 221-227.

Peel, M. (2000). Nobody cares': The challenge of isolation in school to university transition. *Journal of Institutional Research*, *9*(1), 22-34.

Poyrazli, S. (2015). Psychological Symptoms and Concerns Experienced by International Students: Outreach Implications for CounselingCenters. *Journal of International Students*, *5*(3), 306-312.

Reynolds, A. L., & Constantine, M. G. (2007). Cultural adjustment difficulties and career development of international college students. *Journal of Career Assessment*, *15*(3), 338-350.

Robertson, M., Line, M., Jones, S., & Thomas, S. (2000). International Students, Learning Environments and Perceptions: A case study using the Delphi technique. *Higher Education Research & Development*, *19*(1), 89-102.

Sharma, G. (2017). Pros and cons of different sampling techniques. *International Journal of Applied Research*, *3*(7), 749-752.

Sherry, M., Thomas, P., & Chui, W. H. (2009). International students: A vulnerable student population. *Higher education*, *60*(1), 33-46.

Trice, A. G. (2007). Faculty Perspectives regarding Graduate International Students' Isolation from Host National Students. *International Education Journal*, *8*(1), 108-117.

Turula, A. (2002). Language Anxiety and Classroom Dynamics: A Study of Adult Learners. *English Teaching Forum*, 28-33. Retrieved from https://americanenglish.state.gov/files/ae/resource_files/02-40-2-g.pdf

Umbach, P. D., &Wawrzynski, M. R. (2005). Faculty do matter: The role of college faculty in student learning and engagement. *Research in Higher Education*, *46*(2), 153-184.

University of Nebraska-Lincoln. (2016). Feelings of isolation. Retrieved from http://www.unl.edu/mentoring/feelings-isolation

Vighnarajah, & Alias, N. A. (2018). Special Issue on the Assessment of Diversity in Higher Education.*International Journal of Teacher Education and Professional Development*, *1*(2), 1-3.

Vighnarajah & Lim, S. Y. J. (2018). Assessment of diversity through student isolation: Qualitative investigation of academic, social and emotional isolation. *International Journal of Teacher Education and Professional Development*, *1*(2), 1-13.

Vighnarajah, Wong, S. L., & Bakar, K. A. (2008). The shift in the role of teachers in the learning process. *European Journal of Social Sciences, 7*(2), 33-41.

Vighnarajah, Wong, S. L., & Bakar, K. A. (2009). Qualitative findings of students' perception on practice of self-regulated strategies in online community discussion. *Computers & education, 53*(1), 94-103.

Vighnarajah, Hassan, F.A., Aziz, N.A., & Ooi, S.L. (2016). Profiling information-seeking behaviour of distance learning students in Wawasan Open University. *Asian Association of Open Universities Journal, 11*(2), 122-135.

Zhao, C. M., &Kuh, G. D. (2004). Adding value: Learning communities and student engagement. *Research in higher education, 45*(2), 115-138.

An exploratory study using weblogs as a learning and assessment tool in a Public Relations course

Sheila Yvonne Jayasainan *

School of Communication, Faculty of Arts & Social Sciences, Taylor's University

Abstract. The study investigates the usefulness of weblogs in evaluating the understanding and performance of Public Relations (PR) learners. The study also explored PR leaners' reactions towards using weblogs as a learning and assessment tool for their performance. Fifty undergraduate students from a private institution of higher learning in the Klang Valley participated in a semester-long weblog project. The task required students to complete eight weekly-writing reflection entries on the principles and practices of PR onto their individual weblogs. The results indicate that PR leaners' understanding of the weekly topics improved with the use of weblogs. Moreover, 98% of the participants considered their blogging experience in PR-related topics positive. The majority of students believed that the weekly-writing reflections using weblogs constituted a good tool to showcase their learning process and help them reflect on their learning progress in a weblog-based environment. 96% welcomed the idea of employing blogs as an assessment tool. These results carry pedagogical implications in the use of technology for performance and assessment of PR learners.

* Corresponding author: SheilaYvonne.Jayasainan@taylors.edu.my

1 Introduction

The inclusion of technology in teaching and learning has changed the instructional strategies in classrooms. The passive teacher-centred learning approach has now shifted focus to a more student-centred learning, whereby students take up an active role in their learning process. Millennials (or Gen-Y; those born from the mid-1990s to early 2000s) have never known a world without technology. These millennials students have grown up in an environment that possesses unprecedented levels of media saturation and technology. Thus, their attitude towards education are unique. They rely on technology as an essential component to every aspect of their lives. Therefore, there's a growing perception amongst education leaders that the use of technology in the classroom is an effective way to connect and engage with these millennials (Sidek & Yunus, 2012). As such, many higher learning institutions are finding new means to traditional methods in order to bridge the learning efficiency gaps of millennials.

The challenge to education leaders is to modernise teaching methods and create an educational experience that enables students to apply what they have learned in the classroom to the world around them. This trend is also being influenced by the educational teaching and learning methods in Malaysia. The Malaysian Government is encouraging and emphasising the need to incorporate multimedia technology into the curriculum (Mat, 2000 as cited in Neo & Neo, 2010). This would facilitate the development of innovative modals that would enable higher learning institutions to better cater to the needs of the diverse types of students in today's university environment (Sidek & Yunus, 2012).

Smith (2009 as cited in Sidek & Yunus, 2012) states there has been an explosion of experimentation with weblogs or blogging in the classroom in the recent years. Educators believe there's a possibility of engaging students' interest using this technology. Weblogs, a popular form of personal web-publishing allow individuals to publish their thoughts, reflections and commentaries in the form of individual posts of varying lengths on a web page (Sharma & Xie, 2008). Weblogs offer two advantages for reflection: whereby it supports students' ability to record and revisit experience which is crucial for reflective learning; and it engage students' in articulating their thoughts and feeling (by writing posts) which leads to students' understanding of content and subsequently encourages self-learning (Sharma & Xie, 2008).

Although empirical research on weblogs use in higher learning institutions are relatively scarce, informal accounts from practitioners and researchers indicate that weblogs can be an important individual conversational learning space (Fiedler, 2003 as cited in Sharma & Xie, 2008). Bartlett-Bragg (2003) found weblogs useful as it allows students the freedom of commenting, for publishing unconnected pieces of knowledge and as a space to record their self-learning. Williams & Jacob's study found weblogs helpful in higher education whereby it has the potential to provide students with high level of independence while simultaneously providing opportunity for deeper interaction and exchange of information with peers (2004).

The examples cited above are amongst the few studies that cite empirical data in support of students' perceptions of weblogs, however none of them describes in detail the effectiveness of weblogs in evaluating PR learners' understanding and their perception of using weblogs as an assessment tool.

Thus, the purpose of this paper is to gauge PR learners' perception of and experiences with using weblogs as part of their learning process in a Malaysian higher institution classroom. By incorporating this student-centred learning approach coupled with using weblogs to provide a separate student-owned space, the paper hopes to assess the effectiveness of this approach to learning in a Malaysian higher institution classroom.

2 Literature Review

2.1 The rise of technology in the classroom

Multimedia technology is an important component in enhancing the motivation and learning process of students (Neo & Neo, 2010). With the availability of multimedia and Web 2.0 applications, students are able to utilise various applications such as a web-based learning and weblogs as tools in their learning. Therefore, educators have tried to incorporate these online technologies into the classroom environments to increase student satisfaction and learning. Currently, many educators are incorporating weblogs as part of their teaching and learning process as it allows students to publish and share information.

Weblog (or blog in short) is one of the most appealing Web 2.0 tools as it allows users to create personal journals and resource sites to share. The power of a blog lies in its' simplicity and the range of content it can profile on the web without a lot of technical understanding. In the classroom environment, blogging is found to be useful in improving students' critical thinking as the asynchronous mode of communication in weblogs allow students to think deeper before submitting their posts as it is viewed by a wider audience (Wang, 2005 as cited in Wang & Woo, 2008). Besides, the use of weblogs in the classroom provides a safe and comfortable environment for students particularly introverted one to express their ideas freely and creatively.

In addition, students have the freedom to take ownership over their content; they can post, edit, delete and be creative by adding videos, images, infographics and others. By having the freedom to control, it allows students to explore beyond the norm and take more responsibility for their self-learning.

Other studies, however have shown mixed results. Students' weblog usage will not improve their learning without applying suitable content, scaffolding and strategies in the learning environment (Ellison & Wu, 2008 as cited in Top, 2012). Furthermore, students did not consider peer comments valuable for their learning (Ellison & Wu, 2008 as cited in Halic, Lee, Paulus & Spence, 2010). Instead, reading other students' blogs is significantly more helpful than reading comments on one's own blog (Ellison & Wu, 2008 as cited in Halic et al, 2010).

2.2 Student-centred Learning

Student-centred learning is a dominant educational learning approach, whereby students learn by actively participating in learning activities. They have greater freedom and control over choice of subject matter, learning methods and pace of study (Gibbs, 1992 as cited in Neo & Neo, 2010). This suggest that learners can decide on their own learning approaches. Therefore, researchers have defined student-centred learning as active learning; learning of "doing" rather than of passive watching or listening (Neo & Neo, 2010). In other words, students assume a high level of responsibility in their learning process and not solely depend on the educator's direction.

This is in-line with the constructivists claim that knowledge is constructed by students themselves rather than deriving it from educators (Jonassen, 1991 as cited in Wang & Woo, 2008). Social constructivists further argued that knowledge is collaboratively constructed in a social context mediated by discourse; learning is then fostered through interactive and social activities such as information sharing, negotiation and discussion (Wang & Woo, 2008). Thus, weblog basically provides an avenue for students to publish information and obtain feedback from others.

Therefore, with the advent of technology in classroom and Web 2.0 applications, student-centred learning approaches are being implemented in various classroom environments especially in higher education in Malaysia.

2.3 Using weblogs to engage students in reflection

Many educators are using weblogs to promote interactivity and self-reflection in their students learning process. Blogs provide the space for individuals to express their opinions whereby bloggers are given a sense of personal empowerment and belonging to the learning community at large (Farmer, Yue & Brooks, 2008 as cited in Neo & Neo, 2010).

Weblogs provides a way of capturing the learning process. Students are able to externalise their thoughts through writing or visual representation in an ongoing manner. Posts in a weblog are organised in a reverse chronological order. By tracking the blog posting chronologically can help to identify how one constructs meanings over time in various subject matters (Fiedler, 2009 as cited in Wang & Woo, 2008).

Interactivity is a major construct of a technology enhanced learning environment (Mollam, 2003 as cited in Wang & Woo, 2008). The use of weblogs in the classroom to a certain extent promote learner-content interaction as the process of writing can naturally help students to think and reflect. When used in a learning environment, weblogs reflect students' understanding of a subject matter and the meaning associated with it (Linddell, Hubbard & Werner, 2000 as cited in Wang & Woo, 2008). Reflections is said to also have a positive effect on students' academic and personal development (Linddell, Hubbard & Werner, 2000 as cited in Wang & Woo, 2008).

The use of weblogs as part of a subject-matter offers students an effective way to publish and reflect on the content in a discursive manner. Thus, allowing students to make self-improvement progressively. In this learning process, the role of the educator is more likely to be a reviewer who reads the weblog content and gives comments and suggestions for improvement. In addition, by receiving feedback from readers, students can make improvements in their content.

In a study by Strampel & Oliver (2007) states that students perceived blogging as a mean of offering them the ability to evaluate their own learning, to revise their thinking and to reflect. This evidence suggests that by writing about the learning, students were better able to see and understand their thinking and change their conceptions when necessary – both important elements in higher levels of reflection and cognitive processing (Strampel, 2007).

Xie et al (2008 as cited in Halic et al, 2010) reported positive findings related to the effect of time on students' level of reflection: as time lapsed, the reflective thinking scores of students who continued blogging (either with or without feedback) increased.

3 Methodology

3.1 Background

A survey research design was selected to investigate students' perception of and experiences with using weblog in an undergraduate class in a Public Relations course at a private institution of higher learning in the Klang Valley. Students *(n=65)*; those taking the course as a core subject and those taking the course as an elective subject were assigned to engage in weblog reflections related to Public Relations (PR) concepts covered in the weekly lectures. During the course of their semester, each student was required to create an individual weblog and upload one post of 500 words per week for a period of eight consecutive weeks of a 14-week semester. Weebly was selected as a hosting provider for the weblog because it provides a much more reliable access and allows for creativity in the weblog design.

One of the reason weblogs was chosen as a requirement of this course was to encourage students to create an online portfolio of written work and also as a means of reflecting on their understanding of the weekly PR concepts taught during lectures. By using weblogs, these students were able to practice and enhance their writing skills, which is a crucial skill for a Public Relations student and future practitioner. Furthermore, the use of weblogs was also to build students' creative prowess with the use of images, videos, infographics etc. to support their writing reflections. This was the first time in this course that weblogs were used as part of the learning and assessment process.

3.2 Participants

The participants in this study were undergraduate students (n=50), 32 females and 18 males. The gender distribution of this sample was slightly uneven but comparable to the distribution of the overall student population in this Public Relations course where approximately 40% are male students and 60% are female students. The participants' ages ranged from 18 to 24 years. Most of the participants were in their Semester 3 (Year 2, Semester 1) of their undergraduate Public Relations degree.

3.3 Instrument and Procedure

The instrument included five items to record students' perception on their experiences of using weblogs as part of their learning and assessment process. The survey was administered online using Socrative Exit Ticket. Both open and close ended questions were asked to gauge students' perception and experiences of using weblogs in the classroom.

Responses recorded from students were anonymous. Data collection occurred during the last week of the 14-week semester. There was no course credit was given for completing this survey. The survey yielded a response rate of 98%. Once the students completed their responses on the survey, the data was analysed.

Besides close ended questions, students were also asked several open-ended questions. These questions were used to garner feedback directly from the students themselves. Their comments will be shown together with the results of the survey. Their feedback was not edited or summarized. Figure 1 and Figure 2 are samples of these students' weblogs.

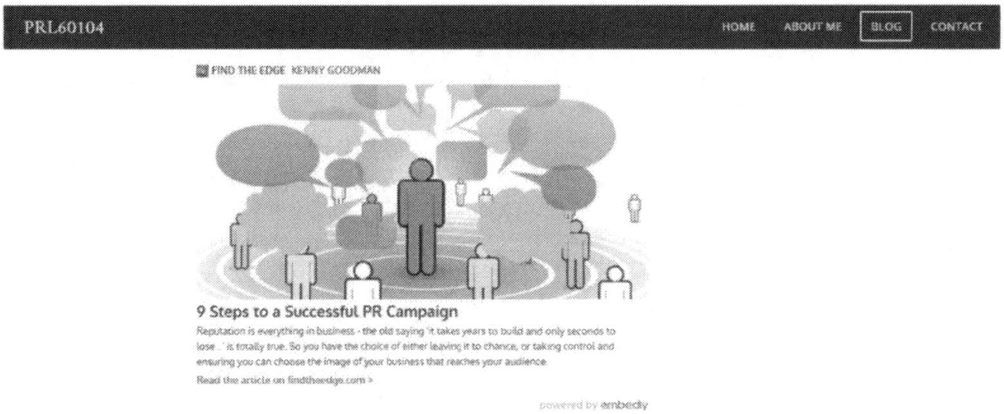

Figure 1. Sample of students' weblog

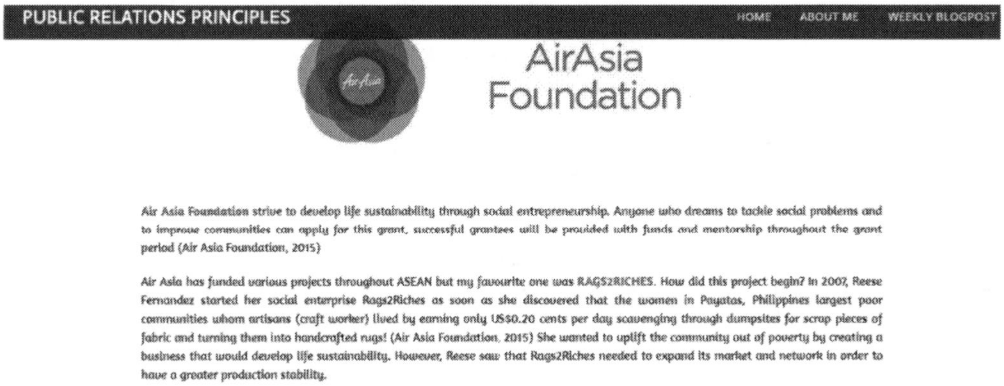

Figure 2. Another sample of students' weblog

4 Results and Discussion

4.1 Students' Experience and Satisfaction with using Weblogs for Learning and Reflection

The majority of the participants in this study reported that their weblog experience was positive in this Public Relations course and it enhance their overall learning in particular helping them to think about PR concepts outsides of the classroom.

This is similar to the views of Churchill (2009 as cited in Halic et al, 2010), the current findings revealed that 98% had a positive experience with the weblog-supported learning environment and acknowledge the learning potential of weblog as a discussion medium. Students found that weblog offered them the opportunity to reflect on PR-course related concepts outside of the classroom and to share their different perspectives on the course material discussed during lectures. Some of the qualitative feedback from the participants are as follows:

"I've learnt and understand more about Public Relations itself".

"I understand the vital aspects for a successful PR campaign".

"I learned how to manage my time doing the blog every week and also understand about the Public Relations industry even more".

"I understand audience engagement through the differences of media as a PR Practitioner".

"I have learnt on how to analyze CSR campaigns and how to determine whether it was a successful one or not. Another thing that I learnt from this module is that, the structure of PR and what a PR practitioner can do in an organization. I have also learnt how important PR is to the world".

"I learned how to write better and more professionally from a PR standpoint and learned the importance of new media and how it affects the PR field and how I must be more adaptable".

These results are consistent with some qualitative findings on students' experiences of blogging as carried out by Paulus, Payne et al (2009 as cited in Halic et al, 2010) which found that students when blogging in an unstructured environment without discussion prompts or active facilitation by instructors offered multiple perspectives on the topic.

Furthermore, a previous study carried out by Sharma & Xie (2008) found that blogging stimulated students thinking and learning by offering alternative perspectives related to the course content. This is similar to the current results where students carefully examine the current course content taught in lectures and extended their discussion outside the classroom in their weekly blogging activity. Overall, these activities engage critical thinking and reflection for students.

4.2 Students' perception of the advantages and disadvantages with using Weblogs for Learning and Reflection

The results showed that students view the use of weblog in their PR-related course as positive and beneficial to their learning and reflection needs. Their weekly reflections have helped them remember what they were taught throughout the semester and provided them recall information on activities already performed.

When asked to list some of the advantages of using weblogs for their course, some of the feedback by students include:

"It helped me in reviewing the lessons".

"I get to practice what I've learn on that topic and enhance the understanding of the topics".

"It indoctrinates discipline in the student and serves as a revision to the students".

"It helps me see my progress every week and let me know my understanding level about the topics as well".

"Sharpening writing skills & understanding of module".

"It keeps me reading new articles and facts consistently and allows me to do research and extra readings. I have learnt a lot of things other than what has been taught in the lectures".

"It is beneficial as it refreshes my memory on each topic that we have learnt every week. It also helps us improve in our writing".

Besides, students were also asked their feedback on the disadvantages of using weblogs for their learning and reflection. Computer expertise emerged as a predictor of perceived learning. Neither previous experience with weblogs nor age or language used had any significant impact on perceived learning.

Computer expertise or technical competence which had previously been identified as a critical factor for success in online environments in a study by Soong et al, 2001 (as cited in Halic et al, 2010) supports the current finding that students' computer expertise influenced their perception of learning. Some of their feedback include:

"The website is slightly hard to use".
"I don't really know how to design a website...."

"I would prefer Wix over Weebly because Weebly hates me".

Besides the lack of computer or technical expertise, other disadvantages from the findings as indicated by the students are weblogging was time consuming and required them to conduct research prior to writing. Their feedback includes:

"It is time consuming and require a lot of effort in researching".

"Having trouble to do it weekly and having to research".

"Time needed for research".

"Time consuming as need to do it every week".

"The least beneficial part would be the duration of research needed for the task".

"Highly opinionated to my own perspective".

4.3 Students' perception of using Weblogs as an Assessment Tool for a PR-course

Since this is the first time weblog is used in the assessment of this PR course in an undergraduate level, the aspect of assessment was analysed. Students were asked for their feedback if the use of weblogs as an assessment tool was appropriate in this course. The findings revealed that 96% of participants felt that using weblogs for grading was appropriate and beneficial to them in this PR-course. Furthermore, students felt that using weblogs allowed them to have control of their assessment progress, to see improvement in their work and it was acceptable for them to make mistakes and learn from it.

Some of their qualitative feedback include:

"....it helped me in learning the modules easier".

"Yes it helped with my critical thinking and polished up my research skills".

"Yes. It had help me improve my writing as a PR student...".

"It's a more friendly assessment for our demographic compared to exams".

"I can see improvement on my writing skills from the first week to the last week of reflection. There is an increase in my writing expression and I try to use new vocabularies in my reflection.

The impact of weblog effectiveness on the understanding of PR-related concepts has had a positive effect on the overall final grades of the participants. It is evident that their performance in this PR-related subject has improved from the previous semester whereby weblogs was not use as part of their assessment. The passing rate at the end of the semester for this subject was 94.74% with only two failures out of 65 students; showing a sharp increase of 15.34% pass rate from the previous semester [the overall pass rate from the previous semester was 79.4%].

5 Conclusion

From the results of the research, the overall students experience and perception of using weblog in their learning process in a PR-related course in an undergraduate classroom proved to be effective in this study. The students were able to comfortably integrate weblogs into their learning process which had not being done in previous classes. The use of weblog in the classroom has helped them understand the subject-matter better and challenged them in their research and critical thinking skills; allowing them to give their best in their weekly reflections.

This study showed that the use of weblogs in a Malaysian undergraduate classroom was an effective instructional tool which could be used to enhance and engage students actively in their learning process. Hence, this technology-enhanced learning mode has a positive impact on students' learning especially these technology-savvy millennials and can serve as a viable option in a student-centred higher learning classroom.

Nevertheless, one of the limitations of this study is that it did not attempt to make comparisons between students who used weblogs and those who did not; even though such an analysis could provide further insights into students' perception of weblog effectiveness to support learning. Future research could investigate into students' attitudes towards peer feedback in a blended learning environment.

6 References

Bartlett-Bragg, A. (2003, December). Blogging to learn. *The Knowledge Tree 4*.

Halic, O., Lee, D., Paulus, T., & Spence, M. (2010). To blog or not to blog: Student perceptions of blog effectiveness for learning in a college-level course. *Internet & Higher Education, 13*, 206-213.

Neo, T.K. & Neo, M. (2010). A Study using web-logs or blogs as a tool for student-centred learning in a computer graphics course: A Malaysian perspective. *The International Journal of Learning*, *17*(9), 77-86.

Sharma, P. & Xie, Y. (2008). Student experiences of using weblogs: An exploratory study. *Journal of Asynchronous Learning Networks*, *12*(3-4), 137-156.

Sidek, E.A.R. & Yunus, M.M. (2012). Students' experiences on using blogs as learning journals. *Procedia – Social & Behavioural Sciences*, *67*, 135-143.

Strampel, K. & Oliver, R.G. (2007). Using blogs in higher education: A pilot study on how online journaling can be effective in language learning. *Proceedings Ascilite Conference Melbourne*. Retrieved from http://www.ascilite.org.au/conferences

Top, E. (2012). Blogging as a social medium in undergraduate course: Sense of community best predictor of perceived learning. *Internet &Higher Education, 15*, 24-28.

Wang, Q. & Woo, H.L. (2008). Affordances and innovative uses of weblogs for teaching and learning. In Kobayashi, R (Ed), *New Educational Technology* (pp.183-199). NY: Nova Science Publishers.

Williams, J.B. & Jacob, J. (2004). Exploring the use of blogs as learning spaces in the higher education sector. *Australasian Journal of Educational Technology, 20*(2), 232-247.

AUTHOR INDEX

EDP Sciences – Web of Conferences
17, Avenue du Hoggar
Parc d'Activité de Courtabouf
BP 112
F-91944 Les Ulis Cedex A
France

ISBN 978-1-5108-7306-3

CURRAN ASSOCIATES INC.
proceedings.com

Applied Nuclear Physics Conference (ANPC 2021)

EPJ Web of Conferences Volume 261 (2022)

Prague, Czech Republic
12 – 17 September 2021

edp sciences